# Geometric Function Theory and Related Topics

Contemporary Mathematics

820

# Geometric Function Theory and Related Topics

29th International Conference
on Finite or Infinite Dimensional Complex Analysis
and Applications
August 21–25, 2023
Pondicherry University, Puducherry, India

Sudeb Mitra
Rakesh Kumar Parmar
Saminathan Ponnusamy
Toshiyuki Sugawa
Editors

American Mathematical Society

2020 *Mathematics Subject Classification*. Primary 32G15, 30C62, 37F31, 37F20, 37F44, 30F45, 37E10, 33C20, 51M10.

---

**Library of Congress Cataloging-in-Publication Data**

Names: International Conference on Finite or Infinite Dimensional Complex Analysis and Applications (29th : 2025 : Puducherry, India), author. | Mitra, Sudeb, editor.
Title: Geometric function theory and related topics : 29th International Conference on Finite or Infinite Dimensional Complex Analysis and Applications, August 21–25, 2023, Pondicherry University, Puducherry, India / Sudeb Mitra, Rakesh Kumar Parmar, Saminathan Ponnusamy, Toshiyuki Sugawa, editors.
Description: Providence, Rhode Island : American Mathematical Society, 2025. | Series: Contemporary mathematics, 0271-4132 ; volume 820 | Includes bibliographical references.
Identifiers: LCCN 2024041979 | ISBN 9781470476342 (paperback) | ISBN 9781470479459 (ebook)
Subjects: LCSH: Geometric function theory–Congresses. | AMS: Several complex variables and analytic spaces – Deformations of analytic structures – Moduli of Riemann surfaces, Teichmüller theory. | Functions of a complex variable – Geometric function theory – Quasiconformal mappings in the plane. | Dynamical systems and ergodic theory – Complex dynamical systems – Combinatorics and topology. | Functions of a complex variable – Riemann surfaces – Conformal metrics (hyperbolic, Poincaré, distance functions). | Dynamical systems and ergodic theory – Complex dynamical systems – Holomorphic families of dynamical systems; the Mandelbrot set; bifurcations. | Dynamical systems and ergodic theory – Low-dimensional dynamical systems – Maps of the circle. | Special functions – Hypergeometric functions – Generalized hypergeometric series. | Geometry – Real and complex geometry – Hyperbolic and elliptic geometries (general) and generalizations.
Classification: LCC QA360 .I56 2025 | DDC 516/.3–dc23/eng/20250207
LC record available at https://lccn.loc.gov/2024041979

Contemporary Mathematics ISSN: 0271-4132 (print); ISSN: 1098-3627 (online)

DOI: https://doi.org/10.1090/conm/820

---



∞ The paper used in this book is acid-free and falls within the guidelines
established to ensure permanence and durability.
Visit the AMS home page at `https://www.ams.org/`

10 9 8 7 6 5 4 3 2 1 30 29 28 27 26 25

# Contents

# Preface

The papers in this volume contain the proceedings of the 29th International Conference on Finite or Infinite Dimensional Complex Analysis and Applications (ICFIDCAA) held at Pondicherry University, India, August 21–25, 2023. This conference was first started by Professor Joji Kajiwara of Kyushu University, Japan, as the chair of the Korean-Japanese Colloquium on Finite or Infinite Dimensional Complex Analysis at Pusan National University, South Korea, in July 1993. Subsequently, this conference blossomed into one of the most significant conferences covering a very broad spectrum of topics in complex analysis (in the spirit of the twentieth century's premier geometric function-theorist Lars V. Ahlfors). This conference has been held at several Asian countries, such as South Korea, Japan, China, Hong Kong, Macau, Vietnam, Thailand, Russia, and India.

Keeping to the philosophy of this conference, the present volume contains papers on a wide range of topics in geometric function theory, such as conformal and quasiconformal mappings, Teichmüller spaces, topics from classical complex analysis, intersections with several aspects of functional analysis. The reader will find survey presentations, aimed at a broad mathematical audience, accounts of recent developments, and the exposition of new trends in these areas, and also original research papers. It contains contributions from many leading experts and also from several postdoctoral fellows and graduate students. We hope that this volume will be a good source of reference for professional researchers working in various areas of complex analysis. It should also be suitable for graduate students who aspire to learn these subjects and pursue their own research.

We want to thank all speakers at this conference and all contributors to this volume. We are very grateful to the anonymous referees for their careful work. Very special thanks to Sergei Gelfand, Linda Joyce, Michael Loss, Bridget Manzella, Arlene O'Sean, and Christine Thivierge of the AMS Editorial Division for their generous help and valuable advice.

Sudeb Mitra
Rakesh Kumar Parmar
Saminathan Ponnusamy
Toshiyuki Sugawa
Editors

Contemporary Mathematics
Volume **820**, 2025
https://doi.org/10.1090/conm/820/16406

# Paths of analytic circle diffeomorphisms

Kuntal Banerjee and Anubrato Bhattacharyya

ABSTRACT. We present here two results on the smooth paths of analytic circle diffeomorphsms. The first result is due to Herman, which says that if $[0,1] \ni t \mapsto f_t$ is a smooth path of analytic circle diffeomorphisms with the condition that the rotation number of $f_0$ is different from the rotation number of $f_1$, then the set of points $t \in [0,1]$ where the map $f_t$ is Lipschitz conjugate to an irrational rotation has measure strictly positive. Next we consider the standard family of analytic circle homeomorphisms of the form $\mathbb{R}/\mathbb{Z} \ni x \mapsto x + a + \frac{b}{2\pi} \sin 2\pi x \mod 1$ for $a, b$ in a subset of $\mathbb{R}$, a parameter space suitable for our purpose. We show that the tongue of rotation number $p/q$ is path connected using some quasiconformal techniques introduced by Dezotti. Finally we connect the two results in the context of the standard family and we raise some open questions.

## 1. Introduction

The circle is one of the simplest one dimensional objects with several useful properties as we observe in many examples. The study of self maps on the circle is a classical topic in Mathematics. Poincaré was the first to study the circle homeomorphisms in [**P**], when he was considering the solutions to ordinary differential equations defined on the torus. The simplest circle homeomorphisms are the rotations. It turns out that the perturbations of rotations by a non linear function have some amazing properties. We can do this perturbation for a parametrized family of analytic circle diffeomorphisms. The family of maps $\mathbb{R}/\mathbb{Z} \ni x \mapsto x + a + \dfrac{b}{2\pi}\sin(2\pi x) \mod 1$ will fall into this category for parameters $a, b$ satisfying $a \in \mathbb{R}, 0 \le b < 1$. This family is called the standard family. One primary advantage of working with this family is the fact that the complexified maps are symmetric (see Definition 4.3) with respect to the unit circle. Each complexified map has two symmetric critical points, with symmetric critical orbits. This makes the study of the dynamics of the family much simpler, just like any unicritical family, e.g. the quadratic family $z \mapsto z^2 + c, c \in \mathbb{C}$. For some parameters,

2020 *Mathematics Subject Classification.* Primary 37E10; Secondary 37C15, 37C05.

*Key words and phrases.* Analytic circle diffeomorphism, rotation number, quasiconformal homeomorphisms.

The first author was supported partly by the Department of Science and Technology (DST), Govt. of India, under the Scheme DST FIST [File No. SR/FST/MS-I/2019/41].

The research of second author was funded by UGC [NTA Ref. No. 201610319430], Govt. of India.

the corresponding circle diffeomorphisms in the standard family exhibit regions of expansions and contractions simultaneously, which is another important aspect.

The aims of this article are two-fold. In the beginning we present a result of Herman in a simplified manner on the measure of a certain subset of points for a path of analytic circle diffeomorphisms. In addition we use quasiconformal techniques similar to the work of Dezotti to show that the rational tongues in the standard family are path connected. Furthermore we analyze Herman's result in the context of the standard family and raise some open questions.

## 2. Some background

We consider the circle as $\mathbb{T} := \mathbb{R}/\mathbb{Z}$. For every orientation preserving circle homeomorphism $f : \mathbb{T} \to \mathbb{T}$ there is a lift $F : \mathbb{R} \to \mathbb{R}$, which is an increasing homeomorphism such that $\Pi \circ F = f \circ \Pi$, where $\Pi : \mathbb{R} \to \mathbb{T}$ is the canonical projection. Conversely for every increasing homeomorphism $F : \mathbb{R} \to \mathbb{R}$ with $F(x+1) = F(x) + 1$ for any $x \in \mathbb{R}$, we have an orientation preserving circle homeomorphism $f : \mathbb{T} \to \mathbb{T}$ induced by $F$, given by $[x] \mapsto [F(x)]$, where $[x]$ denotes $x$ mod 1. Poincaré ([**P**]) showed that when we consider an orientation preserving circle homeomorphism $f : \mathbb{T} \to \mathbb{T}$, the average angle of rotation under the action of $f$ is a unique quantity called the rotation number of $f$. Similarly we can consider the average amount of translation of every point under the action of the lifted map $F : \mathbb{R} \to \mathbb{R}$, and this is called the translation number.

DEFINITION 2.1. The *translation number* of $F : \mathbb{R} \to \mathbb{R}$ is defined as

$$\mathrm{Trans}(F) := \lim_{n \to +\infty} \frac{F^n - \mathrm{Id}}{n}.$$

The *rotation number* of $f$ is the quantity $\rho(f) := \mathrm{Trans}(F)$ mod 1.

As the translation and rotation numbers are reals, so they can either be rational or irrational. The dynamics of the circle homeomorphism is dictated by the nature of these numbers and the following results will justify this statement. Every time we write $p/q$ for a rational number, we implicitly assume that $p$ and $q$ are coprime.

PROPOSITION 2.1 (Poincaré [**P**]). *The rotation number $\rho(f) \in \mathbb{Q}/\mathbb{Z}$ if and only if $f : \mathbb{T} \to \mathbb{T}$ has a periodic point. More precisely, the translation number $\mathrm{Trans}(F) = p/q \in \mathbb{Q}$ if and only if there is a point $a \in \mathbb{R}$ such that $F^q(a) = a + p$.*

Note that $G := F^q - \mathrm{Id} - p$ vanishes on the whole $F$-orbit of $a$, in particular on the $q$-set $\{a, F(a), \ldots, F^{(q-1)}(a)\}$ whose image in $\mathbb{T}$ is a cycle of $f$. We say that such a cycle has rotation number $p/q$.

PROPOSITION 2.2 (Poincaré [**P**]). *If $\rho(f) = \alpha \in \mathbb{R} \setminus \mathbb{Q}$, then $f : \mathbb{T} \to \mathbb{T}$ is semi-conjugate to the rotation $\mathbb{T} \ni [x] \mapsto [x + \alpha] \in \mathbb{T}$.*

The above result does not appear in this form in Poincaré's work, rather it can be proved from his observations. And the semiconjugacy may be obtained as follows. The sequence of maps

$$\frac{1}{N} \sum_{k=0}^{N-1} (F^k - k\alpha)$$

converges to a non-decreasing continuous surjective map $\Phi_F : \mathbb{R} \to \mathbb{R}$, as $N \to +\infty$, which satisfies

$$\Phi_F(x+1) = \Phi_F(x) + 1 \quad \text{and} \quad \Phi_F \circ F(x) = \Phi_F(x) + \alpha.$$

By the following result of Denjoy, the semiconjugacy becomes an actual conjugacy under sufficient regularity condition of the starting homeomorphism, particularly when we consider analytic circle diffeomorphism. In other words $\Phi_F : \mathbb{R} \to \mathbb{R}$ is an increasing homeomorphism when $F$ is an analytic diffeomorphism.

THEOREM 2.2 (Denjoy [**Den**]). *If $\rho(f) = \alpha \in \mathbb{R} \setminus \mathbb{Q}$ and if $f$ is a $C^2$ diffeomorphism, then $f : \mathbb{T} \to \mathbb{T}$ is conjugate to the rotation by angle $\alpha$.*

REMARK 2.3. The rotation number of a circle homeomorphism is invariant under topological conjugacy, even if two orientation preserving circle homeomorphisms are semi-conjugate, they have the same rotation number.

Now we formally introduce the two-parameter family of analytic circle diffeomorphisms, which can be seen as perturbations of rotations. The family is called the standard family or the Arnold family. This is a two-parameter family which is defined for two real parameters $a$ and $b$ as follows

$$f_{a,b} : \mathbb{T} \to \mathbb{T}, \quad f_{a,b}(x) = \big(x + a + \frac{b}{2\pi}\sin(2\pi x)\big) \bmod 1.$$

For our purpose, the parameter space for the standard family is the subset $\mathcal{P} = \{(a,b) : a \in \mathbb{R}, 0 \le b < 1\}$ of parameters $(a,b)$ when the map $f_{a,b}$ is a diffeomorphism. If $f_{a,b}$ is a diffeomorphism then so is $f_{a,-b}$; as well $f_{a,b}(x) = f_{a,-b}(x+1/2)$. For this reason we just concentrate on the part of the parameters for which $b \ge 0$. The Arnold tongues or simply the tongues are the level sets for the map $(a,b) \mapsto \rho(f_{a,b})$ on $\mathcal{P}$.

DEFINITION 2.4. For a fixed $\theta \in \mathbb{R}$, the tongue of rotation number $\theta$ or the $\theta$ tongue is denoted as $T_\theta$ and it is defined by

$$T_\theta := \{(a,b) \in \mathcal{P} : \rho(f_{a,b}) = \theta\}.$$

Depending on the rationality of the rotation number, the tongues exhibit different characteristics. We call the $\theta$ tongue $T_\theta$ an irrational tongue when $\theta$ is irrational and we call it a rational tongue when $\theta$ is rational. It is known that the tongues with irrational rotation numbers are Lipschitz curves and the tongues with rational rotation numbers are closures of components of the hyperbolic set $\mathcal{H} = \{(a,b) \in \mathbb{R} \times [0,1) : f_{a,b}$ has an attracting cycle$\}$ within the parameter space $\mathcal{P}$.

## 3. Herman's result on smooth paths of analytic circle diffeomorphisms

Let $\mathfrak{F} = \{f_t : \mathbb{T} \to \mathbb{T}\}_{t\in[0,1]}$ be a $C^1$ path of analytic circle diffeomorphisms. We consider the rotation number function on this path defined as $t \mapsto \rho(f_t)$. By Denjoy's theorem 2.2, we know that if $\rho(f_t)$ is irrational then $f_t$ is topologically conjugate to the rotation $R_{\rho(f_t)} : [x] \mapsto [x + \rho(f_t)]$ where $[x] = x \bmod 1, x \in \mathbb{R}$. We have something more specific from Herman's work.

THEOREM 3.1 (Herman [**H1**, **Y**]). *There is a set $A \subset \mathbb{R} \setminus \mathbb{Q}$ of full measure and if $\rho(f_t) \in A$ then the analytic circle diffeomorphism $f_t$ is conjugate to the irrational rotation $R_{\rho(f_t)}$ with the conjugating map being analytic.*

Thus when we work with $\mathfrak{F}$, one of our aims will be to analyse the sets of points where the rotation number is irrational and the given map is conjugate to the rotation with that rotation number in a nice way.

DEFINITION 3.2. We call a Lipschitz function with Lipschitz constant $k$ as a $k$-Lipschitz function. A circle diffeomorphism $f : \mathbb{T} \to \mathbb{T}$ is called $k$-Lipschitz conjugate to the rotation $R_{\rho(f)}$ if the conjugating map is $k$-Lipschitz.

DEFINITION 3.3. We denote the Lebesgue measure on $\mathbb{R}$ as $m$ and define

$$L(\mathfrak{F}) = \{t \in [0,1] : f_t \text{ is Lipschitz conjugate to the irrational rotation } R_{\rho(f_t)}\}.$$

We denote $M(\mathfrak{F}) = m(L(\mathfrak{F}))$.

Using the fact that $A$ has full measure and some elementary results we try to prove the following. We define the function $\mathfrak{h} : [0,1] \to \mathbb{R}$ by $\mathfrak{h}(t) = \rho(f_t)$.

THEOREM 3.4 (Herman [**H2**]). *Let $\mathfrak{F} = \{f_t : \mathbb{T} \to \mathbb{T}\}_{t\in[0,1]}$ be a $C^1$ path of analytic circle diffeomorphisms with $\rho(f_0) \neq \rho(f_1)$, then $M(\mathfrak{F}) > 0$. Moreover $\mathfrak{h} : [0,1] \to \mathbb{R}$ is absolutely continuous.*

Before starting the main proof we will prove some lemmas.

LEMMA 3.5. *Suppose $f_{t_0} = R_{\rho(f_{t_0})} : [x] \mapsto [x + \rho(f_{t_0})]$ where $[x] = x \bmod 1$. Then*

$$|\mathfrak{h}(t) - \mathfrak{h}(t_0)| = |\rho(f_t) - \rho(f_{t_0})| \leq \|f_t - R_{\rho(f_{t_0})}\|_\infty.$$

PROOF. Suppose $\rho(f_{t_0}) = \theta_0$ and $F_{t_0} : x \mapsto x + \theta_0$ is a lift of $f_{t_0}$. Assume that $F_t : x \mapsto x + \theta_0 + \phi_t(x)$ a lift of $f_t$, where $\phi_t : \mathbb{R} \to \mathbb{R}$ is a periodic function with period 1. Let $M_t = \|f_t - R_{\rho(f_{t_0})}\|_\infty = \sup_{x\in\mathbb{T}} |f_t(x) - R_{\rho(f_{t_0})}(x)| = \sup_{x\in\mathbb{R}} |\phi_t(x)|$. Then

$$-M_t \leq F_t(x) - x - \theta_0 \leq M_t, \Rightarrow -nM_t \leq {F_t}^n(x) - x - n\theta_0 \leq nM_t.$$

Hence

$$\Rightarrow -M_t \leq \frac{{F_t}^n(x) - x}{n} - \theta_0 \leq M_t$$

Therefore $|\mathfrak{h}(t) - \mathfrak{h}(t_0)| = |\rho(f_t) - \rho(f_{t_0})| \leq \|f_t - R_{\rho(f_{t_0})}\|_\infty$. □

We say that $C > 0$ is the Lipschitz constant for the family $\mathfrak{F}$ at $t_0$ if $\|f_t - f_{t_0}\|_\infty \leq C|t - t_0|$.

LEMMA 3.6. *Let $h \circ f_{t_0} \circ h^{-1} = R_{\rho(f_{t_0})}$ with $h$ being $k$-Lipschitz and $C$ be the Lipschitz constant for the family $\mathfrak{F}$ at $t_0$ then*

$$|\mathfrak{h}(t) - \mathfrak{h}(t_0)| = |\rho(f_t) - \rho(f_{t_0})| \leq k\|f_t - f_{t_0}\|_\infty \leq kC|t - t_0|.$$

PROOF. Define a new family $\mathfrak{G} = \{g_t := h \circ f_t \circ h^{-1}\}_{t\in[0,1]}$. Since rotation number is conjugation invariant $|\mathfrak{h}(t) - \mathfrak{h}(t_0)| = |\rho(f_t) - \rho(f_{t_0})| = |\rho(g_t) - \rho(g_{t_0})|$. Now by Lemma 3.5

$$|\rho(g_t) - \rho(g_{t_0})| \leq \|g_t - g_{t_0}\|_\infty.$$

As $h$ is a homeomorphism,

$$\begin{aligned}|\rho(g_t) - \rho(g_{t_0})| &\leq \sup_{x\in\mathbb{T}} |g_t(h(x)) - g_{t_0}(h(x))| \\ &= \sup_{x\in\mathbb{T}} |h(f_t(x)) - h(f_{t_0}(x))| \\ &\leq k\|f_t - f_{t_0}\|_\infty \\ &\leq kC|t - t_0|.\end{aligned}$$

□

We omit the proof of the following simple observation.

PROPOSITION 3.1. *If a continuous map $f : \mathbb{R} \to \mathbb{R}$ is $k$-Lipschitz on $S \subset \mathbb{R}$ then $m(f(S)) \leq km(S)$.*

PROOF OF THEOREM 3.4. To begin the proof, for each $k \in \mathbb{N}$ we consider the sets

$$D_k = \{t \in [0,1] : h_t \circ f_t \circ h_t^{-1} = R_{\rho(f_t)}, \rho(f_t) \in \mathbb{R} \setminus \mathbb{Q}, h_t \text{ is } k\text{-Lipschitz}\}.$$

Note that $D_k$ defines an increasing sequence of sets. Clearly $\cup_k D_k = L(\mathfrak{F})$. Assume $J = \{\rho(f_t) : t \in [0,1]\} = \mathfrak{h}([0,1])$; this is a closed interval with non-empty interior as $\rho(f_0) \neq \rho(f_1)$. By Theorem 3.1 we have $\mathfrak{h}(\cup_k D_k) \supset A \cap J$. Therefore $m(\mathfrak{h}(\cup_k D_k)) \geq m(A \cap J) = m(J) > 0$. This means $m(\mathfrak{h}(D_k)) > 0$ for some $k$. Moreover if $C = \sup_{x\in\mathbb{T}} |\partial f_t(x)/\partial t|$ then $C > 0$ is the Lipschitz constant for this family $\mathfrak{F}$ at $t_0 \in D_k$ and $\mathfrak{h}$ is $kC$-Lipschitz on $D_k$. Hence by Proposition 3.1 we have $m(\mathfrak{h}(D_k)) \leq kCm(D_k)$ and thus $m(D_k) > 0$. And $D_k \subset L(\mathfrak{F})$ implies $M(\mathfrak{F}) > 0$.

The absolute continuity is remained to be proved now. Suppose $N \subset [0,1]$ is of measure 0. We have to show that $m(\mathfrak{h}(N)) = 0$. We write $\mathfrak{h}(N) = \mathfrak{h}((N \cap \cup_k D_k) \cup (N \cap (\cup_k D_k)^c)) = \mathfrak{h}(N \cap \cup_k D_k) \cup \mathfrak{h}(N \cap (\cup_k D_k)^c)$. Since $\mathfrak{h}(\cup_k D_k) \supset A \cap J$ and $A$ is of full measure, $m(\mathfrak{h}(\cup_k D_k)^c) = 0$, thus $m(\mathfrak{h}(N \cap (\cup_k D_k)^c)) = 0$. Therefore $m(\mathfrak{h}(N)) = m(\mathfrak{h}(N \cap \cup_k D_k)) = m(\cup_k \mathfrak{h}(N \cap D_k))$. But $m(N \cap D_k) = 0$ and $\mathfrak{h}$ is $kC$-Lipschitz on $D_k$. Hence by Proposition 3.1 we are done. □

## 4. Paths within rational tongues in the standard family

It is known [**EKT**] that Arnold tongues in the standard family are connected. The tongues corresponding to irrational rotation numbers are Lipschitz curves [**EKT, B**] and tongues corresponding to rational rotation numbers are open. Here, we will imitate Dezotti's proof of the connectedness of tongues in the double standard family to give an alternate proof of the connectedness of rational tongues in the standard family. The technique [**Dez**] will in fact prove path connectedness.

Recall the expression of a circle map in the standard family

$$f_{a,b} : \mathbb{T} \to \mathbb{T}, \quad f_{a,b}(x) = \big(x + a + \frac{b}{2\pi}\sin(2\pi x)\big) \bmod 1,$$

where $a$ and $b$ are real parameters. We are interested in studying the parameter space $\mathcal{P} = \{(a,b) : a \in \mathbb{R}, 0 \leq b < 1\}$. As $f_{a+1,b} \equiv f_{a,b}$, it is enough to concentrate on the subset $\{(a,b) : a \in \mathbb{T}, 0 \leq b < 1\}$.

A map $f_{a,b}$ in the standard family corresponds to a circle diffeomorphism $g_{a,b} : \mathbb{S}^1 \to \mathbb{S}^1$ after conjugating by the exponential map. Such a diffeomorphism extends holomorphically to an anlytic map $g_{a,b} : \mathbb{C}^* \to \mathbb{C}^*$ by the following formula

$$g_{a,b}(z) = e^{2\pi i a} z \exp\Big(bz - \frac{b}{z}\Big).$$

The family of maps $g_{a,b}$ is called the complexified standard family.

We will now state our main result of the section.

THEOREM 4.1. *In the parameter space $\mathcal{P}$ of the standard family, for any point $(a, b)$ in the rational tongue $T_{p/q}$ of rotation number $p/q$, there is a unique point $(a_0, 1)$ on the boundary of the tongue $T_{p/q}$ such that we can find an analytic path joining $(a, b)$ to $(a_0, 1)$.*

The unique point $(a_0, 1)$ corresponds to a super-attracting cycle. As a corollary we prove the following.

COROLLARY 4.2. *Tongues with rational rotation number in the standard family are path connected.*

We start with a parameter $(a, b) \in T_{p/q}$, for which $f_{a,b}$ has an attacting cycle of period $q$. We use quasiconformal tools to construct a path which is parametrized by the multiplier $t$ in the following manner: for each $t \in (0, 1)$ the path $t \mapsto (a_t, b_t)$ is such that $g_{a,b}$ is conjugate to $g_{a_t,b_t}$ with multiplier $t$, via a quasiconformal homeomorphism $\Phi_t$ which varies analytically with $t$ and as $t \to 0$, $(a_t, b_t) \to (a_0, 1)$, the unique point corresponding to a super-attacting cycle as mentioned in Theorem 4.1. The scheme of the proof is essentially the same as Dezotti's work in [**Dez**].

DEFINITION 4.3 (Symmetry with respect to Circle). A function $g : \mathbb{C}^* \to \mathbb{C}^*$ is symmetric with respect to the unit circle $\mathbb{S}^1 = e^{2\pi i \mathbb{T}} \subset \mathbb{C}^*$, in the sense that $g(1/\overline{z}) = 1/\overline{g(z)}$. A domain $\Omega$ is called symmetric with respect to $\mathbb{S}^1$ when $z \in \Omega$ if and only if $\dfrac{1}{\overline{z}} \in \Omega$.

For the quasiconformal tools, we refer [**Ahl, BF**] to the reader for a detailed overview.

### 4.1. Properties of $g_{a,b}$.

LEMMA 4.4. *For $a \in \mathbb{R}/\mathbb{Z}, 0 \le b < 1$ the map $g_{a,b}(z) = e^{2\pi i a} z \exp\Big(bz - \dfrac{b}{z}\Big)$ has only two critical points at $\dfrac{-1 \pm \sqrt{1 - 4b^2}}{2b}$ and it has essential singularities at $0$ and $\infty$. For $a \in \mathbb{R}/\mathbb{Z}, 0 \le b < 1$, $g_{a,b}$ is orientation preserving restricted to $\mathbb{S}^1$ with critical points away from the unit circle.*

PROOF. The first part is obvious. Since $g_{a,b} : \mathbb{S}^1 \to \mathbb{S}^1$ is conjugate to $f_{a,b} : \mathbb{T} \to \mathbb{T}$ as dynamical systems and $f_{a,b}$ is orientation preserving iff $b \in [0, 1)$, we conclude the last part. ☐

LEMMA 4.5. *For $a \in \mathbb{R}/\mathbb{Z}, 0 \le b < 1$, $g_{a,b}$ has only $0$ and $\infty$ as asymptotic values.*

PROOF. Take a curve $\eta(t)$ so that $\lim_{t\to\infty} g(\eta(t))$ exists where $\eta(t)$ converges to $0$ or $\infty$ as $t \to \infty$. Taking $\eta(t) = \pm t$ we see that $0, \infty$ are asymptotic values. With out loss of generality assume that $\eta(t) \to \infty$. In such a case the $ze^{bz}$ factor of $g_{a,b}$ determines the asymptotics. Now assume $\alpha$ is an asymptotic value other than

$0, \infty$. Let us call $\eta(t) = x(t) + iy(t)$ and observe that $|e^{\eta(t)}| = e^{x(t)}$. Now $\alpha \neq \infty$ implies that $x(t) \to -\infty$ and $\alpha \neq 0$ means $y(t)$ is unbounded. But $x(t) \to -\infty$ means that argument of $\eta$ is bounded for large $t$, which gives a contradiction upon analytically continuing $\log(g_{a,b})$ along $\eta$. □

We state the following Proposition with a sketchy proof. The first part of the proposition is a straight forward calculation, while the later part follows from Lemma 4.5 and the first part of the Proposition 4.1. For further details readers should refer to Proposition 2.5 [**Dez**].

PROPOSITION 4.1. *If* $a \in \mathbb{R}/\mathbb{Z}, 0 \leq b < 1$ *then* $g_{a,b}(z) = e^{2\pi i a} z \exp\Big(bz - \dfrac{b}{z}\Big)$ *is symmetric with respect to the unit circle and if* $g_{a,b}$ *has an attracting cycle on* $\mathbb{S}^1$ *then it attracts both critical points and there is no other attracting cycle.*

PROOF. Let $\psi$ be the linearizing map of an attracting cycle of $g_{a,b}$ of period $n$ which lies on $\mathbb{S}^1$ with multiplier $\lambda$. Let $R$ be the maximal radius of disc centered at 0 for which $\psi$ has inverse. Observe that $g^n_{a,b}$ and $g_{a,b}$ have same asymptotic values as $0, \infty$ do not belong $\mathbb{C}^*$, the domain of definition of $g_{a,b}$. Define $B_r := \psi^{-1}(D(0,r))$. Let $r$ be such that $\dfrac{r}{|\lambda|} > R$. If $B_r$ does not contain any critical points or asymptotic values then $g^n_{a.b} : g^{-n}_{a,b}(B_r) \to B_r$ is a covering map. This gives us a contradiction using maximality of $R$. Then it follows that $B_r$ contains a critical point or an asymptotic value. Since the asymptotic values are not in the domain, $B_r$ contains a critical point of $g^n_{a,b}$. Rest of the Proposition follows from the symmetry. □

Suppose $z_1, \ldots, z_q$ is an attracting $q$-cycle of $g_{a,b}$ on $\mathbb{S}^1$, then $z_j$ is the distinguished attracting periodic point if the immediate basin of $z_j$ contains both the critical points.

**4.2. Finding the super-attracting parameter.** We will require some lemmas to show that given any tongue with rational rotation number in the standard family there is a unique parameter $(a, 1)$ so that the corresponding complexified map has a super-attracting cycle. These are minor modifications of results from [**Dez**].

LEMMA 4.6. *Let* $F_{a,b}(x) = x + a + \dfrac{b}{2\pi}\sin(2\pi x)$, *be the lift of* $f_{a,b}$. *Then* $F^q_{a+1,b}(x) = F^q_{a,b}(x) + q$ *for any* $x \in \mathbb{R}$ *and any* $q \in \mathbb{N}$.

PROOF. This can be proved by induction when we use that $F^n_{a,b}(x+1) = F^n_{a,b}(x) + 1$ for all $x \in \mathbb{R}, n \in \mathbb{N}$ and that $F_{a+1,b}(x) = F_{a,b}(x) + 1$ for all $x \in \mathbb{R}$. □

LEMMA 4.7. *Let* $q \in \mathbb{N}$, *then the mapping from* $\mathbb{T}$ *to itself given by* $a \mapsto f^q_{a,1}(1/2)$ *satisfies the following properties.*

- *The map is strictly increasing.*
- *The map has degree* $q$.

PROOF. This follows directly from the fact that $\dfrac{\partial}{\partial a} f^q_{a,1}(1/2) > 0$, which is a calculation similar to the calculation in Lemma 2.6 of [**MR2**]. The last part follows from Lemma 4.6 above. □

We recall the following well known fact as a Lemma without proof as it follows from Proposition 2.1.

LEMMA 4.8. *If $(a,b) \in T_{p/q}$ and say $y$ is a periodic point of period $q$ of $f_{a,b}$. Let $\Pi(x) = y$ for some $x \in \mathbb{R}$. Then $F^q_{a,b}(x) = x + p$, and conversely if there exists some $x \in \mathbb{R}$ such that, $F^q_{a,b}(x) = x + p$, then $\rho(f_{a,b}) = \dfrac{p}{q}$.*

All this discussion in this sub-section leads to the following proposition.

PROPOSITION 4.2. *For each tongue $T_{p/q}$ corresponding to rational rotation number $p/q$, there is a unique point $(a_0, 1)$ on the boundary of the tongue so that $f_{a_0,1}$ has a super-attracting cycle.*

PROOF. Given any $p \in \{0, 1, \ldots, q-1\}$, from Lemma 4.7 it follows that there exists a unique $(a_0, 1)$ so that $F^q_{a_0,1}(1/2) = 1/2 + p$. Then using Lemma 4.8 we conclude that $1/2$ is a periodic point of period $q$ making the cycle super-attracting. □

**4.3. Deformation.** We state the next three lemmas without proofs, which follow from straightforward computations as studied in [**Dez**].

LEMMA 4.9. *Let $f$ be a holomorphic function defined on a neighbourhood of $\mathbb{S}^1$ which is symmetric with respect to the unit circle and has an attracting cycle $\{x, f(x), \ldots, f^{q-1}(x)\}$ of period $q$ lying on on $\mathbb{S}^1$ with multiplier $\lambda$. Then $\lambda \in \mathbb{R}$. Furthermore, if the linearizing map $\varphi$ defined in some neighbourhood of $x$ satisfies $\varphi'(x) = ix$ then $\varphi\left(\dfrac{1}{\overline{z}}\right) = \overline{\varphi(z)}$ and the pre-image of any disc centered at $0$ is also symmetric with respect to $\mathbb{S}^1$.*

It follows from the lemma that if $D_R$ is a disc on which $\varphi^{-1}$ is well defined then $\varphi^{-1}(D_R)$ is symmetric with respect to $\mathbb{S}^1$. The following lemma will be crucial for our deformation scheme.

LEMMA 4.10. *Let $\alpha > 0$ and*

$$\chi : \mathbb{C}^* \to \mathbb{C}^*$$

$$z \mapsto |z|^\alpha z$$

*Then we have the following.*

- *$\chi$ is real analytic diffeomorphism with $\chi(re^{i\theta}) = r^{\alpha+1}e^{i\theta}$ and $\chi(zz') = \chi(z)\chi(z')$.*
- *$\mu_\chi := \dfrac{\partial\chi/\partial\overline{z}}{\partial\chi/\partial z} = \dfrac{\alpha/2}{1+\alpha/2}\dfrac{z}{\overline{z}}$. In particular $|\mu_\chi|$ is constant and $\|\mu_\chi\|_\infty = \left|\dfrac{\alpha/2}{1+\alpha/2}\right|$.*
- *if $\alpha = \dfrac{\log r}{\log R} - 1$, then $\chi$ sends the disk $D_R$ (of radius $R$) onto the disk $D_r$; moreover if $R, r < 1$ then*

$$|\mu_\chi| = \frac{|1 - \log r/\log R|}{1 + \log r/\log R} < 1.$$

*We have the commutative diagram*

$$\begin{array}{ccc} D_R & \xrightarrow{\chi} & D_r \\ {\scriptstyle z\mapsto\lambda z}\downarrow & & \downarrow{\scriptstyle z\mapsto tz} \\ D_R & \xrightarrow{\chi} & D_r \end{array}$$

*with* $t = \chi(\lambda) = \lambda^{1+\alpha}$. *In particular,* $0 < t < 1$.

LEMMA 4.11. *Referring to the notations in 4.9, 4.10,* $\varphi : U \to D_R$ *and* $\chi : D_R \to D_r$. *Then the composition* $\varphi \circ \chi$ *induces a Beltrami form* $\sigma_t$ *on* $U$*, depending analytically on* $t$*, which is invariant by* $g$*:*

$$\sigma_t = \mu_t \frac{d\overline{z}}{dz} = \frac{\alpha/2}{1+\alpha/2} \cdot \frac{\varphi(z)}{\overline{\varphi(z)}} \cdot \frac{\overline{\varphi'(z)}}{\varphi'(z)} \cdot \frac{d\overline{z}}{dz}$$

*with* $\mu_t\Big(\dfrac{1}{\overline{z}}\Big) = \overline{\mu_t(z)}$ *for* $z \in U$.

Let us now summarize the deformation process using the scheme Dezotti used to prove the connectedness of tongues in the double standard family.

PROPOSITION 4.3. *Let* $(a, b) \in \mathbb{T} \times [0, 1)$ *such that* $g_{a,b}$ *has an attracting cycle on the circle with multiplier* $\lambda$ *and rotation number* $p/q$. *Then for* $t \in (0, 1)$ *there exists quasiconformal homeomorphism* $\varphi_t : \mathbb{P}^1 \to \mathbb{P}^1$ *fixing* $0$ *and* $\infty$*, depending analytically on* $t$ *such that* $\varphi_\lambda = id_{\mathbb{P}^1}$ *and* $\varphi_t$ *conjugates* $g_{a,b}$ *to some* $g_{a_t,b_t}$ *which has the same rotation number* $p/q$ *and multiplier* $t$.

PROOF. Start with a map $g := g_{a,b}$ with $(a, b)$ in some tongue of rotation number $p/q$. Then by 4.11, the linearizing map $\varphi$ around the distinguished attracting periodic point $x$ is symmetric with respect to $\mathbb{S}^1$. For a given $t \in (0, 1)$ one can have a Beltrami form $\sigma_t$ on the domain of linearization as in Lemma 4.11. Then one propagates this form on the whole basin of attraction via the map $g$ and extends it outside as 0. Using Lemma 4.11 we can see that this extended form $\sigma_t$ varies analytically with $t$. The Measurable Riemann Mapping theorem guarantees that there exists quasiconfromal homeomorphisms $\Phi_t$ fixing $0, x, \infty$ such that $t \mapsto \Phi_t(z)$ is analytic for any fixed $z$. Moreover we can choose $\Phi_t$ to be symmetric with respect to $\mathbb{S}^1$ because of the symmetry of $\sigma_t$ from lemma 4.11. Now if we conjugate $g_{a,b}$ via $\Phi_t$, we get a holomorphic map with multiplier $t$ where the holomorphicity follows from the Weyl's lemma. That the deformed map is also conjugate to a map in the complexified standard family is the content of the following lemma.

LEMMA 4.12. *For* $(a, b) \in \mathbb{T} \times [0, 1)$*, Let* $g_{a,b}(z) := e^{2\pi i a} z \exp\Big(bz - \dfrac{b}{z}\Big)$ *be a map in the complexified standard family and* $\Phi_t$ *as in the proof of* 4.3. *Then there exists* $(\alpha_t, \beta_t) \in \mathbb{T} \times \mathbb{D}$ *such that* $\Phi_t \circ g_{a,b} \circ \Phi_t^{-1} = \tilde{g}_{\alpha_t,\beta_t} := e^{2\pi i \alpha_t} z \exp\Big(\beta_t z - \dfrac{\overline{\beta_t}}{z}\Big)$. *Moreover there exists a unique rotation* $R_{\theta(t)}$ *such that* $R^{-1}_{\theta(t)} \circ \tilde{g}_{\alpha_t,\beta_t} \circ R_{\theta(t)} = g_{a_t,b_t}$ *for some* $(a_t, b_t) \in \mathbb{T} \times [0, 1)$ *with multiplier* $t$.

PROOF. Let $\psi, \varphi$ be orientation preserving quasiconformal homeomorphisms of Riemann sphere $\mathbb{P}^1$ fixing $0, \infty$. Let $V$ be the set of critical values of $g_{a,b}$, which contains two points; $\psi(V)$ is the set of critical values of $h := \psi \circ g_{a,b} \circ \varphi$. This means there is an isotopy $\psi_s : \mathbb{P}^1 \to \mathbb{P}^1$ between $\psi_0 = \psi$ and a quasiconformal homeomorphism $\psi_1$ fixing $0, \infty$ and each point in $V$. Then the isotopy can be lifted to an isotopy starting from $\varphi_0 = \varphi$. The final map is quasiconformal as it is locally composition of conformal or quasiconformal maps.

Since $\psi = \Phi_t, \varphi = \Phi_t^{-1}$ are orientation preserving homeomorphisms, they induce identity map on the first integral homology group of $\mathbb{C}^*$. Let $[\gamma]$ be the generator of the integral homology group, where $[\gamma]$ is class of curves of winding number 1 around 0 in the anticlockwise direction. Then

$$\frac{1}{2\pi i}\int_\gamma \frac{h'(z)}{h(z)}\,dz = \frac{1}{2\pi i}\int_\gamma \frac{g'_{a,b}(z)}{g_{a,b}(z)}\,dz.$$

The second integral depends continuously on $(a,b) \in \mathbb{T} \times [0,1)$ which means it is constant. Plugging in $(a,b) = (0,0)$ we see that the value is 1. Which means that $h(z) = ze^{u(z)}$ for some $u : \mathbb{C}^* \to \mathbb{C}^*$ holomorphic. Because a quasiconformal map is Hölder continuous, we get that $h$ has finite order of growth near $\infty$, hence also near 0 using a similar argument. This means $u$ has a finite Laurent series. Suppose the lowest exponent of the Laurent series of $u$ with non zero coefficient is $-p$. Because $h$ has only two critical points, differentiating $h$ gives $1 + zu'(z) = \frac{1}{z^p}(z^p + z^{p+1}u'(z))$, which has to have 2 zeroes. Giving us that $u(z) = \frac{a_{-1}}{z} + a_0 + a_1 z$. The specific form of $h$ follows from the symmetry of $g_{a,b}$ and $\Phi_t$ which makes $h$ symmetric with respect to $\mathbb{S}^1$. And finally we can see that the rotation $R_{\theta(t)} : z \mapsto \frac{|\beta_t|}{\beta_t} z$ conjugates $\tilde{g}_{\alpha_t,\beta_t}$ to $g_{a_t,b_t}$ for some $(a_t, b_t) \in \mathbb{T} \times [0,1)$. □

In the following part, we shall refer to our notation in Lemma 4.12. The deformation process yields that the following diagram is commutative, where $\tilde{f}_{\alpha_t,\beta_t}$ corresponds to $\tilde{g}_{\alpha_t,\beta_t}$ which is obtained after conjugating $g_{a,b}$ by $\Phi_t$. Note that $(\alpha_t, \beta_t) \in \mathbb{T} \times \mathbb{D}$.

$$\begin{array}{ccc} \mathbb{R}/\mathbb{Z} & \xrightarrow{\Psi_t^{-1}} & \mathbb{R}/\mathbb{Z} \\ {\scriptstyle f_{a,b}}\downarrow & & \downarrow{\scriptstyle \tilde{f}_{\alpha_t,\beta_t}} \\ \mathbb{R}/\mathbb{Z} & \xrightarrow{\Psi_t^{-1}} & \mathbb{R}/\mathbb{Z} \end{array}$$

In the diagram $\Psi_t$ is defined by $E \circ \Psi_t = \Phi_{t_{|\mathbb{S}^1}} \circ E$, where $E(x) = e^{2\pi i x}$ is the canonical identification of $\mathbb{R}/\mathbb{Z}$ onto $\mathbb{S}^1 \subset \mathbb{C}$. Observe that $E$ is a homeomorphism implying that $f_{a,b}$ and $\tilde{f}_{\alpha_t,\beta_t}$ are topologically conjugate. Further $\tilde{f}_{\alpha_t,\beta_t}$ is conjugate to $f_{a_t,b_t}$ by rotation $R_{\theta(t)} : z \mapsto \frac{|\beta_t|}{\beta_t} z$, with $(a_t, b_t) \in \mathbb{T} \times [0,1)$. Since rotation number is invariant under conjugacy, $f_{a_t,b_t}$ has same rotation number as in $f_{a,b}$. This finishes our proof of Proposition 4.3. □

### 4.4. Constructing the path.

PROPOSITION 4.4. *Let $(a,b)$ be a parameter in the $T_{p/q}$(tongue corresponding to rotation number $p/q$) such that the attracting cycle of $f_{a,b}$ has multiplier $\lambda$. Referring to the notations in 4.3, the map $\gamma : (0,1) \to T_{p/q}; t \mapsto (a_t, b_t)$; is analytic, with $\gamma(\lambda) = (a,b)$.*

PROOF. For any fixed $z$, $t \mapsto R_{\theta(t)}^{-1} \circ \Phi_t(z)$ is analytic. The critical points of $g_{a,b}$ are $\dfrac{-1 \pm \sqrt{1-4b^2}}{2b}$. Let $\omega$ be a critical point of $g_{a,b}$, then $R_{\theta(t)}^{-1} \circ \Phi_t(\omega)$ is the critical point of $g_{a_t,b_t}$. Since $t \mapsto R_{\theta(t)}^{-1} \circ \Phi_t(\omega)$ is analytic and $R_{\theta(t)}^{-1} \circ \Phi_t(\omega)$ can be expressed in terms of $b_t$, hence $b_t$ can be expressed as analytic function of $t$. Now observe that critical values of $g_{a_t,b_t}$ are given by, $g_{a_t,b_t} \circ R_{\theta(t)}^{-1} \circ \Phi_t(\omega) = R_{\theta(t)} \circ \Phi_t(g_{a,b}(\omega))$. From this identity we can derive an analytic expression for $a_t$. □

LEMMA 4.13. *Referring to the notations in 4.4, $\gamma(t) = (a_t, b_t) \to (a_0, 1)$ as $t \to 0$ where $(a_0, 1)$ is the unique point in the boundary of the rational tongue $T_{p/q}$ guaranteed by 4.2.*

PROOF. Clearly, $b_t \leq 1$, as if $b > 1$ then $f_{a,b}$ is not orientation preserving which cannot be topologically conjugate to an orientation preserving map. Because $f'_{a,b}(x) \geq (1-b)$ for all $(a,b,x)$ and $t \mapsto b_t$ is analytic, it follows that $b_t \to 1$ as $t \to 0$ since $t \mapsto \left(g^q_{a_t,b_t}\right)'(z_t) = t$ is continuous where $z_t$ is any point on the attracting cycle.

CLAIM 4.14. If there exists $t_n \in (0,1), t_n \to 0$ such that $\gamma(t_n) \to (a,1)$ for some point $(a,1)$ in the boundary of $T_{p/q}$ then, $g_{a,1}$ has a super-attracting cycle of period $q$.

PROOF. Recall from above that $t \mapsto \left(g^q_{a_t,b_t}\right)'(z_t) = t$ is continuous where $z_t$ is any point on the attracting cycle. Since the multiplier moves continuously with $t$, we have our claim. □

Let $\Gamma$ be the set of all limit points of $\gamma$ as $t \to 0$. Then $\Gamma := \cap_{n \geq 1} \overline{\gamma((0,1/n))}$. $\Gamma$ is connected in $\mathbb{T} \times \{1\}$. Say $\Gamma$ contains some interval $(a_1, a_2) \times \{1\}$ for some $a_1 < a_2 \in \mathbb{T}$. But in that case $\forall a \in (a_1, a_2), g_{a,1}$ has a super-attracting $q$-cycle. This violates the uniqueness of Proposition 4.2. □

**4.5. Proof of Theorem 4.1.** We have all the necessary preparations to prove Theorem 4.1, our main theorem and Corollary 4.2.

PROOF OF THEOREM 4.1. The proof is trivial combining Proposition 4.4 and Lemma 4.13. □

PROOF OF COROLLARY 4.2. Let $(a,b)$ and $(c,d)$ are two different points in the same tongue $T_{p/q}$. By Theorem 4.1, there exists paths $\gamma_1$ and $\gamma_2$ joining $(a,b)$ and $(c,d)$ to the unique super-attracting point $(a_0,1)$ as $t \to 0$. That means for $t$ very close to 0 both $\gamma_1(t)$ and $\gamma_2(t)$ are in a ball around $(a_0,1)$ inside $\mathbb{T} \times [0,1]$. Such a ball looks like the lower half of a ball in the plane, which is convex. Then for such $t$ we will just join $\gamma_1(t)$ and $\gamma_2(t)$ by a straight line to give us a path connecting $(a,b)$ and $(c,d)$. □

## 5. Final remarks

Let $F : \mathbb{R} \to \mathbb{R}$ be analytic with positive derivative and assume $F - \mathrm{Id}$ is of period 1. For $t \in \mathbb{R}$, set $F_t(x) = F(x) + t$. For $C \geq 1$ denote by $T_C$ the set of $t \in \mathbb{R}$ for which there is a $C$-Lipschitz conjugacy $\Phi_t : \mathbb{R} \to \mathbb{R}$ such that

$$\Phi_t(F_t(x)) = \Phi_t(x) + \mathrm{Trans}(F_t).$$

It is not difficult to see that the function $t \mapsto \text{Trans}(F_t)$ is $C$-Lipschitz on $T_C$. From the work of Arnold [**Arn**] and Herman [**H1**, **Y**] if $\text{Trans}(F_t)$ is a Diophantine number, the conjugacy $\Phi_t$ is analytic. It follows easily that the set of $t \in \mathbb{R}$ for which $\text{Trans}(F_t) \in \mathbb{R} \setminus \mathbb{Q}$ has positive Lebesgue measure if we follow the techniques of Theorem 3.4.

Considering the lift of a map in the standard family $F_{a,b} : x \mapsto x + a + b/2\pi \sin(2\pi x)$ with $(a, b) \in \mathbb{R} \times [0, 1)$, define the set

$$\mu(b) = m\{a \in [0, 1] \mid \text{Trans}(F_{a,b}) \in \mathbb{R} \setminus \mathbb{Q}\}$$

for a fixed $b \in [0, 1)$ where $m$ is the Lebesgue measure on $\mathbb{R}$. One can study the properties of the function $\mu$. It is known that $\mu(b) \to 1$ as $b \to 0$. The most simple question in this context is the following.

CONJECTURE 5.1. *The function $\mu$ is continuous on the interval* $[0, 1)$.

One can also investigate at which rate $\mu(b)$ goes to 1 as $b \to 0$. Note that the sum of the widths of the rational tongues can be obtained by $1 - \mu(b)$. It is proved in [**B**] that the width of $T_{p/q}$ behaves like $\dfrac{2}{\pi q}|C_{p/q}|b^q$ as $b \to 0$. Buff and Chéritat consider a similar problem in their work [**BC**], however they do not find any uniform bound for the constant $|C_{p/q}|$. If one can understand the behaviour of this sum of the rational widths, we can have a better understanding of the function $\Phi_{\text{trunc}}$ found at [**BC**]. We see that quasiconformal techniques provide an insight on the properties of hyperbolic components i.e. rational tongues in the parameter space of the standard family. Another relevant question could be to determine the parametrization of the hyperbolic components of the standard family like the well known parametrization of the hyperbolic components of the quadratic family $z \mapsto z^2 + c, c \in \mathbb{C}$.

## Acknowledgment

The authors appreciate the valuable comments and suggestions of the referee, which have enhanced the presentation of the article. They would also like to thank Sudeb Mitra for the invitation to write this article and for his insightful suggestions and comments.

## References

[Ahl] L. V. Ahlfors, *Lectures on quasiconformal mappings*, 2nd ed., University Lecture Series, vol. 38, American Mathematical Society, Providence, RI, 2006. With supplemental chapters by C. J. Earle, I. Kra, M. Shishikura and J. H. Hubbard, DOI 10.1090/ulect/038. MR2241787

[Arn] V. I. Arnol′d, *Small denominators. I. Mapping the circle onto itself* (Russian), Izv. Akad. Nauk SSSR Ser. Mat. **25** (1961), 21–86. MR140699

[B] K. Banerjee, *Boundaries of the Arnol'd tongues and the standard family*, arXiv:1402.4756, 2014, 30 pages.

[BC] X. Buff and A. Chéritat, *The Brjuno function continuously estimates the size of quadratic Siegel disks*, Ann. of Math. (2) **164** (2006), no. 1, 265–312, DOI 10.4007/annals.2006.164.265. MR2233849

[BF] B. Branner and N. Fagella, *Quasiconformal surgery in holomorphic dynamics*, Cambridge Studies in Advanced Mathematics, vol. 141, Cambridge University Press, Cambridge, 2014. With contributions by Xavier Buff, Shaun Bullett, Adam L. Epstein, Peter Haïssinsky, Christian Henriksen, Carsten L. Petersen, Kevin M. Pilgrim, Tan Lei and Michael Yampolsky. MR3445628

[Den] A. Denjoy, *Sur les courbes deéfinies par les équations différentielles à la surface du tore*, J. Math. Pures Appl. (9) **11** (1932), 333-375.

[Dez] A. Dezotti, *Connectedness of the Arnold tongues for double standard maps*, Proc. Amer. Math. Soc. **138** (2010), no. 10, 3569–3583, DOI 10.1090/S0002-9939-10-10355-4. MR2661556

[EKT] A. Epstein, L. Keen, and C. Tresser, *The set of maps $F_{a,b}\colon x \mapsto x + a + (b/2\pi)\sin(2\pi x)$ with any given rotation interval is contractible*, Comm. Math. Phys. **173** (1995), no. 2, 313–333. MR1355627

[H1] M.-R. Herman, *Sur la conjugaison différentiable des difféomorphismes du cercle à des rotations* (French), Inst. Hautes Études Sci. Publ. Math. **49** (1979), 5–233. MR538680

[H2] M.-R. Herman, *Mesure de Lebesgue et nombre de rotation* (French), Geometry and topology (Proc. III Latin Amer. School of Math., Inst. Mat. Pura Aplicada CNPq, Rio de Janeiro, 1976), Lecture Notes in Math., Vol. 597, Springer, Berlin-New York, 1977, pp. 271–293. MR458480

[P] H. Poincaré, *Mémoire sur les courbes définies par une équation differrentielle III*, J. Math. Pures Appl. **4**, (1885), 167-244.

[MR1] M. Misiurewicz and A. Rodrigues, *Double standard maps*, Comm. Math. Phys. **273** (2007), no. 1, 37–65, DOI 10.1007/s00220-007-0223-5. MR2308749

[MR2] M. Misiurewicz and A. Rodrigues, *On the tip of the tongue*, J. Fixed Point Theory Appl. **3** (2008), no. 1, 131–141, DOI 10.1007/s11784-008-0052-y. MR2402913

[Y] J.-C. Yoccoz, *Analytic linearization of circle diffeomorphisms*, Dynamical systems and small divisors (Cetraro, 1998), Lecture Notes in Math., vol. 1784, Springer, Berlin, 2002, pp. 125–173, DOI 10.1007/978-3-540-47928-4_3. MR1924912

Department of Mathematics, Presidency University, 86/1, College Street, Kolkata - 700073, West Bengal, India.

*Email address*: kbanerjee.maths@presiuniv.ac.in

Department of Mathematics, Presidency University, 86/1, College Street, Kolkata - 700073, West Bengal, India.

*Email address*: anubrato02@gmail.com

Contemporary Mathematics
Volume **820**, 2025
https://doi.org/10.1090/conm/820/16407

# Certain subclass of harmonic functions associated with univalent functions

Prachi Prajna Dash and Jugal Kishore Prajapat

ABSTRACT. In this paper, we define a subclass of sense-preserving harmonic functions associated with a class of analytic functions satisfying a differential inequality. We then establish a close relation between both subclasses. Further, we obtain some characteristic properties including radius properties, convolution, coefficient estimates and their properties, growth estimates, and convex combination for the functions in the defined subclass. Finally, we obtain conditions for some special functions as well as harmonic univalent polynomials to belong to the defined subclass of harmonic functions.

## 1. Introduction

A complex-valued harmonic function in a domain $D \subset \mathbb{C}$ is defined as a function $f(z) = u(z) + iv(z)$, which is twice continuously differentiable and $\Delta f(z) = 4f_{z\overline{z}}(z) = 0$ in $D$. The function $f(z)$ has a canonical representation $f(z) = h(z) + \overline{g(z)}$ when $D$ is simply connected, where $h(z)$ and $g(z)$ are analytic functions in $D$ and are termed as analytic and co-analytic parts of $f(z)$, respectively. It is well known that $f(z)$ is locally univalent in $D$ if, and only if, the Jacobian $J_f(z) \neq 0$ and $f(z)$ is sense-preserving if $J_f(z) > 0$ in $D$, where $J_f(z) = |h'(z)|^2 - |g'(z)|^2$. Let $\mathcal{H}$ be the class of complex-valued harmonic functions $f(z)$ defined in the open unit disk $\mathbb{D} = \{z \in \mathbb{C} : |z| < 1\}$, normalized by the condition $f(0) = 0 = f_z(0) - 1$. As a result, $h(z)$ and $g(z)$ of $f(z)$ have power series representations as

$$h(z) = z + \sum_{n=2}^{\infty} \alpha_n z^n \quad \text{and} \quad g(z) = \sum_{n=1}^{\infty} \beta_n z^n. \tag{1.1}$$

Let $\mathcal{H}^0$ be a subclass of $\mathcal{H}$, defined as $\mathcal{H}^0 = \{f(z) \in \mathcal{H} : f_{\overline{z}}(0) = 0\}$ in $\mathbb{D}$. Thus, for $f(z) \in \mathcal{H}^0$, (1.1) will become

$$h(z) = z + \sum_{n=2}^{\infty} \alpha_n z^n \quad \text{and} \quad g(z) = \sum_{n=2}^{\infty} \beta_n z^n. \tag{1.2}$$

Let $\mathcal{S}_{\mathcal{H}}$ be another subclass of $\mathcal{H}$ containing all univalent functions of $\mathcal{H}$ and let $\mathcal{S}_{\mathcal{H}}^0$ be a subclass of $\mathcal{H}^0$ containing all univalent functions of $\mathcal{H}^0$ in $\mathbb{D}$. Clearly, $\mathcal{S}_{\mathcal{H}}^0 \subset \mathcal{S}_{\mathcal{H}}$. Clunie and Sheil-Small [**1**] stuided the class of harmonic univalent functions. These articles [**3, 10, 11, 14, 15**] provide basic information about harmonic functions as well as some more recent studies.

2020 *Mathematics Subject Classification.* Primary 30C45, 30C50.

A starlike domain with respect to a point in it is defined to be that which contains all line segments connecting any point in it to the respected point in it. It is simply called a starlike domain if the respected point is the origin. Further, a convex domain is defined to be that which is starlike with respect to every point in it. A harmonic starlike function $f(z) \in \mathcal{H}$ is defined to be that which maps $\mathbb{D}$ to a starlike domain. The class of starlike function in $\mathcal{H}$ is denoted by $\mathcal{S}^*_{\mathcal{H}}$. Similarly, a harmonic convex function $f(z) \in \mathcal{H}$ is defined to be that which maps $\mathbb{D}$ to a convex domain, and the collection of such functions is denoted by $\mathcal{K}_{\mathcal{H}}$. A close-to-convex domain is defined to be that which complement can be written as disjoint union of non-crossing half lines. A harmonic close-to-convex function $f(z) \in \mathcal{H}$ is defined to be that which maps $\mathbb{D}$ to a close-to-convex domain, and the collection of such functions is denoted by $\mathcal{C}_{\mathcal{H}}$. Let $\mathcal{K}^0_{\mathcal{H}}$, $\mathcal{S}^{*0}_{\mathcal{H}}$ and $\mathcal{C}^0_{\mathcal{H}}$ be the subclasses of $\mathcal{K}_{\mathcal{H}}$, $\mathcal{S}^*_{\mathcal{H}}$ and $\mathcal{C}_{\mathcal{H}}$ respectively, where $f_{\overline{z}}(0) = 0$.

**1.1. Stable analytic and stable harmonic functions.** For all $|\zeta| = 1$, if the analytic functions $F_\zeta(z) = h(z) + \zeta g(z)$ are convex, starlike, close-to-convex and univalent on $\mathbb{D}$ then the function $F(z) = h(z) + g(z)$ is said to be stable analytic convex, stable analytic starlike, stable analytic close-to-convex and stable analytic univalent on $\mathbb{D}$, respectively. In similar, if the harmonic functions $f_\zeta(z) = h(z) + \zeta\overline{g(z)}$ with $|\zeta| = 1$ are convex, starlike, close-to-convex and univalent on $\mathbb{D}$ then the function $f(z) = h(z) + \overline{g(z)}$ is said to be stable harmonic convex, stable harmonic starlike, stable harmonic close-to-convex and stable harmonic univalent on $\mathbb{D}$, respectively. More information regarding stable harmonic functions is available in [**4**]. The following is one useful result of stable analytic and stable harmonic functions.

LEMMA 1.1 ([**4**]). *A sense-preserving harmonic function $f(z) = h(z) + \overline{g(z)}$ on $\mathbb{D}$ is stable harmonic univalent (respectively, convex), if and only if the analytic function $F(z) = h(z) + g(z)$ is stable analytic univalent (respectively, convex) on $\mathbb{D}$. Same result holds for $f(z)$ to be stable harmonic starlike on $\mathbb{D}$ when the functions $f_\zeta(z) = h(z) + \zeta\overline{g(z)}$ fix the origin on $\mathbb{D}$ for all $|\zeta| = 1$.*

The convolution (or Hadamard product) of two analytic functions $F_1(z) = \sum_{n=0}^{\infty} a_n z^n$ and $F_2(z) = \sum_{n=0}^{\infty} b_n z^n$ is defined by

$$(F_1 * F_2)(z) = \sum_{n=0}^{\infty} a_n\, b_n\, z^n.$$

Following the above, the convolution for two harmonic functions $f_1(z) = h_1(z) + \overline{g_1(z)}$ and $f_2(z) = h_2(z) + \overline{g_2(z)}$ in $\mathcal{H}$ is defined as

$$(f_1 * f_2)(z) = (h_1 * h_2)(z) + \overline{(g_1 * g_2)(z)}.$$

Clunie and Sheil-Small [**1**] also proved a result for the convolution between a harmonic and an analytic function which states, if $f(z)$ is harmonic convex and $F(z)$ is analytic convex, then $f(z) * (F(z) + \gamma\overline{F(z)})$ is a harmonic close-to-convex function where $|\gamma| < 1$.

Recently, Ponnusamy et al. [**13**] (see also [**5**]) have studied the following subclass of $\mathcal{H}$ :

$$\begin{aligned} C^2_{\mathcal{H}} &= \left\{f(z) = h(z) + \overline{g(z)} \in \mathcal{H} : |h'(z) - 1| < 1 - |g'(z)|\right\} \\ &\subseteq C^1_{\mathcal{H}} = \left\{f(z) = h(z) + \overline{g(z)} \in \mathcal{H} : \Re(h'(z)) > \Re(g'(z))\right\}, \end{aligned}$$

and proved that its elements are close-to-convex in $\mathbb{D}$. Motivated by this study, we define the following subclass of $\mathcal{H}^0$ :

DEFINITION 1.2. For $\lambda > 0$

$$\Omega^0_{\mathcal{H}}(\lambda) := \left\{f(z) = h(z) + \overline{g(z)} \in \mathcal{H}^0 : |h(z) - zh'(z)| < \lambda - |g(z) - zg'(z)|,\ z \in \mathbb{D}\right\}.$$

We study many examples of functions in $\Omega^0_{\mathcal{H}}(\lambda)$ in the next section. Now, we consider the subclass of $\mathcal{A}$ defined by us in 2023 [**2**].

DEFINITION 1.3. For $\lambda > 0$

$$\Omega(\lambda) := \{F(z) \in \mathcal{A} : |F(z) - zF'(z)| < \lambda,\ z \in \mathbb{D}\}.$$

The class $\Omega(\lambda)$ is non-empty as the identity function satisfies the Definition 1.3. Note that $\Omega(\lambda_1) \subseteq \Omega(\lambda_2)$ when $\lambda_1 \leq \lambda_2$, and $\Omega(1/2) = \Omega$ is the class defined by Peng and Zhong [**9**] that is closed under convolution. The functions in $\Omega$ are univalent and starlike in $\mathbb{D}$, and convex are in $\mathbb{D}_{1/2}$ (see [**7, 9**]). Obradović and Peng [**8**] gave certain sufficient conditions for functions belongs to the class $\Omega$, also Wani and Swaminathan [**16**] obtained certain sufficient conditions and subordination properties for functions belongs to the class $\Omega$. Thus, the functions in $\Omega(\lambda)$ are univalent and starlike in $\mathbb{D}$, and are convex in $\mathbb{D}_{1/2}$ for $\lambda \in (0, 1/2]$. We [**2**] observed that the function

$$F(z) = z + \frac{\lambda z^2}{2} + \frac{\lambda z^3}{4}, \quad \lambda > 0,\ z \in \mathbb{D}, \tag{1.3}$$

belongs to the class $\Omega(\lambda)$. The following results we obtained in [**2**] for the class $\Omega(\lambda)$ are being used in our main results.

LEMMA 1.4. *The radius of starlikeness of $\Omega(\lambda)$ is $1/2\lambda$. Further, if $F(z) \in \Omega(\lambda)$, then*

$$|z| - \lambda|z|^2 \leq |F(z)| \leq |z| + \lambda|z|^2, \tag{1.4}$$

$$1 - 2\lambda|z| \leq |F'(z)| \leq 1 + 2\lambda|z|. \tag{1.5}$$

*Equality occurs in both the inequalities, if and only if $F(z) = z + \lambda\eta z^2$ for $z \in \mathbb{D}$ and $z \neq 0$, where $|\eta| = 1$.*

LEMMA 1.5. *Let $F(z) \in \Omega(\lambda)$. Then $F(z)$ is univalent in $|z| < 1/2\lambda$, and convex in $|z| < 1/4\lambda$.*

LEMMA 1.6. *Let $F(z) \in \mathcal{A}$. If $|F''(z)| \leq 2\lambda$ then $F(z) \in \Omega(\lambda)$. The number $2\lambda$ is the best possible.*

LEMMA 1.7. *Let $F(z) \in \mathcal{A}$. If*

$$|z^2F''(z) + zF'(z) - F(z)| \leq 3\lambda,$$

*then $F(z) \in \Omega(\lambda)$. The number $3\lambda$ is best possible.*

LEMMA 1.8. *Let $F_1(z)$, $F_2(z) \in \Omega(\lambda)$. Then $(F_1 * F_2)(z) \in \Omega(\lambda)$, if $\lambda \leq 1$.*

In the next section, we prove our main results for class $\Omega^0_{\mathcal{H}}(\lambda)$. Here we establish a close relation between $\Omega^0_{\mathcal{H}}(\lambda)$ and $\Omega(\lambda)$. Further, we obtain some characteristic properties including radius properties, convolution, coefficient estimates and their properties, growth estimates, and convex combination for the functions in $\Omega^0_{\mathcal{H}}(\lambda)$. At last, we produce conditions for some special functions as well as harmonic univalent polynomials to belong to class $\Omega^0_{\mathcal{H}}(\lambda)$.

## 2. Main results

### 2.1. Association between $\Omega^0_{\mathcal{H}}(\lambda)$ and $\Omega(\lambda)$.

THEOREM 2.1. *For a harmonic function $f(z) = h(z) + \overline{g(z)} \in \Omega^0_{\mathcal{H}}(\lambda)$, there exist an analytic function $F_\zeta(z) = h(z)+\zeta g(z) \in \Omega(\lambda)$ and vice versa, where $|\zeta| = 1$.*

PROOF. Let $f(z) = h(z) + \overline{g(z)} \in \Omega^0_{\mathcal{H}}(\lambda)$ and $F_\zeta(z) = h(z) + \zeta g(z)$, for $z \in \mathbb{D}$, where $|\zeta| = 1$. Using Definition 1.2, and for each $\zeta$, we can write

$$\begin{aligned}\left|F_\zeta(z) - zF'_\zeta(z)\right| &= |h(z) + \zeta g(z) - zh'(z) - z\zeta g'(z)| \\ &\leq |h(z) - zh'(z)| + |g(z) - zg'(z)| < \lambda,\end{aligned}$$

for any $\lambda > 0$ and $z \in \mathbb{D}$. This implies $F_\zeta(z) \in \Omega(\lambda)$.

Conversely, let $F_\zeta(z) = h(z)+\zeta g(z) \in \Omega(\lambda)$ for any $\zeta$, where $|\zeta| < 1$. Thus, for $z \in \mathbb{D}$

$$\left|F_\zeta(z) - zF'_\zeta(z)\right| = |h(z) + \zeta g(z) - zh'(z) - z\zeta g'(z)| < \lambda.$$

Now, for an appropriate choice of $\zeta$, we obtain

$$|h(z) - zh'(z)| + |g(z) - zg'(z)| < \lambda,$$

for any $\lambda > 0$ and $z \in \mathbb{D}$. Therefore, $f(z) = h(z) + \overline{g(z)} \in \Omega^0_{\mathcal{H}}(\lambda)$. These complete the proof. □

Using Lemmas 1.4 and 1.5 for a function $F_\zeta(z) = h(z)+\zeta g(z) \in \Omega(\lambda)$ ($|\zeta| = 1$), and in the view of Theorem 2.1 along-with Lemma 1.1, we obtain the following results on radius properties of $\Omega^0_{\mathcal{H}}(\lambda)$.

COROLLARY 2.1. *Let $f(z) = h(z)+\overline{g(z)} \in \Omega^0_{\mathcal{H}}(\lambda)$. Then $f(z)$ is stable harmonic univalent (respectively, starlike) in $|z| < 1/2\lambda$. In particular, a harmonic function $f(z) \in \Omega^0_{\mathcal{H}}(\lambda)$ is univalent (respectively, starlike) in $|z| < 1/2\lambda$.*

COROLLARY 2.2. *Let $f(z) = h(z)+\overline{g(z)} \in \Omega^0_{\mathcal{H}}(\lambda)$. Then $f(z)$ is stable harmonic convex in $|z| < 1/4\lambda$. In particular, a harmonic function $f(z) \in \Omega^0_{\mathcal{H}}(\lambda)$ is convex in $|z| < 1/4\lambda$.*

COROLLARY 2.3. *Let $f(z) = h(z)+\overline{g(z)} \in \Omega^0_{\mathcal{H}}(\lambda)$. Then $f(z)$ is stable harmonic univalent (respectively, starlike) in $\mathbb{D}$, and is stable harmonic convex in $\mathbb{D}_{1/2}$, if $\lambda \leq 1/2$. In particular, a harmonic function $f(z) \in \Omega^0_{\mathcal{H}}(\lambda)$ is univalent (respectively, starlike) in $\mathbb{D}$, and is convex in $\mathbb{D}_{1/2}$, when $\lambda \leq 1/2$.*

Using Lemma 1.6 for a function $F_\zeta(z) = h(z) + \zeta g(z) \in \mathcal{A}$, where $|\zeta| = 1$, and Theorem 2.1, we yield the following result.

COROLLARY 2.4. *Let $F_\zeta(z) = h(z) + \zeta g(z) \in \mathcal{A}$, where $|\zeta| = 1$. If $|F''_\zeta(z)| \leq 2\lambda$ then $f(z) = h(z) + \overline{g(z)} \in \Omega^0_{\mathcal{H}}(\lambda)$. The number $2\lambda$ is the best possible.*

In similar, using Lemma 1.7 for a function $F_\zeta(z) = h(z) + \zeta g(z) \in \mathcal{A}$, where $|\zeta| = 1$, and Theorem 2.1, we yield the following result.

COROLLARY 2.5. *Let $F_\zeta(z) = h(z) + \zeta g(z) \in \mathcal{A}$, where $|\zeta| = 1$. If*

$$|z^2 F_\zeta''(z) + zF_\zeta'(z) - F_\zeta(z)| \leq 3\lambda,$$

*then $f(z) = h(z) + \overline{g(z)} \in \Omega_{\mathcal{H}}^0(\lambda)$. The number $3\lambda$ is best possible.*

We yield the following result for convolution of functions in $\Omega_{\mathcal{H}}^0(\lambda)$ using Lemma 1.8 and Theorem 2.1.

COROLLARY 2.6. *Let $f_1(z),\ f_2(z) \in \Omega_{\mathcal{H}}^0(\lambda)$. Then $(f_1 * f_2)(z) \in \Omega_{\mathcal{H}}^0(\lambda)$, if $\lambda \leq 1$.*

REMARK 2.2. Following Corollary 2.6, if $f_2(z) = F(z) + \zeta\overline{F(z)}$, where $F(z)$ is a convex function in $\mathcal{A}$ for $|\zeta| = 1$, then $f_1(z) * (F(z) + \zeta\overline{F(z)}) \in \Omega_{\mathcal{H}}^0(\lambda)$, for $\lambda \leq 1$.

**2.2. Coefficient estimates and their properties.**

THEOREM 2.3. *For any $f(z) = h(z) + \overline{g(z)} \in \Omega_{\mathcal{H}}^0(\lambda)$ with the form as in (1.2) of $h(z)$ and $g(z)$, the following coefficient bounds hold true.*

$$|\alpha_n| \leq \frac{\lambda}{n-1} \quad \text{and} \quad |\beta_n| \leq \frac{\lambda}{n-1},$$

*for $n \in \mathbb{N} \setminus \{1\}$. Sharpness of both bounds holds.*

PROOF. Let $f(z) \in \Omega_{\mathcal{H}}^0(\lambda)$. Note that, $zh'(z) - h(z)$ is analytic in $\mathbb{D}$ and from (1.2), we obtain

$$zh'(z) - h(z) = \sum_{n=2}^{\infty} (n-1)\alpha_n z^n. \tag{2.1}$$

Using Cauchy's integral formula, for $z = re^{i\theta} \in \mathbb{D}$, we obtain

$$(n-1)|\alpha_n| = \left| \frac{1}{2\pi i} \int_{|z|=r} \frac{zh'(z) - h(z)}{z^{n+1}} dz \right| \leq \frac{1}{2\pi} \int_0^{2\pi} \frac{|re^{i\theta}h'(re^{i\theta}) - h(re^{i\theta})|}{r^n} d\theta,$$

for $n \in \mathbb{N} \setminus \{1\}$. Now using Definition 1.2, for $n \in \mathbb{N} \setminus \{1\}$

$$\begin{aligned}(n-1)|\alpha_n|r^n &< \frac{1}{2\pi} \int_0^{2\pi} \left(\lambda - |re^{i\theta}g'(re^{i\theta}) - g(re^{i\theta})|\right) d\theta \\ &< \lambda - \left| \frac{1}{2\pi} \int_0^{2\pi} \left(re^{i\theta}g'(re^{i\theta}) - g(re^{i\theta})\right) d\theta \right| = \lambda.\end{aligned}$$

Thus, for $n \in \mathbb{N} \setminus \{1\}$

$$|\alpha_n| < \frac{\lambda}{n-1},$$

as $r \to 1^-$. In similar, considering $zg'(z) - g(z)$ in place of $zh'(z) - h(z)$ and vice versa following from (2.1), for $n \in \mathbb{N} \setminus \{1\}$ we obtain

$$|\beta_n| < \frac{\lambda}{n-1}.$$

Sharpness of $|\alpha_n|$ holds for the function $f_\alpha(z)$ in $\Omega^0_{\mathcal{H}}(\lambda)$, where

$$f_\alpha(z) = z + \frac{\lambda}{n-1} z^n,$$

where $h(z) = z + \dfrac{\lambda}{n-1} z^n$ and $g(z) = 0$, for $n \in \mathbb{N} \setminus \{1\}$. Further, sharpness of $|\beta_n|$ holds for the function $f_\beta(z)$ in $\Omega^0_{\mathcal{H}}(\lambda)$, where

$$f_\beta(z) = z - \frac{\lambda}{n-1} \overline{z}^n,$$

where $h(z) = z$ and $g(z) = \dfrac{-\lambda}{n-1} z^n$, for $n \in \mathbb{N} \setminus \{1\}$. This completes the proof. $\square$

The following result provides a sufficient condition for a function to belong to class $\Omega^0_{\mathcal{H}}(\lambda)$.

THEOREM 2.4. *For any $f(z) = h(z) + \overline{g(z)} \in \mathcal{H}^0$, it also belongs to $\Omega^0_{\mathcal{H}}(\lambda)$ if*

$$\sum_{n=2}^{\infty} (n-1)(|\alpha_n| + |\beta_n|) < \lambda. \tag{2.2}$$

PROOF. Let $f(z) \in \mathcal{H}^0$. Now

$$\begin{aligned} |zh'(z) - h(z)| &\le \sum_{n=2}^{\infty} (n-1)|\alpha_n||z|^n \le \sum_{n=2}^{\infty} (n-1)|\alpha_n| \\ &< \lambda - \sum_{n=2}^{\infty} (n-1)|\beta_n| \le \lambda - \sum_{n=2}^{\infty} (n-1)|\beta_n||z|^n \\ &= \lambda - |zg'(z) - g(z)|, \end{aligned}$$

which is true if (2.2) holds true. Thus, $f(z) \in \Omega^0_{\mathcal{H}}(\lambda)$, if and only if (2.2) holds true. This completes the proof. $\square$

**2.3. Growth estimates, convex combination, and characteristic properties.**

THEOREM 2.5. *For any $f(z) = h(z) + \overline{g(z)} \in \Omega^0_{\mathcal{H}}(\lambda)$, we have*

$$|z| - \lambda|z|^2 \le |f(z)| \le |z| + \lambda|z|^2, \tag{2.3}$$

$$1 - 2\lambda|z| \le |f'(z)| \le 1 + 2\lambda|z|. \tag{2.4}$$

*Here equality occurs for the function $f_3(z) = z + \lambda\eta z^2 \in \Omega^0_{\mathcal{H}}(\lambda)$, where $z \in \mathbb{D}$, $z \neq 0$ and $|\eta| = 1$.*

PROOF. Let $f(z) = h(z) + \overline{g(z)} \in \Omega^0_{\mathcal{H}}(\lambda)$ and hence using Theorem 2.1, we have $F_\zeta(z) = h(z) + \zeta g(z) \in \Omega(\lambda)$, for all $|\zeta| = 1$. Using Lemma 1.4, we obtain

$$\begin{aligned} |z| - \lambda|z|^2 &\le |h(z) + \zeta g(z)| \le |z| + \lambda|z|^2, \\ 1 - 2\lambda|z| &\le |h'(z) + \zeta g'(z)| \le 1 + 2\lambda|z|. \end{aligned}$$

Now, for appropriate choices of $\zeta$, we can write

$$|z| - \lambda|z|^2 \le |h(z)| - |g(z)| \quad \text{and} \quad |h(z)| + |g(z)| \le |z| + \lambda|z|^2, \tag{2.5}$$

$$1 - 2\lambda|z| \le |h'(z)| - |g'(z)| \quad \text{and} \quad |h'(z)| + |g'(z)| \le 1 + 2\lambda|z|, \tag{2.6}$$

which further implies

$$|z| - \lambda|z|^2 \le |h(z) + \overline{g(z)}| \quad \text{and} \quad |h(z) + \overline{g(z)}| \le |z| + \lambda|z|^2,$$
$$1 - 2\lambda|z| \le |h'(z) + \overline{g'(z)}| \quad \text{and} \quad |h'(z) + \overline{g'(z)}| \le 1 + 2\lambda|z|,$$

which is equivalent to (2.3) and (2.4), respectively. As the function $f_3(z) = z + \lambda\eta z^2 \in \Omega^0_{\mathcal{H}}(\lambda)$, for $z \in \mathbb{D}$, $z \neq 0$ and $|\eta| = 1$, equality occurs for appropriate choices of $\eta$. This completes the proof. □

THEOREM 2.6. *The class $\Omega^0_{\mathcal{H}}(\lambda)$ is closed under convex combination.*

PROOF. Let $f_k(z) = h_k(z) + \overline{g_k(z)} \in \Omega^0_{\mathcal{H}}(\lambda)$ for $k = 1, 2, 3, \ldots, n$, and $\sum_{k=1}^n t_k = 1$, where $t_k \in [0, 1]$. Note that

$$f(z) = \sum_{k=1}^{n} t_k f_k(z) = h(z) + g(z),$$

is a convex combination of $f_k$'s, where $h(z) = \sum_{k=1}^n t_k h_k(z)$ and $g(z) = \sum_{k=1}^n t_k g_k(z)$. Thus, both $h(z)$ and $g(z)$ are in class $\mathcal{A}$. Now

$$\begin{aligned}|h(z) - zh'(z)| &= \left|\sum_{k=1}^{n} t_k h_k(z) - z\sum_{k=1}^{n} t_k h_k'(z)\right| \le \sum_{k=1}^{n} t_k\,|h_k(z) - zh_k'(z)| \\ &< \lambda - \left|\sum_{k=1}^{n} t_k g_k(z) - zg_k'(z)\right| = \lambda - |g(z) - zg'(z)|\,.\end{aligned}$$

Thus, $f(z) \in \Omega^0_{\mathcal{H}}(\lambda)$. □

THEOREM 2.7. *For any $f(z) = h(z) + \overline{g(z)} \in \Omega^0_{\mathcal{H}}(\lambda)$, we have*

$$J_f(z) \le (1 + 2\lambda|z|)^2.$$

*Here equality occurs for the function $f_3(z) = z + \lambda\eta z^2$, where $z \in \mathbb{D}$, $z \neq 0$, and $|\eta| = 1$.*

PROOF. Let $f(z) \in \Omega^0_{\mathcal{H}}(\lambda)$. Now

$$|h(z) - zh'(z)| < \lambda - |g(z) - zg'(z)| \le \lambda.$$

Thus, $h(z) \in \Omega(\lambda)$. Hence using Lemma 1.4, we obtain

$$|h'(z)| \le 1 + 2\lambda|z|.$$

Now

$$J_f(z) = |h'(z)|^2 - |g'(z)|^2 \le |h'(z)|^2 \le (1 + 2\lambda|z|)^2.$$

Consider the function of $\Omega^0_{\mathcal{H}}(\lambda)$, $f_3(z) = h_3(z) + \overline{g_3(z)} = z + \lambda\eta z^2$, for $z \in \mathbb{D}$, $z \neq 0$, and $|\eta| = 1$. Clearly, $h_3(z) = z + \lambda\eta z^2$. This implies

$$|h_3'(z)| = |1 + 2\lambda\eta z| = 1 + 2\lambda|z|,$$

for an appropriate choice of $\eta$. This completes the proof. □

THEOREM 2.8. *Each $f(z) = h(z) + \overline{g(z)} \in \Omega^0_{\mathcal{H}}(\lambda)$ maps $\mathbb{D}$ onto a domain bounded by a rectifiable Jordan curve, when $\lambda \le 1/2$.*

PROOF. Let $f(z) \in \Omega^0_{\mathcal{H}}(\lambda)$ and let $z_1, z_2 \in \mathbb{D}$ such that $z_1 \neq z_2$ and $|z_1| \geq |z_2|$. For the line segment $[z_1, z_2]$,

$$|f(z_1) - f(z_2)| = \left| \int_{[z_1,z_2]} \frac{\partial f}{\partial \tau} d\tau + \frac{\partial f}{\partial \overline{\tau}} d\overline{\tau} \right| \tag{2.7}$$
$$\leq \int_{[z_1,z_2]} (\, |h'(\tau)| + |g'(\tau)| \,) \, |d\tau|.$$

Using (2.6) (from Theorem 2.5) in (2.7), we obtain

$$\begin{aligned} |f(z_1) - f(z_2)| &\leq \int_{|z_2|}^{|z_1|} (1 + 2\lambda\kappa) d\kappa \\ &= (|z_1| - |z_2|) + \lambda(|z_1|^2 - |z_2|^2) \\ &= (|z_1| - |z_2|)\,(1 + \lambda(|z_1| + |z_2|)) \\ &\leq (|z_1| - |z_2|)\,(1 + 2\lambda) \leq (1 + 2\lambda)|z_1 - z_2| < \epsilon, \end{aligned} \tag{2.8}$$

for any $\epsilon > 0$, such that $|z_1 - z_2| < \delta := \dfrac{\epsilon}{1 + 2\lambda}$, for any $\delta > 0$. Therefore, $f(z)$ is uniformly continuous on $\mathbb{D}$ and consequently can be extended continuously on the boundary of $\mathbb{D}$.

Let $C$ be a curve defined by $f(e^{i\theta})$ for $\theta \in [0, 2\pi]$, and let $0 = \theta_0 < \theta_1 < \cdots < \theta_n = 2\pi$ be a partition of $[0, 2\pi]$. Now using (2.8), we obtain

$$\begin{aligned} \sum_{k=1}^{n} |f(e^{i\theta_k}) - f(e^{i\theta_{k-1}})| &\leq (1 + 2\lambda) \sum_{k=1}^{n} |e^{i\theta_k} - e^{i\theta_{k-1}}| \\ &\leq (1 + 2\lambda)\left(|e^{i\theta_0}| + 2(|e^{i\theta_1}| + \cdots + |e^{i\theta_{n-1}}|) + |e^{i\theta_n}|\right) \\ &\leq (1 + 2\lambda)2n. \end{aligned}$$

Thus, $C$ is rectifiable for a finite $\lambda$.

Following the proof of [**7**, Lemma 3.2], we have $\Re(F'_\zeta(z)) \geq 1 - |z| > 0$, $z \in \mathbb{D}$ for $F_\zeta(z) \in \Omega$. Thus, for $F_\zeta(z) \in \Omega(\lambda)$, $\Re(F'_\zeta(z)) > 0$ for all $z \in \mathbb{D}$ if $\lambda \leq 1/2$. Therefore by the association (Theorem 2.1) and from the proof of [**6**, Theorem 3], it is concluded that $F_\zeta(z)$ is univalent on the boundary of $\mathbb{D}$. Now, let us consider the points, $z_1 \neq z_2$ and let $f(z_1) = f(z_2)$. The latter implies

$$h(z_1) - h(z_2) = \overline{g(z_1) - g(z_2)}.$$

CASE I. Let $h(z_1) = h(z_2)$. Then $g(z_1) = g(z_2)$, and since $F_\zeta(z)$ is univalent, $z_1 = z_2$.

CASE II. Let $h(z_1) \neq h(z_2)$, and let $\theta = \arg\{h(z_1) - h(z_2)\} \in [0, 2\pi)$. Then,

$$e^{-i\theta}\,(h(z_1) - h(z_2)) = \overline{e^{i\theta}\,(g(z_1) - g(z_2))} \in \mathbb{R}^+.$$

Taking conjugates on both sides, we obtain

$$(h(z_1) - h(z_2)) = e^{2i\theta}\,(g(z_1) - g(z_2)),$$

that implies $F_\zeta(z_1) = F_\zeta(z_2)$, where $\zeta = e^{2i\theta}$. Hence, $z_1 = z_2$. Thus, $f(z)$ is univalent on the boundary of $\mathbb{D}$. These all complete the proof. $\square$

REMARK 2.9. From (2.8) of Theorem 2.8 and for $\lambda \leq 1/2$, it can be seen that functions in $\Omega^0_{\mathcal{H}}(\lambda)$ map $\mathbb{D}$ onto a domain which is contained in a disk of radius 2 with center at the origin. Further, the area $f(\mathbb{D})$ is bounded by $4n$, where $n \in \mathbb{N}$ is finite.

**2.4. Producing conditions of belongingness to $\Omega^0_{\mathcal{H}}(\lambda)$.** Here, we produce conditions for functions to belong the class $\Omega^0_{\mathcal{H}}(\lambda)$ whose co-analytic parts include Gaussian hypergeometric functions, defined by

$$ {}_2F_1(a,b;c;z) = F(a,b;c;z) = \sum_{n=0}^{\infty} \frac{(a)_n(b)_n}{n!\,(c)_n} z^n, \quad z \in \mathbb{D}, \tag{2.9} $$

for $a$, $b$, $c \in \mathbb{C}$, and $c \notin \mathbb{Z}^- \cup \{0\}$. Here, $(x)_n = x(x+1)(x+2)\ldots(x+n-1)$, $n \in \mathbb{N}$ is the well-known Pochhammer symbol, with $(x)_0 = 1$. The series (2.9) converges absolutely in $\mathbb{D}$, and converges on $\overline{\mathbb{D}}$ if $\Re(c-a-b) > 0$. The well-known Gauss formula [**17**] is defined by

$$ F(a,b;c;1) = \sum_{n=0}^{\infty} \frac{(a)_n(b)_n}{n!\,(c)_n} = \frac{\Gamma(c)\Gamma(c-a-b)}{\Gamma(c-a)\Gamma(c-b)}, \tag{2.10} $$

for $\Re(c-a-b) > 0$ and the following lemma are useful for our next result.

LEMMA 2.10 ([**12**]). *For any $a$, $b$, $c \in \mathbb{R}^+$, if $c-a-b-1>0$, then*

$$ \sum_{n=0}^{\infty} \frac{(n+1)(a)_n(b_n)}{n!\,(c)_n} = \left(\frac{ab}{c-a-b-1} + 1\right) F(a,b;c;1). $$

THEOREM 2.11. *For any $a$, $b$, $c \in \mathbb{R}^+$ and $\eta \in \mathbb{D}$, satisfying $c-a-b-1>0$, the following results hold.*

(i) *$f_4(z) \in \Omega^0_{\mathcal{H}}(\lambda)$, if*

$$ F(a,b;c;1) < \frac{\lambda}{|\eta|}. \tag{2.11} $$

(ii) *$f_5(z) \in \Omega^0_{\mathcal{H}}(\lambda)$, if*

$$ \left(\frac{ab}{c-a-b-1}\right) F(a,b;c;1) < \frac{\lambda}{|\eta|}. \tag{2.12} $$

(iii) *$f_6(z) \in \Omega^0_{\mathcal{H}}(\lambda)$, if*

$$ \left(\frac{ab}{c-a-b-1} + 1\right) F(a,b;c;1) < \frac{\lambda}{|\eta|}. \tag{2.13} $$

*Here,*

$$\begin{aligned} f_4(z) &= z+\overline{\eta z\int_0^z F(a,b;c;\tau)d\tau},\ f_5(z) \\ &= z+\overline{\eta z\,(F(a,b;c;z)-1)},\ f_6(z) \\ &= z+\overline{\eta z^2 F(a,b;c;z)}. \end{aligned}$$

PROOF. (i) Let $f_4(z) = z+\overline{\eta z\int_0^z F(a,b;c;\tau)d\tau} = z+\overline{\sum_{n=2}^{\infty} I_n\, z^n}$, where

$$I_n = \eta\,\frac{(a)_{n-2}(b)_{n-2}}{(c)_{n-2}(n-1)!}, \qquad n\in\mathbb{N}\setminus\{1\}.$$

Now

$$\begin{aligned} \sum_{n=2}^{\infty}(n-1)|I_n| &= |\eta|\sum_{n=2}^{\infty}\frac{(a)_{n-2}(b)_{n-2}}{(c)_{n-2}(n-2)!} \\ &= |\eta|\sum_{n=0}^{\infty}\frac{(a)_n(b)_n}{n!\,(c)_n} \\ &= |\eta|\,F(a,b;c;1)<\lambda, \end{aligned}$$

if (2.11) holds true. Therefore, from Theorem 2.4, we conclude that $f_4(z)\in\Omega^0_{\mathcal{H}}(\lambda)$.

(ii) Now let $f_5(z) = z+\overline{\eta z\,(F(a,b;c;z)-1)} = z+\overline{\sum_{n=2}^{\infty} J_n\, z^n}$, where

$$J_n = \eta\,\frac{(a)_{n-1}(b)_{n-1}}{(c)_{n-1}(n-1)!}, \qquad n\in\mathbb{N}\setminus\{1\}.$$

Now

$$\begin{aligned} \sum_{n=2}^{\infty}(n-1)|J_n| &= |\eta|\sum_{n=2}^{\infty}\frac{(a)_{n-1}(b)_{n-1}}{(c)_{n-1}(n-2)!} \\ &= |\eta|\sum_{n=0}^{\infty}\frac{(a)_{n+1}(b)_{n+1}}{n!\,(c)_{n+1}} \\ &= |\eta|\,\frac{ab}{c}\sum_{n=0}^{\infty}\frac{(a+1)_n(b+1)_n}{n!\,(c+1)_n} \\ &= |\eta|\,\frac{ab}{c}\,\frac{\Gamma(c+1)\Gamma(c-a-b-1)}{\Gamma(c-a)\Gamma(c-b)} \\ &= |\eta|\left(\frac{ab}{c-a-b-1}\right)F(a,b;c;1)<\lambda, \end{aligned}$$

if (2.12) holds true. Therefore, from Theorem 2.4, we conclude that $f_5(z)\in\Omega^0_{\mathcal{H}}(\lambda)$.

(iii) Further, let $f_6(z) = z+\overline{\eta z^2 F(a,b;c;z)} = z+\overline{\sum_{n=2}^{\infty} H_n\, z^n}$, where

$$H_n = \eta\,\frac{(a)_{n-2}(b)_{n-2}}{(c)_{n-2}(n-2)!}, \qquad n\in\mathbb{N}\setminus\{1\}.$$

Using Lemma 2.10, we obtain

$$\begin{aligned}\sum_{n=2}^{\infty}(n-1)|H_n| &= |\eta|\sum_{n=2}^{\infty}\frac{(n-1)(a)_{n-2}(b)_{n-2}}{(c)_{n-2}(n-2)!}\\ &= |\eta|\sum_{n=0}^{\infty}\frac{(n+1)(a)_n(b)_n}{n!\,(c)_n}\\ &= |\eta|\left(\frac{ab}{c-a-b-1}+1\right)F(a,b;c;1)<\lambda,\end{aligned}$$

if (2.13) holds true. Therefore, from Theorem 2.4, we conclude that $f_6(z)\in\Omega^0_{\mathcal{H}}(\lambda)$.

□

Now we produce conditions for harmonic univalent polynomials to belong the class $\Omega^0_{\mathcal{H}}(\lambda)$.

THEOREM 2.12. *For any $s\in\mathbb{N}$ and $c\in\mathbb{R}^+$, the following results hold*

(i) $p_1(z)\in\Omega^0_{\mathcal{H}}(\lambda)$, *if*

$$\Gamma(c)\Gamma(c+2s)<\lambda\left(\Gamma(c+m)\right)^2. \tag{2.14}$$

(ii) $p_2(z)\in\Omega^0_{\mathcal{H}}(\lambda)$, *if*

$$s^2\Gamma(c)\Gamma(c+2s)<\lambda(c+2s-1)\left(\Gamma(c+m)\right)^2. \tag{2.15}$$

(iii) $p_3(z)\in\Omega^0_{\mathcal{H}}(\lambda)$, *if*

$$(c+s^2+2s-1)\Gamma(c)\Gamma(c+2s)<\lambda(c+2s-1)\left(\Gamma(c+m)\right)^2. \tag{2.16}$$

*Here, $p_1(z)$, $p_2(z)$, and $p_3(z)$ are harmonic univalent polynomials, given by*

$$\begin{cases} p_1(z)=z+\eta\overline{\sum_{n=0}^{s}\binom{s}{n}\frac{(s-n+1)_n}{(c)_n}\frac{z^{n+2}}{n+1}},\\ p_2(z)=z+\eta\overline{\sum_{n=0}^{s}\binom{s}{n}\frac{(s-n+1)_n}{(c)_n}z^{n+1}},\\ p_3(z)=z+\eta\overline{\sum_{n=0}^{s}\binom{s}{n}\frac{(s-n+1)_n}{(c)_n}z^{n+2}}.\end{cases} \tag{2.17}$$

PROOF. For any $\mu\in\mathbb{C}\setminus\mathbb{Z}^-$ and $n\in\mathbb{N}\cap\{0\}$, we can write

$$\frac{(-1)^n(-\mu)_n}{n!}=\binom{\mu}{n}=\frac{\Gamma(\mu+1)}{n!\Gamma(\mu-n+1)}.$$

But when this $\mu=s\in\mathbb{N}$, where $s\geq n$, we can write

$$(-s)_n=\frac{(-1)^n s!}{(s-n)!}.$$

Using this relation on $f_4(z)$, $f_5(z)$, and $f_6(z)$ of Theorem 2.11, we obtain the harmonic univalent polynomials $p_1(z)$, $p_2(z)$, and $p_3(z)$, respectively as defined on (2.17). Now putting $a=b=-s$ in the conditions (2.11), (2.12) and (2.13), we obtain respective belongingness conditions (2.14), (2.15) and (2.16), of $p_1(z)$, $p_2(z)$, and $p_3(z)$. This completes the proof. □

## References


[1] J. Clunie and T. Sheil-Small, *Harmonic univalent functions*, Ann. Acad. Sci. Fenn. Ser. A I Math. **9** (1984), 3–25, DOI 10.5186/aasfm.1984.0905. MR752388

[2] P. P. Dash and J. K. Prajapat, On certain analytic functions defined by differential inequality, *Pre-print*, 2023.

[3] P. Duren, *Harmonic mappings in the plane*, Cambridge Tracts in Mathematics, vol. 156, Cambridge University Press, Cambridge, 2004, DOI 10.1017/CBO9780511546600. MR2048384

[4] R. Hernández and M. J. Martín, *Stable geometric properties of analytic and harmonic functions*, Math. Proc. Cambridge Philos. Soc. **155** (2013), no. 2, 343–359, DOI 10.1017/S0305004113000340. MR3091525

[5] D. Kalaj, S. Ponnusamy, and M. Vuorinen, *Radius of close-to-convexity and fully starlikeness of harmonic mappings*, Complex Var. Elliptic Equ. **59** (2014), no. 4, 539–552, DOI 10.1080/17476933.2012.759565. MR3177008

[6] T. H. MacGregor, *A class of univalent functions*, Proc. Amer. Math. Soc. **15** (1964), 311–317, DOI 10.2307/2034061. MR158985

[7] H. Mahzoon and R. Kargar, *Further results for a subclass of univalent functions related with differential inequality*, Indian J. Pure Appl. Math. **52** (2021), no. 1, 205–215, DOI 10.1007/s13226-021-00071-2. MR4285969

[8] M. Obradović and Z. Peng, *Some new results for certain classes of univalent functions*, Bull. Malays. Math. Sci. Soc. **41** (2018), no. 3, 1623–1628, DOI 10.1007/s40840-017-0546-0. MR3818438

[9] Z. Peng and G. Zhong, *Some properties for certain classes of univalent functions defined by differential inequalities*, Acta Math. Sci. Ser. B (Engl. Ed.) **37** (2017), no. 1, 69–78, DOI 10.1016/S0252-9602(16)30116-3. MR3581527

[10] S. Ponnusamy, J. K. Prajapat, and A. Sairam Kaliraj, *Uniformly starlike and uniformly convex harmonic mappings*, J. Anal. **23** (2015), 121–129. MR3491652

[11] S. Ponnusamy and A. Rasila, *Planar harmonic and quasiregular mappings*, Topics in modern function theory, Ramanujan Math. Soc. Lect. Notes Ser., vol. 19, Ramanujan Math. Soc., Mysore, 2013, pp. 267–333. MR3220953

[12] S. Ponnusamy and F. Rønning, *Starlikeness properties for convolutions involving hypergeometric series*, Ann. Univ. Mariae Curie-Skłodowska Sect. A **52** (1998), no. 1, 141–155. Dedicated to Professor Eligiusz J. Złotkiewicz. MR1665537

[13] S. Ponnusamy, H. Yamamoto, and H. Yanagihara, *Variability regions for certain families of harmonic univalent mappings*, Complex Var. Elliptic Equ. **58** (2013), no. 1, 23–34, DOI 10.1080/17476933.2010.551200. MR3011974

[14] J. K. Prajapat and M. Mathi, *Injectivity of harmonic mappings with fixed analytic part*, Proc. Nat. Acad. Sci. India Sect. A **93** (2023), no. 1, 23–29, DOI 10.1007/s40010-022-00781-0. MR4544311

[15] J. K. Prajapat, M. Manivannan, and S. Maharana, *Harmonic mappings with analytic part convex in one direction*, J. Anal. **28** (2020), no. 4, 961–972, DOI 10.1007/s41478-020-00226-0. MR4181909

[16] A. Swaminathan and L. A. Wani, *Sufficient conditions and radii problems for a starlike class involving a differential inequality*, Bull. Korean Math. Soc. **57** (2020), no. 6, 1409–1426, DOI 10.4134/BKMS.b191074. MR4180186

[17] N. M. Temme, *Special functions*, A Wiley-Interscience Publication, John Wiley & Sons, Inc., New York, 1996. An introduction to the classical functions of mathematical physics, DOI 10.1002/9781118032572. MR1376370



Department of Mathematics, Central University of Rajasthan, Kishangarh-305817, India

*Email address*: `prachiprajnadash@gmail.com`

Department of Mathematics, Central University of Rajasthan, Kishangarh-305817, India

*Email address*: `jkprajapat@curaj.ac.in`

Contemporary Mathematics
Volume **820**, 2025
https://doi.org/10.1090/conm/820/16408

# Some geometrical applications of tame quasiconformal motions

Xinlong Dong

ABSTRACT. The concept of *tame quasiconformal motions* was first introduced in a paper by Y. Jiang, S. Mitra, H. Shiga, and Z. Wang [Tohoku Math. J. (2) 70 (2018), no. 4, 607–631]. In this paper, we study some geometrical applications of tame quasiconformal motions.

## 1. Introduction

The concept of quasiconformal motions was first introduced in the paper by Sullivan and Thurston ([**11**]). Theorem 3 of that paper claimed that every quasiconformal motion of any set over an interval can be extended to the Riemann sphere. In the paper [**7**] the authors presented a counterexample to Theorem 3 of [**11**]. In that same paper ([**7**]), they introduced a new concept, called *tame quasiconformal motions*, and showed that Theorem 3 of [**11**] holds for tame quasiconformal motions over any simply connected Hausdorff space. The crucial idea was to show that tame quasiconformal motions have a certain "universal property" that quasiconformal motions (in the sense of Sullivan and Thurston) do not have.

In this paper, we study some geometrical properties of tame quasiconformal motions.

## 2. Statements of the main theorems

Throughout this paper, $\mathbb{C}$ denotes the complex plane, $\widehat{\mathbb{C}} = \mathbb{C} \cup \{\infty\}$ denotes the Riemann sphere, $D$ is a region in $\widehat{\mathbb{C}}$, $E$ is an infinite closed subset of $\widehat{\mathbb{C}}$ and $X$ is a connected Hausdorff space with a base point $x_0$. We shall always assume that $E$ contains the points 0, 1, and $\infty$.

We state the main theorems of this paper. The basic definitions and relevant facts are given in §3.

THEOREM A. *Let $D$ be a region in $\widehat{\mathbb{C}}$ with boundary $E$, and let $\phi$ be a tame quasiconformal motion of $E$ over the simply connected Hausdorff space $X$ with a basepoint $x_0$. For any closed Poincaré geodesic $\gamma$ in $D$, the hyperbolic length of the closed Poincaré geodesic $\phi_x(\gamma)$ in $\phi_x(D)$ is a continuous function of $x$ in $X$.*

2020 *Mathematics Subject Classification.* Primary 32G15, 32Q45; Secondary 30F60.
*Key words and phrases.* Teichmüller spaces, Tame quasiconformal motions.

THEOREM B. *Let $D$ be a region in $\widehat{\mathbb{C}}$ with boundary $E$, and let $\phi$ be a tame quasiconformal motion of $E$ over the simply connected Hausdorff space $X$ with a basepoint $x_0$. For any family $\Gamma$ of curves in $D$, the extremal length of the family $\phi_x(\Gamma)$ in $\phi_x(D)$ is a continuous function of $x$ in $X$.*

THEOREM C. *Let $E$ be the boundary of the non-degenerate ring domain $D$. If $\phi$ is a tame quasiconformal motion of $E$ over the connected Hausdorff space $X$ with a basepoint $x_0$, then the set $\phi_x(D)$ is a non-degenerate ring domain for each $x$ in $X$, and the function $x \mapsto mod(\phi_x(D))$ is continuous in $X$.*

THEOREM D. *Let $E$ be the boundary of the quadrilateral $(D; a, b, c, d)$, and let $\phi$ be a tame quasiconformal motion of $E$ over the connected Hausdorff space $X$ with a basepoint $x_0$. The function $x \mapsto mod(\phi_x(D); \phi(x,a), \phi(x,b), \phi(x,c), \phi(x,d))$ is continuous in $X$.*

## 3. Preliminaries

In this section, we summarize the definitions and some basic properties of the tame quasiconformal motions.

**3.1. Basic definitions.** We begin with some definitions.

DEFINITION 3.1. Let $E$ be a subset of $\widehat{\mathbb{C}}$, and let $X$ be a connected Hausdorff space with a basepoint $x_0$. A motion of $E$ over $X$ is a map $\phi : X \times E \to \widehat{\mathbb{C}}$ satisfying

(i) $\phi(x_0, z) = z$ for all $z \in E$, and

(ii) for all $x \in X$, the map $\phi(x, \cdot) : E \to \widehat{\mathbb{C}}$ is injective.

We say that $X$ is the parameter space of the motion $\phi$.

We will assume that $0, 1$, and $\infty$ belong to $E$ and that the motion is normalized; i.e. $0, 1$, and $\infty$ are fixed points of the map $\phi(x, \cdot)$ for every $x$ in $X$.

Let $E \subset \widehat{E}$, $\phi : X \times E \to \widehat{\mathbb{C}}$ and $\widetilde{\phi} : X \times \widehat{E} \to \widehat{\mathbb{C}}$ be two motions. We say that $\widetilde{\phi}$ extends $\phi$ if $\widetilde{\phi}(x, z) = \phi(x, z)$ for all $(x, z) \in X \times E$.

For any motion $\phi : X \times E \to \widehat{\mathbb{C}}$, $x$ in $X$, and any quadruplet of distinct points $a, b, c, d$ in $E$, let $\phi_x(a, b, c, d)$ denote the cross-ratio of the values $\phi(x, a)$, $\phi(x, b)$, $\phi(x, c)$, and $\phi(x, d)$. We will often write $\phi(x, z)$ as $\phi_x(z)$ for $x$ in $X$ and $z$ in $E$. So we have:

$$\phi_x(a, b, c, d) = \frac{(\phi_x(a) - \phi_x(c))(\phi_x(b) - \phi_x(d))}{(\phi_x(a) - \phi_x(d))(\phi_x(b) - \phi_x(c))}$$

for each $x$ in $X$.

It is obvious that condition (ii) in Definition 3.1 holds if and only if $\phi_x(a, b, c, d)$ is a well-defined point in the thrice-punctured sphere $\widehat{\mathbb{C}} \setminus \{0, 1, \infty\}$ for all $x \in X$ and all quadruplets $a, b, c, d$ of distinct points in $E$.

Let $\rho$ be the Poincaré distance on $\widehat{\mathbb{C}} \setminus \{0, 1, \infty\}$.

DEFINITION 3.2. A quasiconformal motion is a motion $\phi : X \times E \to \widehat{\mathbb{C}}$ of $E$ over $X$ with the following additional property:

(iii) Given any $x$ in $X$ and any $\epsilon > 0$, there exists a neighborhood $U_x$ of $x$ such that for any quadruplet of distinct points $a, b, c, d$ in $E$, we have

$$\rho(\phi_y(a, b, c, d), \phi_{y'}(a, b, c, d)) < \epsilon$$

for all $y, y' \in U_x$.

DEFINITION 3.3. Let $W$ be a connected complex manifold that is modeled on a complex Banach space with a basepoint $x_0$. A holomorphic motion is a motion $\phi : W \times E \to \widehat{\mathbb{C}}$ of $E$ over $W$ such that the map $\phi(\cdot, z) : W \to \widehat{\mathbb{C}}$ is holomorphic for each $z$ in $E$.

REMARK 3.4. Suppose that $\phi : W \times E \to \widehat{\mathbb{C}}$ is a holomorphic motion. Then $\phi$ is also a quasiconformal motion.

See Remark 1 in [**7**].

DEFINITION 3.5. A tame quasiconformal motion is a motion $\phi : X \times E \to \widehat{\mathbb{C}}$ of $E$ over $X$ with the additional property:

(iii) Given any $x \in X$, there exists a quasiconformal map $w : \hat{\mathbb{C}} \to \hat{\mathbb{C}}$, a neighborhood $N(x)$ with a basepoint $x$, and a quasiconformal motion $\psi : N(x) \times \hat{\mathbb{C}} \to \hat{\mathbb{C}}$ over $N(x)$ such that $\phi(y, z) = \psi(y, w(z))$ for all $(y, z) \in N(x) \times E$.

**3.2. Some properties of tame quasiconformal motions.** Recall that a homeomorphism of $\widehat{\mathbb{C}}$ is called normalized if it fixes $0, 1$, and $\infty$.

We use $M(\mathbb{C})$ to denote the open unit ball of the complex Banach space $L^\infty(\mathbb{C})$. Each $\mu$ in $M(\mathbb{C})$ is the Beltrami coefficient of a unique normalized quasiconformal homeomorphism $w^\mu$ of $\widehat{\mathbb{C}}$ onto itself. The basepoint of $M(\mathbb{C})$ is the zero function.

We need the following properties of quasiconformal motions of $\widehat{\mathbb{C}}$, proved in [**10**].

PROPOSITION 1. *A motion $\phi : X \times \widehat{\mathbb{C}} \to \widehat{\mathbb{C}}$ is quasiconformal if and only if it satisfies*

(a) *the map $\phi_x : \widehat{\mathbb{C}} \to \widehat{\mathbb{C}}$ is quasiconformal for each $x$ in $X$, and*
(b) *the map from $X$ to $M(\mathbb{C})$ that sends $x$ to the Beltrami coefficient of $\phi_x$ for each $x$ in $X$ is continuous.*

Part (b) means that the map $x \mapsto \mu_x = \frac{(\phi_x)_{\bar{z}}}{(\phi_x)_z}, x \in X$, is continuous.
The following results were proved in [**7**].

LEMMA 3.6. *A motion $\phi : X \times E \to \widehat{\mathbb{C}}$ is a tame quasiconformal motion if and only if given any $x$ in $X$, there exists a neighborhood $N(x)$, and a continuous map $g_x : N(x) \to M(\mathbb{C})$ such that $\phi(y, z) = w^{g_x(y)}(z)$ for all $(y, z) \in N(x) \times E$.*

PROPOSITION 2. *Let $\phi : W \times E \to \widehat{\mathbb{C}}$ be a holomorphic motion where $W$ is a connected complex Banach manifold with a basepoint $x_0$. Then, $\phi$ is a tame quasiconformal motion.*

See Proposition 6 in [**7**].

**3.3. Teichmüller space of a closed set $E$.** We need some analytic properties of the Teichmüller space of a closed set $E$ in $\widehat{\mathbb{C}}$. These Teichmüller spaces were first studied by G. Lieb in his 1990 Cornell University doctoral dissertation; see [**8**]. Subsequently, they have been extensively studied; the reader is referred to the papers [**3**] and [**5**] for many details.

Let $E \subset \widehat{\mathbb{C}}$ where $E$ is a closed set that has the points $0, 1$, and $\infty$. Two normalized quasiconformal self-mappings $f$ and $g$ of $\widehat{\mathbb{C}}$ are said to be $E$-equivalent if $f^{-1} \circ g$ is isotopic to the identity rel $E$. The Teichmüller space $T(E)$ is the set of all $E$-equivalence classes of normalized quasiconformal self-mappings of $\widehat{\mathbb{C}}$. The base point of $T(E)$ is the $E$-equivalence class of the identity map.

We define the the projection

$$P_E : M(\mathbb{C}) \to T(E)$$

by setting $P_E(\mu)$ equal to the $E$-equivalence class of $w^\mu$, written as $[w^\mu]_E$. Clearly, $P_E$ maps the basepoint of $M(\mathbb{C})$ to the basepoint of $T(E)$.

In his doctoral dissertation (see [**8**]), G. Lieb proved that $T(E)$ is a complex Banach manifold such that the projection map $P_E$ is a holomorphic split submersion. For more details, see [**3**]. Using Lemma 5 and Proposition 7 in [**2**], one can show the following:

PROPOSITION 3. *There is a continuous basepoint preserving map $s$ from $T(E)$ to $M(\mathbb{C})$ such that $P_E \circ s$ is the identity map on $T(E)$.*

See Proposition 7.22 in [**3**] and Proposition 3.3 in [**5**].

DEFINITION 3.7. The map $s$ from $T(E)$ to $M(\mathbb{C})$ is called the Douady-Earle section of $P_E$ for the Teichmüller space $T(E)$.

**3.4. Universal tame quasiconformal motion.** The universal holomorphic motion $\Psi_E$ of $E$ over $T(E)$ is defined as follows:

$$\Psi_E(P_E(\mu), z) = w^\mu(z)$$

for $\mu \in M(\mathbb{C})$ and $z \in E$.

By Proposition 2, $\Psi_E : T(E) \times E \to \widehat{\mathbb{C}}$ is also a tame quasiconformal motion. In [**7**], it was proved that this is the universal tame quasiconformal motion of the closed set $E$ over a simply connected Hausdorff space. Here is the precise statement:

THEOREM 3.8. *Let $\phi : X \times E \to \widehat{\mathbb{C}}$ be a tame quasiconformal motion. If $X$ is a simply connected Hausdorff space with a basepoint $x_0$, there exists a unique basepoint preserving continuous map $f : X \to T(E)$ such that $f^*(\Psi_E) = \phi$.*

See Theorem II in [**7**].

COROLLARY 1. *Let $X$ be a simply connected Hausdorff space with a basepoint $x_0$, and $\phi : X \times E \to \widehat{\mathbb{C}}$ be a tame quasiconformal motion. Then, there exists a quasiconformal motion $\widetilde{\phi} : X \times \widehat{\mathbb{C}} \to \widehat{\mathbb{C}}$ such that $\widetilde{\phi}$ extends $\phi$.*

See Corollary 1 in [**7**].

**3.5. Extremal length of a curve family.** Let $D$ be a region in $\widehat{\mathbb{C}}$. Suppose $\Gamma$ is a family of curves in $D$. Each $\gamma \in \Gamma$ shall be a countable union of open arcs, closed arcs or closed curves, and every closed sub arc shall be rectifiable.

The extremal length $\Lambda(\Gamma)$ of a family $\Gamma$ of curves in a region $D$ in $\widehat{\mathbb{C}}$ is a conformal invariant defined as follows (see [**1**]). Let $\rho \geq 0$ be a measurable function on $D$ such that

$$\iint_D \rho(z)\, dxdy = 1, \quad z = x + iy.$$

Define

$$L(\rho) = \inf_{\gamma\in\Gamma} \int_\gamma \rho(z)\ |dz|.$$

The extremal length $\Lambda(\Gamma)$ is defined as

$$\Lambda(\Gamma) = \sup_\rho L(\rho)^2.$$

This quantity is quasi-invariant under quasiconformal mappings. In fact, if $f : D \to D'$ is a $K$-quasiconformal homeomorphism and $\Gamma' = f(\Gamma)$, then

$$K^{-1}\Lambda(\Gamma) \le \Lambda(\Gamma') \le K\Lambda(\Gamma).$$

In particular, if $f$ is conformal, then $\Lambda(\Gamma') = \Lambda(\Gamma)$.

**3.6. Quadrilateral.** A quadrilateral is a closed Jordan domain $D$ in $\widehat{\mathbb{C}}$ together with four distinct points $p_1, p_2, p_3$, and $p_4$ in positive cyclic order on its boundary. Here the $p_i$ are called the vertices of $D$. We say that an orientation-preserving homeomorphism $f$ maps a quadrilateral $D$ with vertices $p_1, p_2, p_3, p_4$ to another quadrilateral $D'$ with vertices $q_1, q_2, q_3, q_4$ if it is a homeomorphism from the region $D$ to the region $D'$ and maps each $p_i$ to $q_i$.

In a quadrilateral $D$, we call the two subarcs of $\partial D$ connecting $p_1$ to $p_2$ and $p_3$ to $p_4$ as $a$-sides and the two subarcs connecting $p_1$ to $p_4$ and $p_2$ to $p_3$ as $b$-sides.

An example of a quadrilateral is a rectangle $R = [0, a] \times [0, b]$ with four vertices labeled so it has $a$-sides $[0, a] \times \{0\}$ and $[0, a] \times \{b\}$ and $b$-sides $\{0\} \times [0, b]$ and $\{a\} \times [0, b]$.

By the Riemann mapping theorem, there is a conformal map $f$ from any quadrilateral $D$ to a rectangle $R$ mapping $a$-sides to $a$-sides and $b$-sides to $b$-sides. The ratio $mod(D) = a/b$ is called the modulus of $D$ and is a conformal invariant.

**3.7. Annulus.** A topological annulus $A$ is a doubly connected region such that each of the two components of $\widehat{\mathbb{C}} \setminus A$ contains at least two points. An example of a topological annulus is a round annulus $RA(r) = \{z \in \mathbb{C} \mid 1 \le |z| \le r\}$.

For every topological annulus $A$, there is a conformal map $f$ from $A$ to a round annulus $RA(r)$. The number $r$ is a conformal invariant. The modulus of $A$ defined as $mod(A) = (1/2\pi)\log r$.

We call the region $D$ in $\widehat{\mathbb{C}}$ a non-degenerate ring domain if and only if it is conformally equivalent to the round annulus

$$RA(r) = \{z \in \mathbb{C} \mid 1 \le |z| \le r\}$$

for some finite number $r > 1$.

**3.8. Two lemmas.** We will need the following lemmas.

LEMMA 3.9. *Let $X$ be a simply connected Hausdorff space with a basepoint $x_0$, and $\phi : X \times E \to \widehat{\mathbb{C}}$ be a tame quasiconformal motion that extends to a quasiconformal motion $\widetilde{\phi} : X \times \widehat{\mathbb{C}} \to \widehat{\mathbb{C}}$. If $\gamma$ is a closed Poincaré geodesic in a component $D$ of the complement $\widehat{\mathbb{C}} \setminus E$, the homotopy class of $\widetilde{\phi}_x(\gamma)$ in $\widetilde{\phi}_x(D)$ contains a unique closed Poincaré geodesic which we denote by $\phi_x(\gamma)$.*

The above lemma is similar to Corollary and Definition 2.4 in [**3**] and Proposition 4 in [**6**]. The proof is, verbatim, the same.

LEMMA 3.10. *Let $\phi : X \times E \to \widehat{\mathbb{C}}$ be a tame quasiconformal motion of $E$ over $X$. For each $x$ in $X$ there is a quasiconformal mapping $\widetilde{\phi}$ of $\widehat{\mathbb{C}}$ onto itself satisfying*

$$\widetilde{\phi}_x(z) = \phi(x, z)$$

*for all $z \in E$. If $D$ is a component of $\widehat{\mathbb{C}} \setminus E$, then the region $\widetilde{\phi}_x(D)$ depends only on $x, \phi$, and $D$, not on the choice of $\widetilde{\phi}_x$.*

The above lemma is similar to Proposition 2.6 in [**3**]. The proof is, verbatim, the same.

DEFINITION 3.11. Let $\phi$ be a tame quasiconformal motion of $E$ over $X$, let $x \in X$, and let $D$ be a region in $\widehat{\mathbb{C}}$ and $E$ is its boundary (that is, $D$ is a component of $\widehat{\mathbb{C}} \setminus E$). We set

$$\phi_x(E) = \{\phi(x, z) \mid z \in E\},$$

and we define $\phi_x(D)$ to be the component of $\widehat{\mathbb{C}} \setminus \phi_x(E)$ such that $\phi_x(D) = \widetilde{\phi}_x(D)$ for any quasiconformal map $\widetilde{\phi}_x$ of $\widehat{\mathbb{C}}$ onto itself satisfying $\widetilde{\phi}_x(z) = \phi(x, z)$ for all $z \in E$.

## 4. Proof of Theorem A

We need the following well-known lemma, see Lemma 5.1 in [**3**].

LEMMA 4.1. *For each $\mu$ in $M(\mathbb{C})$, let $L^\mu(\gamma)$ be the length of the closed Poincaré geodesic that is homotopic to $w^\mu(\gamma)$ in $w^\mu(D)$. The function $\mu \mapsto L^\mu(\gamma)$ is real analytic on $M(\mathbb{C})$.*

Since $X$ is simply connected, by Corollary 1 there exists a quasiconformal motion $\widetilde{\phi} : X \times \widehat{\mathbb{C}} \to \widehat{\mathbb{C}}$ that extends $\phi$. By Proposition 1, for each $x \in X$, the map $\widetilde{\phi}_x : \widehat{\mathbb{C}} \to \widehat{\mathbb{C}}$ is quasiconformal, and its Beltrami coefficient $\mu_x$ depends continuously on $x$. The map $x \mapsto \mu_x$ is a basepoint preserving continuous map from $X$ to $M(\mathbb{C})$. By Lemma 3.9, the homotopy class of $\widetilde{\phi}_x(\gamma) = w^{\mu_x}(\gamma)$ contains a unique Poincaré geodesic $\phi_x(\gamma)$, whose length we denote by $L^{\mu_x}(\gamma)$. By Lemma 4.1, the function $\mu_x \to L^{\mu_x}(\gamma)$ is real analytic. It follows that the function $x \mapsto L^{\mu_x}(\gamma)$ is continuous on $X$.

## 5. Proof of Theorem B

Since $X$ is simply connected, by Corollary 1 there exists a quasiconformal motion $\widetilde{\phi} : X \times \widehat{\mathbb{C}} \to \widehat{\mathbb{C}}$ that extends $\phi$. By Proposition 1, for each $x \in X$, the map $\widetilde{\phi} : \widehat{\mathbb{C}} \to \widehat{\mathbb{C}}$ is quasiconformal and its Beltrami coefficient $\mu_x$ varies continuously with respect to $x$.

In order to show that $\Lambda(\phi_{x'}(\Gamma)) \to \Lambda(\phi_x(\Gamma))$ whenever $x' \to x$, we consider the composition $\Phi = \widetilde{\phi}_x \circ (\widetilde{\phi}_{x'})^{-1}$ which sends $\phi_{x'}(\Gamma)$ to $\phi_x(\Gamma)$. Let $K(\Phi)$ denote its dilatation. By the quasi-invariance of the extremal length, we have

$$\frac{1}{K(\Phi)}\Lambda(\phi_{x'}(\Gamma)) \le \Lambda(\phi_x(\Gamma)) \le K(\Phi)\Lambda(\phi_{x'}(\Gamma)).$$

As $x' \to x$, $K(\Phi) \to 1$ (by Proposition 1). Hence, $\Lambda(\phi_{x'}(\Gamma)) \to \Lambda(\phi_x(\Gamma))$.

## 6. Proof of Theorem C

(1) By Lemma 3.10 and Definition 3.11, the region $\phi_x(D)$ is well-defined for each $x \in X$. And $\phi_x(D)$ is a non-degenerate ring domain because it is the image of $D$ under a quasiconformal mapping.
(2) To show that $x \mapsto mod(\phi_x(D))$ is continuous in $X$. We need the following lemma.

LEMMA 6.1. *Let $\phi : X \times E \to \widehat{\mathbb{C}}$ be a tame quasiconformal motion. Consider any point $x_1$ in $X$. Let $X_1$ be a simply connected neighborhood of $x_1$ in $X$ and let $x_1$ be the basepoint of $X_1$. Define $\widehat{\phi} : X_1 \times \phi_{x_1}(E) \to \widehat{\mathbb{C}}$ by*

$$\widehat{\phi}(y, \phi(x_1, z)) = \phi(y, z)$$

*for all $(y, z) \in X_1 \times E$. Then $\widehat{\phi} : X_1 \times \phi_{x_1}(E) \to \widehat{\mathbb{C}}$ of $\phi_{x_1}(E)$ over $X_1$ is a tame quasiconformal motion.*

PROOF. Since $\phi : X \times E \to \widehat{\mathbb{C}}$ is a tame quasiconformal motion, by Lemma 3.6 for $x_1 \in X_1 \subset X$, there exists a neighborhood $N(x_1)$ and a continuous map $g_{x_1} : N(x_1) \to M(\mathbb{C})$ such that $\phi(y, z) = w^{g_{x_1}(y)}(z)$ for all $(y, z) \in N(x_1) \times E$. Set $w = w^{g_{x_1}(x_1)}$. Then $w : \widehat{\mathbb{C}} \to \widehat{\mathbb{C}}$ is a quasiconformal map and $\phi(x_1, z) = w(z)$ for all $z \in E$.

Given any $x' \in X_1 \subset X$, there exists a neighborhood $N(x')$ and a continuous map $g_{x'} : N(x') \to M(\mathbb{C})$ such that $\phi(y, z) = w^{g_{x'}(y)}(z)$ for all $(y, z) \in N(x') \times E$. By Proposition 1, there exists a quasiconformal motion $w^{g_{x'}} : N(x') \times \widehat{\mathbb{C}} \to \widehat{\mathbb{C}}$ over $N(x')$ such that

$$\widehat{\phi}(y, \phi(x_1, z)) = \phi(y, z) = w^{g_{x'}(y)}(z)$$

for all $(y, z) \in N(x') \times E$. Let $\phi(x_1, z) = \widehat{z} \in \phi_{x_1}(E)$. Then, we have

$$\widehat{\phi}(y, \widehat{z}) = w^{g_{x'}(y)}(w^{-1}(\widehat{z}))$$

for all $y \in N(x')$. By Definition 3.5, $\widehat{\phi}$ is a tame quasiconformal motion. □

We need the following well-known formula, see Lemma 3.1 in [**3**].

LEMMA 6.2. *If $\gamma$ is a simple closed geodesic in the non-degenerate ring domain $D$, then the length of $\gamma$ equals $\frac{\pi}{2mod(D)}$.*

Choose a point $x_1 \in X$. Let $X_1$ be a simply connected neighborhood of $x_1$ and let $x_1$ be the basepoint of $X_1$. It suffices to show that $x \mapsto mod(\phi_x(D))$ is continuous in $X_1$. By Lemma 6.1, we define the tame quasiconformal motion $\widehat{\phi} : X_1 \times \phi_{x_1}(E) \to \widehat{\mathbb{C}}$ of $\phi_{x_1}(E)$ over $X_1$ by

$$\widehat{\phi}(x, \phi(x_1, z)) = \phi(x, z)$$

for all $(x, z) \in X_1 \times E$. Then $\widehat{\phi}_x(\phi_{x_1}(D)) = \phi_x(D)$ for all $x \in X_1$.

Let $\gamma_1$ be a simple closed Poincaré geodesic in $\phi_{x_1}(D)$. By Theorem A, the length $L(\widehat{\phi}_x(\gamma_1))$ of $\widehat{\phi}_x(\gamma_1)$ in $\phi_x(D)$ is a continuous function of $x$ in $X_1$. By Lemma 6.2, $L(\widehat{\phi}_x(\gamma_1)) = \frac{\pi}{2mod(\phi_x(D))}$. Hence, $mod(\phi_x(D))$ is a continuous function of $x$.

## 7. Proof of Theorem D

We may assume $a = 0, b = 1$, and $c = \infty$ and choose an additional point $d$ in $E$ so that $(D; 0, 1, \infty, d)$ is a quadrilateral. We need the following well-known lemma, see Lemma 4.3 in [**3**].

LEMMA 7.1. *The function* $\mu \to mod(w^\mu(D); 0, 1, \infty, w^\mu(d))$ *is real analytic.*

For $x \in X$, set $g(x) = mod(\phi_x(D); 0, 1, \infty, \phi_x(x, d))$.

Choose a point $x_1 \in X$. Let $X_1$ be a simply connected neighborhood of $x_1$ and let $x_1$ be the basepoint of $X_1$. By Lemma 6.1, we define the tame quasiconformal motion $\widehat{\phi} : X_1 \times \phi_{x_1}(E) \to \widehat{\mathbb{C}}$ of $\phi_{x_1}(E)$ over $X_1$ by

$$\widehat{\phi}(x, \phi(x_1, z)) = \phi(x, z)$$

for all $(x, z) \in X_1 \times E$. Set $E_1 = \phi_{x_1}(E)$, $D_1 = \phi_{x_1}(D)$ and $d_1 = \phi(x_1, d)$. Then,

$$g(x) = mod(\widehat{\phi}_x(D_1); 0, 1, \infty, \widehat{\phi}(x, d_1))$$

for all $x \in X_1$.

Since $X_1$ is simply connected, by Corollary 1 there exists a quasiconformal motion $\widetilde{\widehat{\phi}} : X_1 \times \widehat{\mathbb{C}} \to \widehat{\mathbb{C}}$ that extends $\widehat{\phi}$. By Proposition 1, for each $x \in X_1$, the map $\widetilde{\widehat{\phi}} : \widehat{\mathbb{C}} \to \widehat{\mathbb{C}}$ is quasiconformal and its Beltrami coefficient $\mu_x$ varies continuously with respect to $x$. Therefore,

$$\begin{aligned} g(x) &= mod(\widetilde{\widehat{\phi}}_x(D_1); 0, 1, \infty, \widetilde{\widehat{\phi}}(x, d_1)) \\ &= mod(w^{\mu_x}(D_1); , 0, 1, \infty, w^{\mu_x}(d_1)) \end{aligned}$$

for all $x \in X_1$. By Lemma 7.1, the function

$$\mu_x \mapsto mod(w^{\mu_x}(D_1); , 0, 1, \infty, w^{\mu_x}(d_1))$$

is real analytic. It follows that the function

$$x \mapsto mod(w^{\mu_x}(D_1); , 0, 1, \infty, w^{\mu_x}(d_1))$$

is continuous on $X$.

REMARK 7.2. The concept of "quasiconformal motion" was first introduced in the important paper [**11**]. Quasiconformal motions of an infinite closed set $E$ in $\widehat{\mathbb{C}}$ do not have the universal property; but "tame quasiconformal motions" studied in [**7**] have this universal property; this is Theorem II in [**7**] (Theorem 3.8 in the present paper). Theorems A, B, C and D of our paper are all applications of this universal property.

### Acknowledgment

We thank the referee for his/her careful reading of the paper and for his/her helpful comments.

## References

[1] L. V. Ahlfors, *Lectures on quasiconformal mappings*, 2nd ed., University Lecture Series, vol. 38, American Mathematical Society, Providence, RI, 2006. With supplemental chapters by C. J. Earle, I. Kra, M. Shishikura and J. H. Hubbard, DOI 10.1090/ulect/038. MR2241787

[2] A. Douady and C. J. Earle, *Conformally natural extension of homeomorphisms of the circle*, Acta Math. **157** (1986), no. 1-2, 23–48, DOI 10.1007/BF02392590. MR857678

[3] C. J. Earle and S. Mitra, *Variation of moduli under holomorphic motions*, In the tradition of Ahlfors and Bers (Stony Brook, NY, 1998), Contemp. Math., vol. 256, Amer. Math. Soc., Providence, RI, 2000, pp. 39–67, DOI 10.1090/conm/256/03996. MR1759669

[4] J. H. Hubbard, *Teichmüller theory and applications to geometry, topology, and dynamics; Volume 1*, Matrix Editions, Ithaca, NY, 2006.

[5] Y. Jiang and S. Mitra, *Douady-Earle section, holomorphic motions, and some applications*, Quasiconformal mappings, Riemann surfaces, and Teichmüller spaces, Contemp. Math., vol. 575, Amer. Math. Soc., Providence, RI, 2012, pp. 219–251, DOI 10.1090/conm/575/11400. MR2933902

[6] Y. Jiang and S. Mitra, *Variation of moduli under continuous motions*, In the tradition of Ahlfors-Bers. VII, Contemp. Math., vol. 696, Amer. Math. Soc., Providence, RI, 2017, pp. 205–214, DOI 10.1090/conm/696/14023. MR3715448

[7] Y. Jiang, S. Mitra, H. Shiga, and Z. Wang, *Teichmüller spaces and tame quasiconformal motions*, Tohoku Math. J. (2) **70** (2018), no. 4, 607–631, DOI 10.2748/tmj/1546570827. MR3896139

[8] G. S. Lieb, *Holomorphic motions and Teichmuller space*, ProQuest LLC, Ann Arbor, MI, 1989. Thesis (Ph.D.)–Cornell University. MR2638376

[9] S. Mitra, *Teichmüller spaces and holomorphic motions*, J. Anal. Math. **81** (2000), 1–33, DOI 10.1007/BF02788984. MR1785276

[10] S. Mitra, *Extensions of holomorphic motions to quasiconformal motions*, In the tradition of Ahlfors-Bers. IV, Contemp. Math., vol. 432, Amer. Math. Soc., Providence, RI, 2007, pp. 199–208, DOI 10.1090/conm/432/08309. MR2342817

[11] D. P. Sullivan and W. P. Thurston, *Extending holomorphic motions*, Acta Math. **157** (1986), no. 3-4, 243–257, DOI 10.1007/BF02392594. MR857674

Department of Mathematics and Computer Science, Kingsborough Community College of the City University of New York, Brooklyn, New York 11235–2398

*Email address*: xinlong.dong@kbcc.cuny.edu

Contemporary Mathematics
Volume **820**, 2025
https://doi.org/10.1090/conm/820/16409

# On compositions of quaternionic Möbius transformations

Krishnendu Gongopadhyay and Abhishek Mukherjee

Abstract. Let $\mathbb{H}$ be the division ring of real quaternions. Let $\mathrm{SL}(2, \mathbb{H})$ be the group of $2 \times 2$ quaternionic matrices with Dieudonné determinant $\det A = |ad - aca^{-1}b| = 1$. The group $\mathrm{PSL}(2, \mathbb{H})$ can be identified with the group of orientation-preserving isometries of the five dimensional real hyperbolic space $\mathbf{H}^5$. We prove results that control type of compositions of elements in $\mathrm{PSL}(2, \mathbb{H})$. As an application we observe that for three generic loxodromic elements $f, g, h$ in $\mathrm{PSL}(2, \mathbb{H})$, the groups $\langle f^{-n} g f^n, h\rangle$ are free for large values of $n$.

## 1. Introduction

Let $\mathbb{H}$ be the division ring of real quaternions. Let $\mathrm{SL}(2, \mathbb{H})$ be the group of $2\times 2$ quaternionic matrices with Dieudonné determinant one. The group $\mathrm{PSL}(2, \mathbb{H}) = \mathrm{SL}(2, \mathbb{H})/\{\pm I\}$ may be identified with the group of orientation-preserving isometries of the five dimensional hyperbolic space $\mathbf{H}^5$ using their action on the boundary sphere $\mathbb{S}^4 \simeq \mathbb{H} \cup \{\infty\}$ by quaternionic linear fractional transformations:

$$\begin{pmatrix} a & b \\ c & d \end{pmatrix} : Z \mapsto (aZ + b)(cZ + d)^{-1}.$$

The group $\mathrm{PSL}(2, \mathbb{H})$ also identifies with the group of orientation-preserving conformal automorphisms of $\mathbb{S}^4$. We refer the reader to [**9**], [**11**] for more details. The group $\mathrm{SL}(2, \mathbb{C})$ may be realized as a subgroup of $\mathrm{SL}(2, \mathbb{H})$, and hence the above action restricts to the usual Möbius transformations when the coefficients are from complex numbers. In analogy with the geometry of $\mathrm{SL}(2, \mathbb{C})$ action on the Riemann sphere, one may ask to understand similar questions for $\mathrm{SL}(2, \mathbb{H})$ action on the four dimensional sphere $\mathbb{S}^4$, cf. [**7**], [**8**].

Given two elements in $\mathrm{PSL}(2, \mathbb{H})$, it is a problem of potential application to investigate the behavior of compositions of different iterations of the elements. In the classical action of $\mathrm{SL}(2, \mathbb{C})$ on the two dimensional sphere, some of the results in this direction are available in [**3**, Section 2] that controls the type of compositions. The aim of this paper is to investigate such results for $\mathrm{PSL}(2, \mathbb{H})$.

---

2020 *Mathematics Subject Classification.* Primary 51M10; Secondary 20H10, 20H25 .

*Key words and phrases.* quaternionic matrices, compositions, reversing symmetries, hyperbolic space.

The first author was partially supported by the grant DST/INT/JSPS/P-323/2020. The second author was partially supported by the NBHM grant: Ref. No. 02011/35/2023/NBHM(R.P)/R & D II/16887.

To state our main results, we recall from [**6**], [**9**] that a parabolic element in $\mathrm{SL}(2, \mathbb{H})$ is conjugate to

$$\begin{pmatrix} \lambda & 1 \\ 0 & \lambda \end{pmatrix}, \ |\lambda| = 1,$$

and up to conjugacy, an elliptic or loxodromic element $A$ is given by

$$A = \begin{pmatrix} \lambda & 0 \\ 0 & \mu \end{pmatrix},$$

where $\lambda, \mu \in \mathbb{C}$, and $A$ is loxodromic if and only if $|\lambda| \neq 1 \neq |\mu|$. If $|\lambda| = |\mu| = 1$ and $\lambda$ is not similar to $\mu$, then $A$ is called a *regular elliptic*.

DEFINITION 1 ([**5**]). Let $A$ be an elliptic or loxodromic element in $\mathrm{PSL}(2, \mathbb{H})$. Then up to conjugacy, it has the above diagonal form. We define the *argument trace* of $A$ by

$$\mathrm{argtr}(A) = \arg(\lambda) + \arg(\mu),$$

and the *absolute trace* of $A$ by:

$$\mathrm{abstr}(A) = |\lambda| + |\mu|.$$

Note that an element of $\mathrm{PSL}(2, \mathbb{H})$ is loxodromic if and only if $\mathrm{abstr}(A) > 2$. Now we state our main results below.

THEOREM 1.1. *Let $f$ be a loxodromic element in* $\mathrm{PSL}(2, \mathbb{H})$ *with attractive fixed point $p^*$ and repulsive fixed point $p_*$. Choose two isometries $g, h$ in* $\mathrm{PSL}(2, \mathbb{H})$ *which are different from $f$.*

(1) *If $g(p^*) \neq p^*, h(p_*) \neq p_*$, then corresponding to a large number $M > 0$, there exists $n_0 \in \mathbb{N}$ such that*

$$\mathrm{abstr}(f^{-n} g f^n h) > M$$

*and for all $n \geq n_0$, $f^{-n} g f^n h$ is loxodromic which does not share a fixed point with $g$ or $h$.*

(2) *If $g(p_*) \neq p_*, h(p^*) \neq p^*$, then corresponding to a large number $M > 0$, there exists $n_0 \in \mathbb{N}$ such that* $\mathrm{abstr}(f^{-n} g f^n h) > M$ *and for all $n \leq -n_0$, $f^{-n} g f^n h$ is loxodromic which does not share a fixed point with $g$ or $h$.*

THEOREM 1.2. *Let $f$ be a non-parabolic element in* $\mathrm{PSL}(2, \mathbb{H})$ *that admits exactly two fixed points on the boundary $\mathbb{S}^4$. Let $g$ be a loxodromic element in* $\mathrm{PSL}(2, \mathbb{H})$. *If $g$ permutes the boundary fixed points of $f$, then $f$ also permutes the fixed points of $g$ if and only if $f$ is either a loxodromic or a regular elliptic, and*

$$\mathrm{abstr}(f) = \mathrm{abstr}(g).$$

We prove the above results in Section 3. Before that we discuss some preliminary materials in Section 2. It is worth noting that Proposition 1.4 is an immediate consequence of Theorem 1.1 and Proposition 1.3. We state Proposition 1.3, which can be proved following standard arguments as in the proof of Lemma 2.1.4 in [**3**]; in our case, Lemma 4.1 and Lemma 4.2 in Section 4 can be used to prove it.

We now state Proposition 1.3 that has been used to deduce the Proposition 1.4 stated below. In the quaternionic set up, we get a quantified bound on the elements using abstr to get a condition for generation of a classical Schottky group by two elements. For stating the result, recall the notion of a 'classical Schottky group'.

Let $k \geq 1$. Suppose that we are given a collection of disjoint closed round balls $B_1^+, \cdots, B_k^+, B_1^-, \cdots, B_k^- \subset \mathbb{S}^4$ and Möbius transformations $\gamma_j \in \mathrm{PSL}(2, \mathbb{H})$ so that $\gamma_j(int(B_j^+)) = ext(B_j^-)$. We assume that for each pair $B_j^+, B_j^-$, there exists a diffeomorphism of $\mathbb{S}^4$ which carries these balls to the round balls. Then $\mathbb{S}^4 - (B_j^+ \cup int(B_j^-))$ will be a fundamental domain for the group $\Gamma$ generated by $\gamma_1, \cdots, \gamma_k$. The group $\Gamma$ is defined to be a Schottky group. It is isomorphic to a free group of rank $k$. It should be noted that every non-trivial element of the group $\Gamma$ is loxodromic, cf. [**12**].

PROPOSITION 1.3. *Consider a loxodromic element $f$ in* $\mathrm{PSL}(2, \mathbb{H})$ *having attractive fixed point $p^*$ and repulsive fixed point $p_*$.*

(1) *For a given point $s \neq p^*, p_*$ and $T > 2$, there exists $\delta > 0$ such that if $g$ is any loxodromic transformation with fixed points $p, q$ with $p$ and $q$ lie in a $\delta-$neighborhood of $s$, and* $\mathrm{abstr}(g) \geq T$, *then the two generator subgroup $\langle f, g\rangle$ will be a classical Schottky group in* $\mathrm{PSL}(2, \mathbb{H})$.
(2) *Given two points $p, q \neq p^*, p_*$, there exists $T > 2$ such that if $g$ is any loxodromic transformation with fixed points $p, q$ having satisfied that* $\mathrm{abstr}(g) \geq T$ *and if the given transformation $f$ satisfies* $\mathrm{abstr}(f) \geq T$, *then the two generator subgroup $\langle f, g\rangle$ produces a classical Schottky group.*

PROPOSITION 1.4. *Suppose $f$ is a loxodromic transformation in* $\mathrm{PSL}(2, \mathbb{H})$ *with attractive fixed point $p^*$ and repulsive fixed point $p_*$. Let $g, h$ be two loxodromic transformations in* $\mathrm{PSL}(2, \mathbb{H})$.

(1) *If $g(p^*) \neq p^*$ and $h(p_*) \neq p_*$, then there exists $N \geq 0$ such that for all $n \geq N$, $f^{-n}gf^n$ and $h$ generate a free group.*
(2) *If $g(p_*) \neq p_*$ and $h(p^*) \neq p^*$, there exists $N \geq 0$ such that for all $n \leq -N$, $f^{-n}gf^n$ and $h$ generate a free group.*

The proposition follows using standard arguments as in the proof of Lemma 2.1.4 in [**3**]. In our case Lemma 4.1 and Lemma 4.2 proven in Section 4 can be used in the arguments.

We must note here that a version of the above two propositions might have been known to experts for all dimensions, and even at more generality, e.g. as consequences of the Tits alternative. However, we have not seen any literature with the viewpoint taken in this paper. We generalize the classical complex number theoretic approach using simple quaternionic analysis for conformal automorphisms of $\mathbb{S}^4$, and this might be of independent interest.

## 2. Preliminaries

**2.1. The quaternions.** Let $\mathbb{H}$ denote the division ring of quaternions. Recall that every element of $\mathbb{H}$ is of the form $a_0 + a_1 i + a_2 j + a_3 k$,where $a_0, a_1, a_2, a_3 \in \mathbb{R}$, and $i, j, k$ satisfy relations: $i^2 = j^2 = k^2 = -ijk = -1$. Any $a \in \mathbb{H}$ can be uniquely written as $a = a_0 + a_1 i + a_2 j + a_3 k$. We define $\Re(a) = a_0$= the real part of $a$ and $\Im(a) = a_1 i + a_2 j + a_3 k$ = the imaginary part of $a$. Also, define the conjugate of $a$ as $\overline{a} = \Re(a) - \Im(a)$. The norm of $a$ is $|a| = \sqrt{a_0^2 + a_1^2 + a_2^2 + a_3^2}$.

Two quaternions $a, b$ are said to be *similar*, denoted $a \backsim b$, if and only if $\Re(a) = \Re(b)$ and $|a| = |b|$. The similarity class of every quaternion $a$ contains a pair of complex conjugates with absolute-value $|a|$ and real part equal to $\Re(a)$. Let $a$ is similar to $re^{i\theta}$, $\theta \in (-\pi, \pi]$. We shall adopt the convention of calling $|\theta|$ as the *argument* of $a$ and will denote it by $\arg(a)$.

Here we refer to the quaternionic right vector space $\mathbb{H}^2$ which is comprising of elements of the form Z$=\begin{pmatrix} z_1 \\ z_2 \end{pmatrix}$, where $z_1, z_2 \in \mathbb{H}$, and equipped with the usual operations.

**2.2. Quaternionic matrices.** Let $\mathrm{M}(2,\mathbb{H})$ denote the group of all $2 \times 2$ quaternionic matrices. For a matrix $A = \begin{pmatrix} a & b \\ c & d \end{pmatrix}$, the Dieudonné determinant of $A$ is given by

$$\det A = |a|^2|d|^2 + |b|^2|c|^2 - 2\,\mathrm{Re}(a\bar{c}d\bar{b}).$$

Assume $a, b, c, d \neq 0$. Let

$$l_{11} = da - dbd^{-1}c \qquad l_{12} = bdb^{-1}a - bc$$
$$l_{21} = cac^{-1}d - cb \qquad l_{22} = ad - aca^{-1}b.$$

THEOREM 2.1 ([**9**]). *Let* $M = \begin{pmatrix} a & b \\ c & d \end{pmatrix} \in \mathrm{M}(2,\mathbb{H})$ *be such that* $\det M \neq 0$. *Then* $M$ *is invertible and*

$$M^{-1} = \begin{pmatrix} d^{\sim} & -b^{\sim} \\ -c^{\sim} & a^{\sim} \end{pmatrix}, \text{ where}$$

$$d^{\sim} = l_{11}^{-1}d, \qquad c^{\sim} = l_{21}^{-1}c, \qquad b^{\sim} = l_{12}^{-1}b, \qquad a^{\sim} = l_{22}^{-1}a.$$

If either of the co-efficients of the matrix $A$ is zero, the above expressions can be modified suitably to give $A^{-1}$. For more details see [**11**].

Now along this line let us recall that the standard lift of an element $z \in \widehat{\mathbb{H}}$ to $\mathbb{H}^2$ is defined by the following:

$$z \longmapsto \mathbf{Z} = \begin{pmatrix} z \\ 1 \end{pmatrix}, \forall z \in \mathbb{H}$$

and

$$\infty \longmapsto \begin{pmatrix} 1 \\ 0 \end{pmatrix}.$$

Now we see that $\mathrm{PSL}(2,\mathbb{H})$ acts on $\widehat{\mathbb{H}}$ by left multiplication on the standard lift followed by the natural right projection given by the following: For $A = \begin{pmatrix} a & b \\ c & d \end{pmatrix} \in \mathrm{PSL}(2,\mathbb{H})$ and $\forall z \in \mathbb{H}$, we have

$$\begin{aligned} A(z) &= \mathbb{P}A\mathbf{Z} \\ &= \mathbb{P}\begin{pmatrix} a & b \\ c & d \end{pmatrix}\begin{pmatrix} z \\ 1 \end{pmatrix} \\ &= (az+b)(cz+d)^{-1} \\ \text{and } A(\infty) &= \mathbb{P}\begin{pmatrix} a & b \\ c & d \end{pmatrix}\begin{pmatrix} 1 \\ 0 \end{pmatrix} = ac^{-1}. \end{aligned}$$

This corresponds to the *quaternionic Möbius transformations* and the group $\mathrm{PSL}(2,\mathbb{H})$ of all quaternionic Möbius transformations is isomorphic to $\mathrm{SL}(2,\mathbb{H})/\{\pm I\}$.

**2.3. Classification of elements of** $\mathrm{PSL}(2, \mathbb{H})$**.** We consider the extended quaternionic plane $\widehat{\mathbb{H}} = \mathbb{H} \cup \{\infty\}$ as the boundary of the hyperbolic space $\mathbf{H}^5$. The group $\mathrm{SL}(2, \mathbb{H})$ acts by the linear fractional transformations on $\widehat{\mathbb{H}}$ and this action can be extended to $\overline{\mathbf{H}}^5$. Given an element $A$ in $\mathrm{PSL}(2, \mathbb{H})$, it has two lifts $A$ and $-A$ in $\mathrm{SL}(2, \mathbb{H})$. We shall often work with the linear lifts for elements of $\mathrm{PSL}(2, \mathbb{H})$ without explicitly referring to those.

Every element $A$ of $\mathrm{PSL}(2, \mathbb{H})$ has a fixed point on the closure of the hyperbolic space $\overline{\mathbf{H}}^5 = \mathbf{H}^5 \cup \widehat{\mathbb{H}}$ and this gives us the usual classification of elliptic, parabolic and loxodromic elements in $\mathrm{PSL}(2, \mathbb{H})$. Further, it follows from the Lefschetz' fixed point theorem that every element of $\mathrm{PSL}(2, \mathbb{H})$ has a fixed point on the conformal boundary. Up to conjugacy, we can take that fixed point to be $\infty$ and hence, every element in $\mathrm{PSL}(2, \mathbb{H})$ is conjugate to an upper-triangular matrix. Using this one may classify the elements algebraically that also reflects their dynamical types, see [**1**], [**2**], [**6**], [**10**].

The action of $\mathrm{SL}(2, \mathbb{H})$ on $\widehat{\mathbb{H}}$ is also induced by a linear action of $\mathrm{SL}(2, \mathbb{H})$ on $\mathbb{H}^2$, and the point $\infty$ projects by the vector $\begin{pmatrix} 1 \\ 0 \end{pmatrix}$. The point 0 corresponds to the vector $\begin{pmatrix} 0 \\ 1 \end{pmatrix}$. For more details see the monograph [**9**].

**2.4. Fixed points and eigenvalues.** Recall that a quaternion $\lambda$ is a *right eigenvalue* of a matrix $A$ in $\mathrm{SL}(2, \mathbb{H})$ if and only if there is a non-zero column vector $v \in \mathbb{H}^2$ such that $Av = v\lambda$, where $v$ is defined to be a *right eigenvector* of $A$. For $u \neq 0$, let $w = vu^{-1}$ which explores that all quaternions similar to $\lambda$ are also right eigenvalues of $A$.

Also let us discuss the relationship between the fixed points of a quaternionic Möbius transformation and the right eigenvectors of the corresponding quaternionic matrix. This is similar to the real and complex cases. Suppose that $v$ is fixed by $A(z) = (az + b)(cz + d)^{-1}$ and $\mathbf{v}$, the standard lift of $v$, is an eigenvector of the corresponding matrix $A$. Then, we have

$$vcv + vd - av - b = 0.$$

Then it is easy to see that

$$\begin{pmatrix} a & b \\ c & d \end{pmatrix} \begin{pmatrix} v \\ 1 \end{pmatrix} = \begin{pmatrix} v \\ 1 \end{pmatrix} (cv + d).$$

By conjugating a matrix in $\mathrm{SL}(2, \mathbb{H})$ to a triangular form, one can prove that every quaternionic matrix in $\mathrm{SL}(2, \mathbb{H})$ must have a right eigenvalue coming from the *quaternionic characteristic polynomial*

$$t^2 - \tau t + \sigma = 0,$$

where $\tau = cac^{-1} + d, \sigma = cac^{-1}d - cb$, for $c \neq 0$.

In view of the above, one may conjugate a matrix $A = \begin{pmatrix} a & b \\ c & d \end{pmatrix}$ of $\mathrm{SL}(2, \mathbb{H})$ to a simplest triangular form whose two right eigenvalues will be respectively $cv + d$ and $a - vc$, where $v$ is a fixed point of $A$. Also corresponding to a right eigenvalue $t$ of the given matrix, whenever $c \neq 0$, the point $v = c^{-1}(t - d)$ is a fixed point of the transformation $A(z) = (az + b)(cz + d)^{-1}$. The upper triangular form for such an $A$ would be

$$\begin{pmatrix} 0 & 1 \\ -1 & v \end{pmatrix}\begin{pmatrix} a & b \\ c & d \end{pmatrix}\begin{pmatrix} v & -1 \\ 1 & 0 \end{pmatrix} = \begin{pmatrix} cv+d & -c \\ 0 & a-vc \end{pmatrix}$$

Evidently the above matrix has right eigen values $t_1 = cv + d$ and $t_2 = a - vc$ as mentioned above. For details see **[9]**, **[10]**.

## 3. Proofs of the Main Theorems

### 3.1. Proof of Theorem 1.1.

(1) Without loss of generality let $f = \begin{pmatrix} \lambda & 0 \\ 0 & \mu \end{pmatrix}$, where $|\lambda\mu| = 1, |\lambda| > 1$ in $\mathrm{SL}(2,\mathbb{H})$. Let $g = \begin{pmatrix} a & b \\ c & d \end{pmatrix}$, and $h = \begin{pmatrix} u & v \\ w & x \end{pmatrix}$ in $\mathrm{SL}(2,\mathbb{H})$ be such that $g(\infty) \neq \infty, h(0) \neq 0$.

We see that $c \neq 0, v \neq 0$. Note that

$$f^{-n}gf^nh = \begin{pmatrix} \lambda^{-n}a\lambda^n u + \lambda^{-n}b\mu^n w & \lambda^{-n}a\lambda^n v + \lambda^{-n}b\mu^n x \\ \mu^{-n}c\lambda^n u + \mu^{-n}d\mu^n w & \mu^{-n}c\lambda^n v + \mu^{-n}d\mu^n x \end{pmatrix}.$$

This gives

$$\begin{aligned} \mathrm{abstr}(f^{-n}gf^nh) &= |\lambda^{-n}a\lambda^n u + \lambda^{-n}b\mu^n w| + |\mu^{-n}c\lambda^n v + \mu^{-n}d\mu^n x| \\ &\geq |\lambda^{-n}a\lambda^n u| - |\lambda^{-n}b\mu^n w| + |\mu^{-n}c\lambda^n v| - |\mu^{-n}d\mu^n x| \\ &= |au| - |\mu|^{2n}|bw| + |\lambda|^{2n}|cv| - |dx| \\ &= (|au| - |dx|) + |\lambda|^{2n}|cv| - |\mu|^{2n}|bw|. \end{aligned}$$

Since $|\mu| < 1$, so $|\mu|^{2n} \to 0$ as $n \to \infty$. That is, for a given large number $M > 0, \exists N_1 \in \mathbb{N}$ such that

$$|\mu|^{2n} < \frac{1}{M}, \forall n \geq N_1.$$

Thus $\forall n \geq N_1$,

$$\begin{aligned} \mathrm{abstr}(f^{-n}gf^nh) &\geq (|au| - |dx| - \frac{1}{M}|bw|) + |\lambda|^{2n}|cv| \\ &\to \infty \text{ as } n \to \infty. \end{aligned}$$

This again shows that for large values of $n$, the isometries $f^{-n}gf^nh$ must be loxodromic. For the second part, we let that $q$ is a fixed point of both $h$ and $f^{-n}gf^nh$. Then we must have $q \neq 0$. Observe that

$$\begin{aligned} f^{-n}gf^nh(q) = q &\iff f^{-n}gf^n(q) = q \\ &\iff gf^n(q) = f^n(q). \end{aligned}$$

That is, $g(\lambda^n q\mu^{-n}) = \lambda^n q\mu^{-n}$. From this we see that

$$\begin{aligned} |g(\lambda^n q\mu^{-n})| &= |\lambda^n q\mu^{-n}| \\ &= |\lambda|^{2n}|q| \to \infty. \end{aligned}$$

This implies that $g(\infty) = \infty$, which contradicts our assumption. Therefore, we have $f^{-n}gf^nh(q) \neq q$.

Now, suppose $q$ is a fixed point of both $g$ and $f^{-n}gf^nh$ for large values of $n$. We claim that $q \neq 0$. If not, then $f^{-n}gf^nh(0) = 0 \implies h(0) = 0$, contradicting our choice of $h$. Further, we must have that $h(q) \neq 0$. If it was not so, then we see that

$$q = f^{-n}gf^nh(q) = f^{-n}gf^n(0) = f^{-n}g(0).$$

Since $g(0) = f^n(q) \to \infty$ as $n \to \infty$, this implies $q = \infty$, which contradicts the choice of $g$. Hence, $h(q) \neq 0$, and therefore, we must have $f^{-n}gf^nh(q) \neq q$, for all large $n$.

(2) This can be proved as above in which case we will observe that $b \neq 0$ and $w \neq 0$. So, in that case we have,

$$\begin{aligned}\operatorname{abstr}(f^{-n}gf^nh) &= |\lambda^{-n}a\lambda^n u + \lambda^{-n}b\mu^n w| + |\mu^{-n}c\lambda^n v + \mu^{-n}d\mu^n x| \\ &\geq |\lambda^{-n}b\mu^n w| - |\lambda^{-n}a\lambda^n u| + |\mu^{-n}d\mu^n x| - |\mu^{-n}c\lambda^n v| \\ &= |\mu|^{2n}|bw| - |\lambda|^{2n}|cv| + (|dx| - |au|).\end{aligned}$$

Since $|\lambda| > 1$, so $|\lambda|^{2n} \to 0$ as $n \to -\infty$ which implies that for a given large number $M > 0, \exists N_0 \in \mathbb{N}$ such that

$$|\lambda|^{2n} < \frac{1}{M}, \forall n \leq -N_0.$$

This shows that $\forall n \leq -N_0$,

$$\begin{aligned}\operatorname{abstr}(f^{-n}gf^nh) &\geq |\mu|^{2n}|bw| - \frac{1}{M}|cv| + (|dx| - |au|) \\ &\to \infty \text{ as } n \to -\infty.\end{aligned}$$

This shows that $f^{-n}gf^nh$ is loxodromic for small values of $n$. The rest follows similarly as above. □

**3.2. Proof of Theorem 1.2.** We consider a non-parabolic transformation $f = \begin{pmatrix}\lambda & 0\\ 0 & \mu\end{pmatrix}$, where $|\lambda||\mu| = 1$ and $\lambda$ is not similar to $\mu$. Let $g = \begin{pmatrix}a & b\\ c & d\end{pmatrix}$ be a loxodromic transformation in $\mathrm{SL}(2, \mathbb{H})$ such that $g(\infty) = 0$ and $g(0) = \infty$. Then $g$ has the form

$$g = \begin{pmatrix}0 & b\\ c & 0\end{pmatrix}, \text{ where } b, c \neq 0,\ |bc| = 1.$$

Now the right eigenvalues of $g$ associated to a given fixed point $z_1$ of $g$ are

$$q_1 = cz_1, q_2 = -z_1c.$$

Observe that if $z_2$ is another fixed point of $g$, then we can express it as

$$z_2 = c^{-1}(q_2 - d) = c^{-1}(-z_1c) = -c^{-1}z_1c.$$

Following this, we can assert that the corresponding eigenvalues are also not similar. Therefore we can transform $g$ into a upper triangular form $\begin{pmatrix}q_1 & -c\\ 0 & q_2\end{pmatrix}$.

Now consider the quaternionic characteristic polynomial for $g$ which is given by

$$t^2 - cb = 0.$$

Then we see that

$$\begin{aligned} |t|^2 &= |bc| = 1 \\ \implies |cz_1|^2 &= 1 \\ \implies |z_1|^2 &= |c|^{-2} = |b|^2, \\ \text{i.e. } |z_1| &= |b|. \end{aligned}$$

This suggests us to represent the fixed points in the quaternionic polar form: $z_1 = |b|e^{I\alpha}, z_2 = |b|e^{-J\alpha}$, where $\alpha = \arg z_1 \in (0, \pi]$ and $I, J$ being purely quaternionic units so that $J = c^{-1}Ic$.

Observe that $|e^{I\alpha}| = |e^{-J\alpha}| = 1$. If we assume further that $f$ permutes the respective fixed points of $g$, then we have that

$$\begin{aligned} \lambda z_1 \mu^{-1} &= z_2 \\ \iff \lambda |b| e^{I\alpha} &= |b| e^{-J\alpha} \mu \\ \iff |\lambda||e^{I\alpha}| &= |\mu||e^{-J\alpha}| \\ \iff |\lambda| &= |\mu|. \end{aligned}$$

Since $|\lambda||\mu| = 1$, this implies $|\lambda| = |\mu| = 1$. This shows that

$$\begin{aligned} \text{abstr}(g) &= |q_1| + |q_2| \\ &= |cz_1| + |cz_2| \\ &= |c||b||e^{I\alpha}| + |c||b||e^{-I\alpha}| \\ &= |\mu| + |\lambda| = \text{abstr}(f) \end{aligned}$$

This completes the proof. □

## 4. Useful lemmas

LEMMA 4.1.

(1) *Let $f$ be a loxodromic element in* $\mathrm{PSL}(2, \mathbb{H})$. *Suppose $g \in \mathrm{PSL}(2, \mathbb{H})$ is any arbitrary element that does not permute the fixed points of $f$. Then for a given $M > 0$, $\exists\, n_0 \in \mathbb{N}$ such that $gf^n$ is loxodromic and* $\text{abstr}(gf^n) > M$, $\forall n \geq n_0$.

(2) *Let $f$ be a loxodromic element in* $\mathrm{PSL}(2, \mathbb{H})$. *Let $g \in \mathrm{PSL}(2, \mathbb{H})$ sends exactly one fixed point of $f$ to the other. Then for a given $M > 0$, $\exists\, n_0 \in \mathbb{N}$ such that $gf^n$ is loxodromic and* $\text{abstr}(gf^n) > M$ *for all*
   (i) *$n \geq n_0$ if $g$ sends the repulsive fixed point of $f$ to the attractive fixed point.*
   (ii) *$n \leq -n_0$ if $g$ sends the attractive fixed point of $f$ to the repulsive fixed point.*

(3) *Let $f$ be a parabolic element in* $\mathrm{PSL}(2, \mathbb{H})$. *Let $g \in \mathrm{PSL}(2, \mathbb{H})$ be any arbitrary element such that $g$ does not have the same fixed point as that of $f$. Then for a given $M > 0$, $\exists n_0 \in \mathbb{N}$ such that $gf^n$ is loxodromic and* $\text{abstr}(gf^n) > M$, *for all $|n| \geq n_0$.*

PROOF.

(1) Let $g = \begin{pmatrix} a & b \\ c & d \end{pmatrix}$ and $f = \begin{pmatrix} \lambda & 0 \\ 0 & \mu \end{pmatrix}$, where $|\lambda| > 1$ and $|\lambda||\mu| = 1$. Then we have that, $\forall n \in \mathbb{N}$,

$$gf^n = \begin{pmatrix} a\lambda^n & b\mu^n \\ c\lambda^n & d\mu^n \end{pmatrix}.$$

Now $\operatorname{abstr}(gf^n) = |a\lambda^n| + |d\mu^n| \geq |a\lambda^n| = |a||\lambda|^n \to \infty$ as $n \to \infty$. Also this shows that for large values of $n$, $gf^n$ is loxodromic.

(2) In view of the hypothesis, assume that $g$ sends 0 to $\infty$. So, we may assume $g = \begin{pmatrix} a & b \\ c & 0 \end{pmatrix}$. Then we have that, $\forall n \in \mathbb{N}$,

$$gf^n = \begin{pmatrix} a\lambda^n & b\mu^n \\ c\lambda^n & 0 \end{pmatrix}.$$

Now $\operatorname{abstr}(gf^n) = |a\lambda^n| + 0 = |a\lambda^n| = |a||\lambda|^n \to \infty$ as $n \to \infty$. Thus, $gf^n$ is loxodromic for large values of $n$.

(3) Suppose that $f = \begin{pmatrix} \lambda & 1 \\ 0 & \lambda \end{pmatrix}$, where $|\lambda| = 1, \lambda \in \mathbb{C}$. We have,

$$\begin{aligned} gf^n &= \begin{pmatrix} a & b \\ c & d \end{pmatrix} \begin{pmatrix} \lambda^n & n\lambda^{n-1} \\ 0 & \lambda^n \end{pmatrix} \\ &= \begin{pmatrix} a\lambda^n & na\lambda^{n-1} + b\lambda^n \\ c\lambda^n & nc\lambda^{n-1} + d\lambda^n \end{pmatrix}. \end{aligned}$$

Then we observe that

$$\begin{aligned} \operatorname{abstr}(gf^n) &= |a\lambda^n| + |nc\lambda^{n-1} + d\lambda^n| \\ &= |a\lambda^n| + |\lambda|^{n-1}|nc + d\lambda| \\ &\geq |a||\lambda|^n + |nc + d\lambda| \\ &\geq |a| + (|n||c| - |d||\lambda|) \\ &= (|a| - |d||\lambda|) + |n||c| \to \infty \text{ as } |n| \to \infty. \end{aligned}$$

Also as in the previous case $gf^n$ is loxodromic for large values of $n$. This completes the proof.

□

LEMMA 4.2. *Let $f$ be a loxodromic element in* $\mathrm{PSL}(2, \mathbb{H})$ *with attractive fixed point $p^*$ and repulsive fixed point $p_*$.*

(1) *For an arbitrary element $g \in \mathrm{PSL}(2, \mathbb{H})$, the fixed points of the sequence $f^n g$ converge to $p^*$ and $g^{-1}(p_*)$ as $n \to \infty$ and also the fixed points of $gf^n$ converge to $g(p^*)$ and $p_*$ as $n \to \infty$.*
(2) *For an arbitrary element $g \in \mathrm{PSL}(2, \mathbb{H})$, the fixed points of the sequence $f^{-n} g$ converge to $p_*$ and $g^{-1}(p^*)$ as $n \to \infty$ and also the fixed points of $gf^{-n}$ converge to $g(p_*)$ and $p^*$ as $n \to \infty$.*

PROOF. Suppose that $f = \begin{pmatrix} \lambda & 0 \\ 0 & \mu \end{pmatrix}$, $|\lambda| > 1$, $g = \begin{pmatrix} a & b \\ c & d \end{pmatrix}$. Assume that $ac \neq 0$. If $p_n$ is a fixed point, then using the results in Section 2.4 we can conjugate the matrix of $f^n g$ to

$$\begin{pmatrix} \mu^n c p_n + \mu^n d & -\mu^n c \\ 0 & \lambda^n a - p_n \mu^n c \end{pmatrix}.$$

Note that $t_1 = \mu^n c p_n + \mu^n d$ and $t_2 = \lambda^n a - p_n \mu^n c$ are two eigen-values of $f^n g$, where $|t_1||t_2| = 1$. Now $p_n = c^{-1}\mu^{-n}(t_1 - \mu^n d) = c^{-1}(\mu^{-n} t_1 - d)$. It should be observed that $|\mu| < 1$ which implies that $|\mu|^{-n} \to \infty$ as $n \to \infty$.

For this, we see that $p_n = \lambda^n a c^{-1} \mu^{-1} - t_2 c^{-1} \mu^{-n}$, and

$$\begin{aligned} |p_n| &\geq |\lambda^n||ac^{-1}||\mu^{-n}| - |t_2||c^{-1}||\mu^{-n}|, \\ &= (|ac^{-1}||\mu|^{-n} - |t_2||c^{-1}|)|\mu|^{-n}, \text{ since } |\mu|^{-1} \geq 1, \\ &\geq (|a| - |t_2|)|c^{-1}||\mu|^{-n} \to \infty \text{ as } n \to \infty. \end{aligned}$$

This shows that the sequence of fixed points $\{p_n\}$ converges to $\infty$. On the other hand, we see that the other fixed point of $f^n g$ will be $q_n = c^{-1}(t_2 - d) = c^{-1}(\lambda^n a - p_n \mu^n c - d)$. This gives

$$|q_n + a^{-1}b| = |c^{-1}(\lambda^n a - p_n \mu^n c - d) + a^{-1}b|.$$

Recall that corresponding to the right eigen-value $t_2$, we must have an eigen-vector $\begin{pmatrix} q_n \\ 1 \end{pmatrix}$ such that $f^n g \begin{pmatrix} q_n \\ 1 \end{pmatrix} = \begin{pmatrix} q_n \\ 1 \end{pmatrix} t_2$. By using the equivalent matrix of $f^n g$, we have, $t_1 q_n - q_n t_2 = \mu^n c$, $\implies t_1 q_n = q_n t_2 + \mu^n c$

$$\implies |q_n| \leq |c|(|t_1| - |t_2|)^{-1}|\mu|^n.$$

Also it is to be noted that,

$$\begin{aligned} \lambda^n a q_n + \lambda^n b &= q_n(\lambda^n a - q_n \mu^n c), \\ \text{and } \mu^n c q_n + \mu^n d &= \lambda^n a - p_n \mu^n c, \\ \implies \lambda^n (a q_n + b) &= q_n \lambda^n (a - \lambda^{-n} p_n \mu^n c \\ \implies q_n + a^{-1} b &= a^{-1} \lambda^{-n} q_n \lambda^n (a - \lambda^{-n} p_n \mu^n c) \\ \implies |q_n + a^{-1} b| &\leq |c|(|t_1| - |t_2|)^2 |1 - a^{-1} \lambda^{-n} p_n \mu^n c|. \end{aligned}$$

Now, $|a^{-1}\lambda^{-n} p_n \mu^n c| \leq |a|^{-1}||\lambda|^{-n}||t_2|$. From this we get

$$\begin{aligned} |q_n + a^{-1}b| &\leq |a|^{-1}||c|(|t_1| - |t_2|)^2 |t_2||\mu|^{2n} \\ &= |a|^{-1}||c|(\frac{|t_1|}{|t_2| - 1})^2 |\mu|^{2n} \\ &\to 0 \text{ as } n \to \infty. \end{aligned}$$

Hence, $q_n \to -a^{-1}b = -\tilde{b}\tilde{a}^{-1} = g^{-1}(0)$. This completes the proof. □

## References


[1] W. Cao, *On the classification of four-dimensional Möbius transformations*, Proc. Edinb. Math. Soc. (2) **50** (2007), no. 1, 49–62, DOI 10.1017/S0013091505000398. MR2294003

[2] B. Foreman, *Conjugacy invariants of* Sl(2, ℍ), Linear Algebra Appl. **381** (2004), 25–35, DOI 10.1016/j.laa.2003.11.002. MR2039798

[3] D. Gallo, M. Kapovich, and A. Marden, *The monodromy groups of Schwarzian equations on closed Riemann surfaces*, Ann. of Math. (2) **151** (2000), no. 2, 625–704, DOI 10.2307/121044. MR1765706

[4] K. Gongopadhyay and A. Mukherjee, *Extremality of quaternionic Jørgensen inequality*, Hiroshima Math. J. **47** (2017), no. 2, 113–137, DOI 10.32917/hmj/1499392822. MR3679886

[5] K. Gongopadhyay, A. Mukherjee, and S. K. Sardar, *Test map and discreteness in* $\mathrm{SL}(2,\mathbb{H})$, Glasg. Math. J. **61** (2019), no. 3, 523–533, DOI 10.1017/s0017089518000332. MR3991354

[6] K. Gongopadhyay, *Algebraic characterization of the isometries of the hyperbolic 5-space*, Geom. Dedicata **144** (2010), 157–170, DOI 10.1007/s10711-009-9394-x. MR2580424

[7] R. Kellerhals, *Volumes in hyperbolic* 5*-space*, Geom. Funct. Anal. **5** (1995), no. 4, 640–667, DOI 10.1007/BF01902056. MR1345017

[8] R. Kellerhals, *Collars in* $\mathrm{PSL}(2,\mathbf{H})$, Ann. Acad. Sci. Fenn. Math. **26** (2001), no. 1, 51–72. MR1816562

[9] J. R. Parker, *Hyperbolic Spaces*, Volume 2 of *Jyväskylä Lectures in Mathematics*, University of Jyväskylä, 2008.

[10] J. R. Parker and I. Short, *Conjugacy classification of quaternionic Möbius transformations*, Comput. Methods Funct. Theory **9** (2009), no. 1, 13–25, DOI 10.1007/BF03321711. MR2478260

[11] J. B. Wilker, *The quaternion formalism for Möbius groups in four or fewer dimensions*, Linear Algebra Appl. **190** (1993), 99–136, DOI 10.1016/0024-3795(93)90222-A. MR1230354

[12] B. Maskit, *Kleinian groups*, Grundlehren der mathematischen Wissenschaften [Fundamental Principles of Mathematical Sciences], vol. 287, Springer-Verlag, Berlin, 1988. MR959135

Indian Institute of Science Education and Research (IISER) Mohali, Knowledge City, Sector 81, SAS Nagar, Punjab 140306, India

*Email address*: krishnendu@iisermohali.ac.in

Department of Mathematics, Kalna College, Kalna, Dist. Burdwan, West Bengal 713409

*Email address*: abhiquaternion@gmail.com, abhishekmukherjee@kalnacollege.ac.in

Contemporary Mathematics
Volume **820**, 2025
https://doi.org/10.1090/conm/820/16410

# Tameness conditions in complex dynamics and in quasiconformal motions

Yunping Jiang

ABSTRACT. Extending mathematical results from finite contexts to infinite ones presents notable challenges. This survey examines our efforts to address this challenge through two illustrative examples. First, we explore complex dynamical systems by extending Thurston's theorem for post-critically finite branched coverings to geometrically finite ones. Second, we delve into quasiconformal motions, extending the full extension theorem of quasiconformal motions from finite subsets of the Riemann sphere over simply connected Hausdorff spaces to infinite subsets. While elegant theorems exist under finite assumptions, they will not hold in infinite frameworks. However, by introducing a manageable "tameness" condition in each context, we revive these theorems, preserving their elegance even in infinite scenarios.

## 1. Tameness condition in complex dynamical systems

Suppose $\mathbb{C}$ denotes the complex plane. Let $\widehat{\mathbb{C}} = \mathbb{C}\cup\{\infty\}$ be the Riemann sphere, also interpreted as the 2-sphere $S^2$ equipped with standard complex structure. A rational map

$$R = \frac{P}{Q}$$

is defined as a quotient of two polynomials $P$ and $Q$. Assume $P$ and $Q$ share no common factors. The degree of $R$ is defined as

$$deg(R) = \max\{deg(P), deg(Q)\}.$$

We further assume that $deg(R) > 1$. A rational map is an orientation-preserving holomorphic branched covering of $S^2$. In general, a continuous branched covering $f$ of $S^2$ is characterized by a set $C_f$ of finite points known as the set of branched points. These points satisfy the condition that the local degree $deg_p f > 1$ for $p \in C_f$ and $\deg_q f = 1$ for $q \in S^2 \setminus C_f$, on which $f$ is locally homeomorphic. When $f = R$ is a rational map, a point $p$ in $C_R$ is referred to as a critical point since $R'(p) = 0$. We only consider orientation-preserving maps. We simply call a continuous branched covering a branched covering.

2020 *Mathematics Subject Classification.* Primary 37F20, 37F44; Secondary 37F34, 37F31.
*Key words and phrases.* rational map, tame geometrically finite branched covering, sub-hyperbolic tame geometrically finite branched covering, tame combinatorial equivalence, Thurston obstruction, canonical obstruction, tame quasiconformal motion.

Suppose $f$ is a branched covering of $S^2$. Let

$$f^n = \underbrace{f \circ \cdots \circ f}_{n}.$$

denote the $n$-composition of $f$. The post-critical set

$$P_f = \overline{\cup_{n \geq 1} f^n(C_f)},$$

emerges as a fundamental set in examining the dynamics of $f$, as it represents the closure of singularities for all inverse branches of $f^{-n}$ for all $n \geq 1$. We classify $f$ as post-critically finite if $\#P_f < \infty$, and post-critically infinite if $\#P_f = \infty$.

A significant inquiry within complex dynamical systems pertains to determining when $f$ can be represented by a rational map $R$ in the term of combinatorial equivalence. Let $f, g : S^2 \to S^2$ be two branched coverings. They are deemed *combinatorially equivalent* if there is a pair of orientation-preserving homeomorphisms $(\phi, \psi)$ of $S^2$ such that the diagram

$$\begin{array}{ccc} S^2 & \xrightarrow{\psi} & S^2 \\ \downarrow f & & \downarrow g \\ S^2 & \xrightarrow{\phi} & S^2 \end{array}$$

commutes, i.e., $\phi \circ f = g \circ \psi$, and $\psi$ is isotopic to $\phi$ rel $P_f$. Here $\psi$ being isotopic to $\phi$ rel $P_f$ entails the existence of a continuous function

$$F(t, x) : [0, 1] \times S^2 \to S^2$$

satisfying

(i) $F_t(\cdot) = F(t, \cdot)$ is an orientation-preserving homeomorphism of $S^2$ for every $0 \leq t \leq 1$;
(ii) $F(t, p) = \phi(p) = \psi(p)$ for all $p \in P_f$ and $0 \leq t \leq 1$; and
(iii) $F(0, x) = \phi(x)$ and $F(1, x) = \psi(x)$ for all $x \in S^2$.

Here, $(\phi, \psi)$ is termed a *combinatorial equivalence between $f$ and $g$*. It is notable that $\phi|P_f = \psi|P_f$ represents a topological conjugacy between $f|P_f$ and $g|P_g$. If $g = R$ is a rational function, we say $f$ can be represented by a rational map.

In 1982, Thurston proposed in [**31**] that a post-critically finite branched covering can be represented by a unique rational map up to conjugation by Möbius transformations, provided it lacks a certain topological hindrance now known as a Thurston obstruction and its orbifold is hyperbolic. Thurston's Theorem was formally introduced in the paper by Douady and Hubbard [**9**] (see also [**12**]).

THEOREM 1. *Let $f$ be a post-critically finite branched covering, and suppose the associated orbifold $\mathcal{O}_f$ is hyperbolic. Then $f$ is combinatorially equivalent to a rational map $R$ if and only if $f$ has no Thurston obstruction. Furthermore, the rational map $R$ is uniquely determined up to conjugation by Möbius transformations.*

Let us first delve into understanding the concept of Thurston obstruction. A simple closed curve in $S^2 - P_f$ is called *essential* if it does not bound a disk in $S^2 - P_f$ and *peripheral* if it encloses a single point of $P_f$.

A *multicurve* $\Gamma = \{\gamma_i\}$ on $S^2 - P_f$ is a finite non-empty collection of disjoint simple closed curves, each being essential, non-peripheral, and non-isotopic one another rel $P_f$. A multicurve $\Gamma$ is termed *$f$-invariant* if for any $\gamma \in \Gamma$, every non-peripheral essential component of $f^{-1}(\gamma)$ is homotopic to an element of $\Gamma$ rel $P_f$.

Given an $f$-invariant multicurve $\Gamma$, it determines a transition matrix $A = A_\Gamma = (a_{\delta\gamma}) : \mathbb{R}^\Gamma \to \mathbb{R}^\Gamma$ using the formula:

$$a_{\delta\gamma} = \sum_\alpha \frac{1}{\deg(f : \alpha \to \gamma)},$$

where the sum is taken over components $\alpha$ of $f^{-1}(\gamma)$ which are homotopic to $\delta$ rel $P_f$. Since every entry in $A$ is non-negative, the Perron-Frobenius Theorem implies that $A$ possesses a non-negative and maximum eigenvalue $\lambda = \lambda_\Gamma$. Then $\Gamma$ is termed a Thurston obstruction if the maximum eigenvalue $\lambda \geq 1$.

Now, let us elucidate the concept of the associated orbifold for a post-critically finite branched covering $f$. For each point $x$ in $S^2$, we define $v_f(x)$ as the least common multiple of local degrees $\deg_y f^n$ for all $n > 0$ and all $y \in S^2$ such that $f^n(y) = x$. It is important to note that $v_f(x) = 1$ if $x$ is not in $P_f$ and that $v_f(x)$ may be $\infty$. The associated orbifold is

$$\mathcal{O}_f = (S^2, P_f, v_f)$$

and the signature of $\mathcal{O}_f$ is the function

$$v_f : P_f \to \mathbb{N} \cup \{\infty\}.$$

We say $\mathcal{O}_f$ is hyperbolic if its signature is not $(2,2,2,2)$ and parabolic if its signature is $(2,2,2,2)$.

Essentially, it is demonstrated in McMullen's paper [**24**] that a rational map does not exhibit Thurston obstruction.

THEOREM 2. *Let $R : \hat{\mathbb{C}} \to \hat{\mathbb{C}}$ be a rational map. Consider $\Gamma$, a $R$-invariant multicurve on $\hat{\mathbb{C}} - P_R$. Then $\lambda \leq 1$. The equality holds only in the following cases:*

- *$R$ is post-critically finite and $\mathcal{O}_R$ is parabolic.*
- *$R$ is post-critically infinite, and $\Gamma$ comprises the essential curves in a finite system of annuli permuted by $R$. These annuli lie in Siegel disks, or Herman rings for $R$, and each annulus serves as a connected component of $\hat{\mathbb{C}} - P_R$.*

An important problem in the realm of complex dynamical systems is to investigate Theorem 1 when $\#P_f = \infty$. In our initial preprint [**5**] around 1996, which we split into two published preprints [**6, 7**] around 2004 and their subsequent publications [**4**] and [**8**] and [**29**] (see also our survey paper [**14**]), we began addressing this inquiry. We first presented a counter-example of a post-critically infinite branched covering, illustrating the infeasibility of a direct generalization of Theorem 1, even within the realm of geometrically finite branched coverings. Subsequently, we introduced a manageable "tameness" condition within the context of geometrically finite cases in [**4**]. Finally, we demonstrated that Theorem 1 can be generalized under this tameness condition, focusing on sub-hyperbolic tame geometrically finite branched coverings under the tame combinatorial equivalence. It is worth noting that the proofs in [**7**] and its subsequently published version [**8**] and the proof in [**29**] (see also [**14**]) differ significantly. The former approach involves reducing a tame geometrically finite branched covering to a post-critically finite form and then applying Theorem 1, whereas the latter provides a direct proof utilizing Teichmüller theory and bounded geometry (see [**14**]). Notably, the latter method also yields a new proof of Theorem 1 (see [**14**]). The motivation for us seeking new proof lies in avoiding [**9**, Lemma 5.2], for which a detailed proof was given in [**20**] later.

DEFINITION 1. A branched covering $f$ of $S^2$ is called geometrically finite if its post-critical set $P_f$ contains infinitely many points, but the limiting set $P'_f$ of its post-critical set contains only finitely many points.

As established in [**4**, Theorem 2.3], this definition is aligned with the definition of a geometrically finite rational map $R$ by considering its Julia set $J_R$, specifically, $\#(P_R \cap J_R) < \infty$.

We provided a counter-example to demonstrate that a direct generalization of Theorem 1 to the geometrically finite case is not feasible. In other words, the absence of Thurston obstruction alone cannot ensure that a geometrically finite branched covering can be represented by a rational map; there exist other topological obstructions. We formulated our counter-example as a theorem in the paper [**4**, Theorem 4.1].

THEOREM 3. *There exists a geometrically finite branched covering $f$ such that $f$ has no Thurston obstruction, but $f$ is not combinatorially equivalent to any rational map.*

REMARK 1. The orbifold associated with a geometrically finite branched covering can be defined similarly and is always hyperbolic.

We review the proof of this theorem in detail since we will encounter a similar counterexample later in the study of the full extension of quasiconformal motions (see Theorem 9). Before proving this theorem, let us introduce some concepts. Suppose $f$ is a geometrically finite branched covering. In this scenario, every point in $P'_f$ is periodic (see [**4**]).

DEFINITION 2. Let $p \in P'_f$ be a periodic point of period $k \geq 1$. Consider a family $\Gamma = \{\gamma_n\}_{n=0}^{\infty}$ of disjoint pairwise simple closed curves in $\widehat{\mathbb{C}} - P_f$. We refer to $\Gamma$ as a *combinatorially $f$-invariant family of curves nested at* $p$ if for all $n \geq 0$, $\gamma_{n+1}$ separates $\gamma_n$ from $p$ and there is a component of $f^{-k}(\gamma_{n+1})$ homotopic to $\gamma_n$ rel $P_f$. A combinatorially $f$-invariant family of curves $\Gamma$ nested at $p \in P'_f$ is termed *shrinking* if for every neighborhood $U$ of $p$, there is a simple closed curve $\beta \subset U - P_f$ such that $\beta$ is homotopic to some $\gamma_n \in \Gamma$ rel $P_f$.

DEFINITION 3. Let $p \in P'_f$ be a periodic point of period $k \geq 1$. We say $f$ is *locally combinatorially attracting* at $p$ if there exists a combinatorial equivalence $(\phi, \psi)$ from $f$ to a branched covering $g : \hat{\mathbb{C}} \to \hat{\mathbb{C}}$ such that $g$ is holomorphic in a neighborhood of the forward orbit of $\phi(p)$ under $g$ and such that $\phi(p)$ is a holomorphically attracting or superattracting fixed point of $g^k$, that is, $0 \leq |(g^k)'(\phi(p))| < 1$.

We have established the following three lemmas in [**4**].

LEMMA 1. *Let $p \in P'_f$ be a periodic point of period $k \geq 1$ with $\deg_p f^k = 1$. Then $f$ is locally combinatorially attracting at $p$ if and only if there is a combinatorially $f$-invariant shrinking family of curves nested at $p$.*

LEMMA 2. *Let $p \in P'_f$ be a fixed point of $f$ with $\deg_p f = 1$. If $f$ is locally combinatorially attracting at $p$, then there is a neighborhood $V$ of $p$ such that any combinatorially $f$-invariant family of curves nested at $p$ in $V$ is shrinking.*

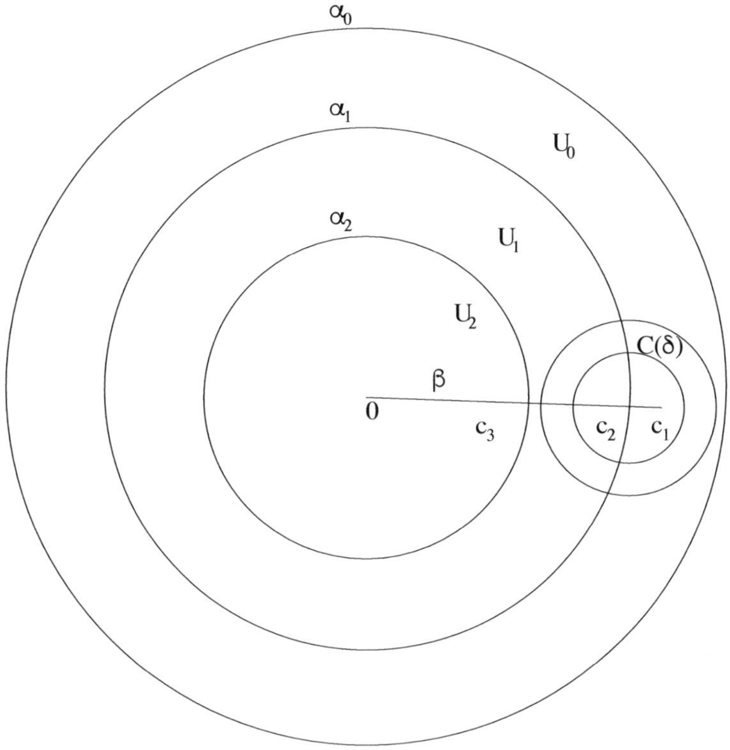


FIGURE 1

LEMMA 3. *Let $p \in P'_f$ be a fixed point with $deg_p f = 1$. If there is only one critical point whose forward orbit converges to $p$ in the spherical distance, then there is at most one combinatorially $f$-invariant shrinking family of curves nested at $p$ up to isotopy rel $P_f$. In other words, if $\Gamma = \{\gamma_n\}_{n=0}^{\infty}$ and $\Gamma' = \{\beta_n\}_{n=0}^{\infty}$ are combinatorially $f$-invariant shrinking families of curves nested at $p$, then there exist integers $k$ and $m \geq 1$ such that $\gamma_{n+k}$ is isotopic to $\beta_n$ rel $P_f$ for all $n \geq m$.*

PROOF OF THEOREM 3. Consider the quadratic polynomial $Q(z) = \lambda z + z^2$, where $0 < |\lambda| < 1$. The point $c = -\lambda/2$ is the unique critical point of $Q$ in the complex plane $\mathbb{C}$. Let $PO = \{c_n = Q^n(c)\}_{n=1}^{\infty}$ be the orbit of the critical value $Q(c)$. Then it converges to 0 as $n \to \infty$, which is the unique attractive fixed point of $Q$ in $\mathbb{C}$. Note that $P_Q = PO \cup \{0, \infty\}$.

There is a combinatorially $Q$-invariant shrinking family of curves $\Gamma = \{\alpha_n\}_{n=0}^{\infty}$ nested at 0 such that for $U_n$, the domain bounded by $\alpha_n$, we have that $Q|U_0$ is univalent, $U_0 \supset P_Q$, and $U_n = Q^n(U_0)$. There is an arc $\beta$ in $U_0$ which connects $c_1$ with 0 and passes over every point $c_n$ such that $Q(\beta) \subset \beta$ and $\beta$ intersects with each $\alpha_n$ ($n > 0$) at one point (see Figure 1).

Denote by $A$ the annulus bounded by $\alpha_0$ and $\alpha_2$. Take $\delta$ to be a Jordan curve in $A$ such that $\delta$ intersects with $\beta$ at one point and $\delta$ bounds a disk $D$ in $A$ containing $c_1$ and $c_2$ (see Figure 1).

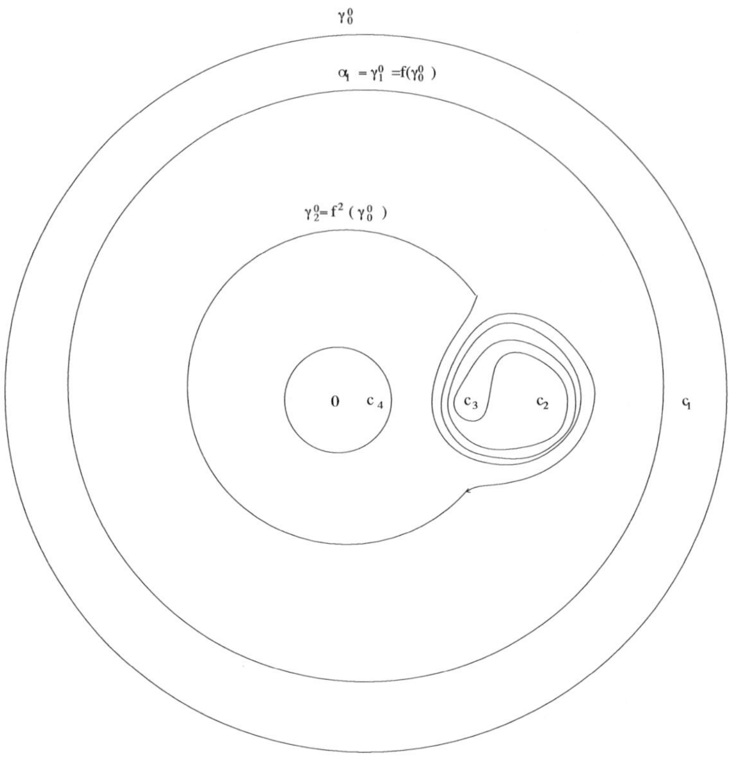


FIGURE 2

Let $\tau_0$ be the Dehn twist along $\delta$ (see Figure 2). Here $\tau_0$ can be picked as a homeomorphism, which is the identity out of an annulus neighborhood $C(\delta)$ of $\delta$ in $A - \{c_1, c_2\}$ and which is $h^{-1} \circ R \circ h$ on $C(\delta)$, where $h$ is a conformal map from $C(\delta)$ to a standard annulus $C = \{z \mid 0 < r < |z| < 1\}$ and $R$ is a homeomorphism from $C$ to itself defined as

$$R(\rho e^{i\theta}) = \rho e^{i(\theta + 2\pi \frac{\rho - r}{1-r})}, \quad r < \rho < 1.$$

Define $\tau_{2n} = Q^{2n} \circ \tau_0 \circ (Q|U_0)^{-2n}$ on $Q^{2n}(C(\delta))$ and the identity elsewhere. Let

$$f_n = Q \circ \tau_0^2 \circ \tau_2^2 \circ \cdots \circ \tau_{2n}^2.$$

For any $x \neq 0$, there is an $n(x) \geq 1$ such that $f_n(x) = f_{n(x)}(x)$ for all $n \geq n(x)$. Define $f(x) = f_n(x)$ for all $n \geq n(x)$. Since $f_n(0) = 0$ for all $n \geq 1$, define $f(0) = 0$. For any $\epsilon > 0$, there is a $\delta > 0$ such that $f_n(\Delta_\delta) \subset \Delta_\epsilon$ for all $n$ large, where $\Delta_\delta = \{z \in \mathbb{C} \mid |z| < \delta\}$ and $\Delta_\epsilon = \{z \in \mathbb{C} \mid |z| < \epsilon\}$. So $f(\Delta_\delta) \subset \Delta_\epsilon$. Thus $f$ is continuous at 0. If we consider the local inverse branch of $f_n^{-1}$ fixing 0, similarly, we can define $f^{-1}$ locally at 0 and fixing 0, such that $f^{-1}$ is continuous at 0. Thus $f$ is locally homeomorphic at 0. Thus $\deg_0 f = 1$. And $f : \hat{\mathbb{C}} \to \hat{\mathbb{C}}$ is a branched covering of degree 2. The sequence $\{f_n\}$ of branched coverings converges to $f$ as $n$ goes to $\infty$ in the maximal norm with respect to the spherical distance. In particular,

$$P_f = P_Q = PO \cup \{0, \infty\}$$

(see Figure 2).

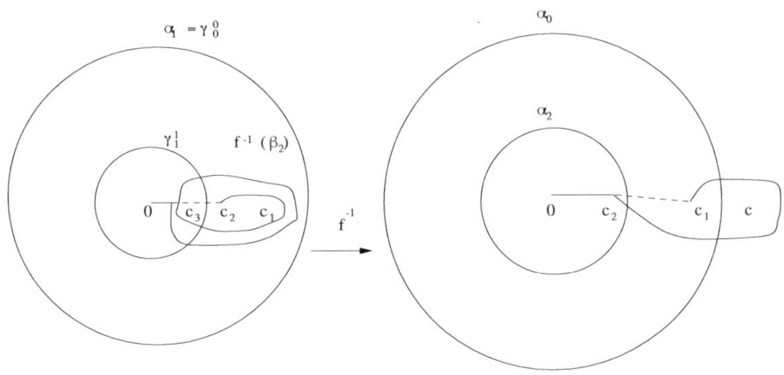


FIGURE 3

Let $\gamma_n^n = \alpha_{2n}$ and $\gamma_{n+k}^n = f^k(\gamma_n^n) \subset U_{2n}$ for $k > 0$. Since $\gamma_{n+1}^n = f(\gamma_n^n) = \alpha_{2n+1}$, we see that $\Gamma^n = \{\gamma_{n+k}^n\}_{k=1}^{\infty}$ is a combinatorially $f$-invariant family of curves nested at 0 for all $n \geq 0$.

Since $f(\gamma_{n+1}^n)$ is not isotopic to $\gamma_{n+1}^{n+1}$ rel $P_f$ (refer to Figure 2), each curve in $\Gamma^n$ is not isotopic to any curve in $\Gamma^{n+1}$ rel $P_f$. If $\Gamma^m$ is shrinking for some $m$, so is $\Gamma^{m+1}$ since $\gamma_{m+1}^{m+1}$ separates $\gamma_m^m$ from 0. Thus $\Gamma^n$ is not shrinking for all $n \geq 0$ by Lemma 3. Because $\gamma_n^n$ converges to 0 in the spherical distance as $n$ goes to $\infty$, $f$ has no combinatorially $f$-invariant shrinking family of curves nested at 0 and hence is not locally combinatorially attracting at 0 by Lemma 1 and Lemma 2. This proves that $f$ cannot be combinatorially equivalent to a rational map.

Now, we show that $f$ has no Thurston obstruction. For every $n \geq 0$, let $f^{-n}$ be the maximal local inverse branch of $f^n$ fixing 0. We prove by induction that $f^{-2n}(U_{2n}) \supset P_f$ for all $n \geq 0$.

Clearly $U_0 \supset P_f$. Assume $f^{-2n+2}(U_{2n-2}) \supset P_f$ for $n \geq 1$. Note that

$$U_{2n-1} \supset U_{2n} \supset \{c_{2n+1}, c_{2n+2}, \cdots\}.$$

Since $f^{-2n+1}(U_{2n-1}) = f^{-2n+2}(U_{2n-2}) \supset P_f$ and $f^{2n-1}$ is a homeomorphism on $f^{-2n+1}(U_{2n-1})$, we have $f^{-2n+1}(U_{2n}) \supset P_f - \{c_1\}$. Let $\beta_n$ be the segment of $\beta$ with endpoints $c_n$ and $c_{n+1}$ for $n \geq 1$. Then $\beta_{n+1} \subset U_n$ for $n \geq 0$. In particular, $\beta_{2n}$ connects $c_{2n}$ and $c_{2n+1}$ in $U_{2n-1}$, and $\beta_{2n+1}$ connects $c_{2n+1}$ with $c_{2n+2}$ in $U_{2n}$. By the definition of $f$, $f^{-1}(\beta_{2n+1}) = \tau_{2(n-1)}^{-2}(\beta_{2n})$ (refer to Figure 3). So $\beta_{2n} \cup f^{-1}(\beta_{2n+1})$ is a loop whose winding number with respect to $c_{2n-1}$ is $\pm 2$.

Now, $f^{-2n+2}(\beta_{2n})$ connects $c_2$ with $c_3$ in $f^{-2n+2}(U_{2n-1})$ and $f^{-2n+1}(\beta_{2n+1})$ connects $c_2$ with $c_3$ in $f^{-2n+1}(U_{2n})$. Their union is also a loop whose winding number with respect to $c_1$ is also $\pm 2$ (see Figure 3).

By the assumption $f^{-2n+2}(U_{2n-2}) \supset P_f$, we know that $f^{-2n+1}(\beta_{2n})$ connects $c_2$ with $c_1$. Since $c_1$ is a critical value of degree 2, the path $f^{-2n}(\beta_{2n+1})$ must connect $c_2$ with $c_1$ in $f^{-2n}(U_{2n})$. That means

$$f^{-2n}(U_{2n}) \supset P_f$$

(see Figure 3) .

Suppose that $\Gamma$ is an $f$-invariant multicurve. Denote by $A = A_\Gamma$ the transition matrix determined by $\Gamma$. For any element $\tau \in \Gamma$, since $\gamma_n^n$ converges to zero in the spherical distance as $n$ goes to $\infty$, there is $m > 0$ such that $\tau$ is outside the closure of $U_{2m}$. Since $f^{-2m}(U_{2m}) \supset P_f$ is a Jordan domain, each component of preimage of $\tau$ under $f^{-2m}$ is either non-essential or peripheral. This shows that $A^{2m}$ is the zero matrix. Hence its spectral radius is 0. This implies that the spectral radius of $A$ is less than one. So $\Gamma$ is not a Thurston obstruction. Therefore, $f$ has no Thurston obstruction. This completes the proof. $\square$

The definition of a tame geometrically finite branched covering, as introduced in [**4**], provides a criterion for ensuring certain desirable properties. Here, we outline the definition and remark provided in the paper:

DEFINITION 4. A geometrically finite branched covering $f$ is termed a tame geometrically finite branched covering if it satisfies the following conditions:

(1) $f$ is holomorphic in a neighborhood of $P'_f$;
(2) each cycle $< p_0, \cdots, p_{k-1} >$ of period $k \geq 1$ in $P'_f$ is either attractive, or super-attractive, or or parabolic, i.e.,,either $0 < |(f^k)'(p_0)| < 1$, or $(f^k)'(p_0) = 0$, or $|(f^k)'(p_0)| = 1$ and $\left((f^k)'(p_0)\right)^q = 1$ for some integer $q \geq 1$; and
(3) each attracting petal associated with a parabolic cycle in $P'_f$ contains a point in the post-critical set $P_f$.

Additionally, if all cycles in $P'_f$ are either attractive or super-attractive, the branched covering $f$ is termed a sub-hyperbolic tame geometrically finite branched covering.

REMARK 2. It is noted that every geometrically finite rational map is a tame geometrically finite branched covering. The term “tameness” was suggested by Clifford Earle in a conversation, replacing the previously used term “semi-rational” in the context of [**4**].

In our work [**4**], we established the following result:

THEOREM 4. *Any tame geometrically finite branched covering $f$ is combinatorially equivalent to a sub-hyperbolic tame geometrically finite branched covering $g$.*

This theorem suggests that our focus can be narrowed to investigating the rational representation problem solely within the realm of tame geometrically finite branched coverings among all sub-hyperbolic tame geometrically finite branched coverings. Additionally, we introduce the concept of tame combinatorial equivalence to refine our analysis further.

DEFINITION 5. Let $f$ and $g$ be two sub-hyperbolic tame geometrically finite branched coverings. They are deemed *tame combinatorially equivalent* if there exist a pair of orientation-preserving homeomorphisms $\phi, \psi : \widehat{\mathbb{C}} \to \widehat{\mathbb{C}}$ satisfying:

(1) $\psi$ is isotopic to $\phi$ rel $P_f$,
(2) $\phi \circ f = g \circ \psi$,
(3) $\phi|U_f = \psi|U_f$ is holomorphic on some open set $U_f \supset P'_f$.

REMARK 3. We clarify our terminology in [**4**], where “CLH-equivalence” denotes “tame combinatorial equivalence,” with “CLH” signifying “combinatorial and locally holomorphical equivalence.”

Furthermore, in our works in [**7**] and subsequently published version [**8**] and in [**29**] (also see our survey paper [**14**]), we establish:

THEOREM 5. *Let $f$ be a sub-hyperbolic tame geometrically finite branched covering. Then $f$ is tame combinatorially equivalent to a rational map $R$ if and only if $f$ has no Thurston obstruction. Furthermore, the rational map $R : \widehat{\mathbb{C}} \to \widehat{\mathbb{C}}$ is uniquely determined up to conjugation by Möbius transformations.*

Additionally, we present the following corollary regarding the "combinatorial equivalence" for tame geometrically finite branched coverings:

COROLLARY 1. *Let $f$ be a tame geometrically finite branched covering. Then $f$ is combinatorially equivalent to a rational map $R$ if and only if $f$ has no Thurston obstructions. However, the rational map $R$ may not be determined uniquely up to conjugation by Möbius transformations.*

A post-critically finite or sub-hyperbolic tame geometrically finite branched covering must possess a Thurston obstruction if it cannot be represented by any rational map. Among various Thurston obstructions, the canonical one (detailed, for example, in [**14**]) stands out as the most compelling. Pilgrim, in [**28**], established the following result for post-critically finite branched coverings:

THEOREM 6. *Let $f$ be a post-critically finite branched covering that is not combinatorially equivalent to a rational map. Then, it must have the canonical obstruction, which is a Thurston obstruction.*

We extended this finding to sub-hyperbolic tame branched coverings in our work [**3**]:

THEOREM 7. *Let $f$ be a sub-hyperbolic tame geometrically finite branched covering that is not tame combinatorially equivalent to a rational map. Then, it must have the canonical obstruction, which is a Thurston obstruction.*

## 2. Tameness condition in quasiconformal motions

In our investigation of quasiconformal motions, we observed a similar phenomenon to the one discussed in the preceding section, as detailed in [**19**]. Here, we offer an overview to compare with our work in the preceding section in complex dynamical systems.

We denote a connected Hausdorff topological space with basepoint $x_0$ by $X$. In the case where $X$ is simply connected, meaning it is path-connected with a trivial fundamental group, we denote it by $V$ with basepoint $v_0$. When $X$ is a complex Banach manifold, we denote it by $W$ with basepoint $w_0$. Let $E$ be a subset of $\widehat{\mathbb{C}}$.

DEFINITION 6. A *motion $\phi$ of $E$ over $X$* is a map $\phi : X \times E \to \widehat{\mathbb{C}}$ satisfying the following conditions:

(i) $\phi(x_0, z) = z$ for all $z \in E$, and
(ii) for all $x \in X$, the map $\phi(x, \cdot) : E \to \widehat{\mathbb{C}}$ is injective.

DEFINITION 7. A motion $\phi : X \times E \to \widehat{\mathbb{C}}$ is a *continuous* motion of $E$ over $X$ if $\phi : X \times E \to \widehat{\mathbb{C}}$ is a continuous map, and for each $x \in X$, the map $\phi(x, \cdot)$ is a homeomorphism of $E$ onto its image, that fixes the points $0, 1$, and $\infty$.

For a given motion $\phi : X \times E \to \widehat{\mathbb{C}}$, we denote two maps as

$$\phi_x(\cdot) = \phi(x, \cdot) : X \to \widehat{\mathbb{C}} \quad \text{and} \quad \phi^z(\cdot) = \phi(\cdot, z) : E \to \widehat{\mathbb{C}}.$$

We will assume that 0, 1, and $\infty$ belong to $E$ and that the motion $\phi$ is *normalized*, meaning the map $\phi_x(\cdot)$ fixes 0, 1, $\infty$ for every $x$ in $X$. Let $E \subset \widehat{E}$, $\phi : X \times E \to \widehat{\mathbb{C}}$ and $\widehat{\phi} : X \times \widehat{E} \to \widehat{\mathbb{C}}$ be two motions. We say that $\widehat{\phi}$ *extends* $\phi$ if $\widehat{\phi}|X \times E = \phi$.

Given a motion $\phi : X \times E \to \widehat{\mathbb{C}}$ and any quadruplet of distinct points $a$, $b$, $c$, $d$ of points in $E$, define $\phi_x(a, b, c, d)$ as the cross-ratio of the values $\phi_x(a)$, $\phi_x(b)$, $\phi_{(x}(c)$, and $\phi_x(d)$ for every $x \in X$, that is,

$$\phi_x(a, b, c, d) = \frac{(\phi_x(a) - \phi_x(c))(\phi_x(b) - \phi_x(d))}{(\phi_x(a) - \phi_x(d))(\phi_x(b) - \phi_x(c))}. \tag{1}$$

Condition (ii) holds if and only if $\phi_x(a, b, c, d)$ is a well-defined point in the thrice-punctured sphere $\mathbb{C}_{0,1} = \widehat{\mathbb{C}} \setminus \{0, 1, \infty\}$ for all $x$ in $X$ and all quadruplets $a, b, c, d$ of distinct points in $E$. Let $\rho$ be the Poincaré distance on $\mathbb{C}_{0,1}$.

DEFINITION 8. A motion $\phi : X \times E \to \widehat{\mathbb{C}}$ is termed a *quasiconformal* motion if, in addition to (i) and (ii) in Definition 6, it also satisfies

(iii) given any $x$ in $X$ and any $\epsilon > 0$, there exists a neighborhood $U_x$ of $x$ such that for any quadruplet of distinct points $a, b, c, d$ in $E$, we have

$$\rho\big(\phi_y(a, b, c, d), \phi_{y'}(a, b, c, d)\big) < \epsilon \quad \text{for all } y \text{ and } y' \text{ in } U_x.$$

DEFINITION 9. A motion $\phi \colon W \times E \to \widehat{\mathbb{C}}$ is a *holomorphic* motion if we can replace (*iii*) as

(iii)′ the map $\phi^z(\cdot) \colon W \to \widehat{\mathbb{C}}$ is holomorphic for each $z$ in $E$.

REMARK 4. For a holomorphic motion $\phi \colon W \times E \to \widehat{\mathbb{C}}$, the map $w \mapsto \phi_w(a, b, c, d)$ from $W$ into $\mathbb{C}_{0,1}$ is holomorphic. Therefore, it is distance-decreasing with respect to the Kobayashi metrics on $W$ and on $\mathbb{C}_{0,1}$. Thus (*iii*)′ implies (*iii*). So, we have that any holomorphic motion is a quasiconformal motion. Moreover, any quasiconformal motion is a continuous motion due to the $\lambda$-lemma first proved in [**22, 23**] (see also [**11, 15**]) for holomorphic motions, and the same method in the proof can be used on the context of quasiconformal motions. For a normalized continuous motion $\phi : X \times \widehat{\mathbb{C}} \to \widehat{\mathbb{C}}$, each $\phi_x(\cdot)$, where $x$ in $X$, is a continuous map from $\widehat{\mathbb{C}}$ to itself that fixes 0, 1, and $\infty$. Since $\phi_x$ is injective and continuous, it is a homeomorphism of $\widehat{\mathbb{C}}$ onto itself by invariance of the domain.

A natural question arises: if $\phi : X \times E \to \widehat{\mathbb{C}}$ is a quasiconformal motion, does there exist a quasiconformal motion $\widehat{\phi} : X \times \widehat{\mathbb{C}} \to \widehat{\mathbb{C}}$ such that $\widehat{\phi}$ extends $\phi$? If the answer is affirmative, we call $\phi$ fully extendable.

When $X$ is not simply connected, the answer, in general, is negative (see [**13**]) without introducing a topological condition called the trivial monodromy condition (see [**2**]). So we pose the same question for a quasiconformal motion $\phi : V \times E \to \widehat{\mathbb{C}}$ for a simply connected parameter space $V$. In this case, when $\#E < \infty$, the answer is always positive.

Let $M(\mathbb{C})$ be the space of all Beltrami coefficients, which is the unit ball of $L^\infty(\mathbb{C})$, defined as:

$$M(\mathbb{C}) = \{\mu \in L^\infty(\mathbb{C}) \mid \|\mu\|_\infty < 1\}.$$

The measurable Riemann mapping theorem (see [**1**]) ensures that for every $\mu \in M(\mathbb{C})$, the Beltrami equation

$$w_{\overline{z}} = \mu w_z$$

has a quasiconformal homeomorphism solution. The normalized quasiconformal homeomorphism solution $w^\mu$, which fixes $0, 1, \infty$, is unique and holomorphically depends on $\mu \in M(\mathbb{C})$. The quasiconformal dilatation of $w^\mu$ is

$$K(w^\mu) = \frac{1 + \|\mu\|_\infty}{1 - \|\mu\|_\infty}.$$

THEOREM 8. *Let $E_n = \{0, 1, \infty, z_1, \cdots, z_n\}$, $z_i \neq 0, 1, \infty$, $z_i \neq z_j$ for all $1 \leq i \neq j \leq n$, be a set containing $n+3$ points in $\widehat{\mathbb{C}}$. Suppose $\phi : V \times E_n \to \widehat{\mathbb{C}}$ is a quasiconformal motion of $E_n$ over $V$. Then, it can be extended to a quasiconformal motion $\widehat{\phi} : V \times \widehat{\mathbb{C}} \to \widehat{\mathbb{C}}$. For each $v \in V$, the map $\widehat{\phi}_v(\cdot) : \widehat{\mathbb{C}} \to \widehat{\mathbb{C}}$ is a quasiconformal homeomorphism of $\widehat{\mathbb{C}}$. Moreover, the Beltrami coefficeint*

$$\mu_v = \frac{\overline{\partial}\widehat{\phi}_v}{\partial\widehat{\phi}_v} : V \to M(\mathbb{C})$$

*is a continuous map and*

$$\phi(v, z) = w^{\mu_v}(z), \quad \forall (v, z) \in V \times E_n.$$

PROOF. We use Teichmüller theory for closed subsets in the Riemann sphere to prove it. So let us first review some facts in Teichmüller theory for closed subsets in the Riemann sphere. Suppose $E$ is a closed subset of $\widehat{\mathbb{C}}$. Two normalized solutions $w^\mu$ and $w^{\widehat{\mu}}$ are said to be isotopic rel $E$ if there is a continuous map $H(t, z) : [0, 1] \times \widehat{\mathbb{C}} \to \widehat{\mathbb{C}}$ satisfying:

(1) for every $0 \leq t \leq 1$, $H_t(\cdot) = H(t, \cdot) : \widehat{\mathbb{C}} \to \widehat{\mathbb{C}}$ is a homeomorphism;
(2) $H(0, z) = z$ and $H(1, z) = ((w^\mu)^{-1} \circ w^{\widehat{\mu}})(z)$ for all $z \in \widehat{\mathbb{C}}$;
(3) $H(t, z) = z$ for all $0 \leq t \leq 1$ and $z \in E$;

The Teichmüller space $T(E)$ consists of all isotopy classes $[w^\mu]_E$ with Teichmüller metric given by:

$$d(\tau, \widehat{\tau}) = \frac{1}{2} \inf_{w^\mu \in \tau, w^{\widehat{\mu}} \in \widehat{\tau}} \log K((w^\mu)^{-1} \circ w^{\widehat{\mu}}).$$

From [**10, 17, 21, 25**], we know that $T(E)$ is a contractible complex Banach manifold and the projection

$$P_E(\mu) = [w^\mu]_E : M(\mathbb{C}) \to T(E)$$

is a holomorphic split submersion. This means that for any point $\tau \in T(E)$, there exists a neighborhood $U_{\tau,E}$ around $\tau$ and a holomorphic map $s_{\tau,E} : U_\tau \to M(\mathbb{C})$ such that $P_E \circ s_{\tau,E} = Id$. We call such a map a local holomorphic section at $\tau$. We will not have a global holomorphic section if $\#E \geq 5$. But we have a continuous section, that is, there exists a global continuous map $S_E : T(E) \to M(\mathbb{C})$ such that $P_E \circ S_E = Id$.

Since $E_n$ is finite, $T(E_n)$ is holomorphically isomorphic to the classical Teichmüller space $Teich(\widehat{\mathbb{C}} \setminus E_n)$ of the punctured Riemann sphere $\widehat{\mathbb{C}} \setminus E_n$ (refer to [**25**, Example 3.1]), which is a simply connected domain in $\mathbb{C}^n$ and is a universal holomorphic cover of

$$Y_n = \{\mathbf{z} = (\zeta_1, \cdots, \zeta_n) \in \mathbb{C}^n \mid \zeta_i \neq \zeta_j, 1 \leq i \neq j \leq n\}$$

(see [**27**, p. 166, Section 2.8.5]). Let $\pi_{E_n} : T(E_n) \to Y_n$ be the universal holomorphic cover map.

For the quasiconformal motion $\phi : V \times E_n \to \widehat{\mathbb{C}}$, we have a continuous map

$$f_0(v) = (\phi^{z_1}(v), \cdots, \phi^{z_n}(v)) : V \to Y_n.$$

Since $V$ is simply connected, we can lift $f_0$ to a continuous map $f_1 : V \to T(E_n)$ such that $\pi_{E_n} \circ f_1 = f_0$. Let $f = S_{E_n} \circ f_1 : V \to M(\mathbb{C})$. Then the quasiconformal motion

$$\widehat{\phi}(v, z) = w^{f(v)}(z) : V \times \widehat{\mathbb{C}} \to \widehat{\mathbb{C}}$$

extends $\phi : V \times E_n \to \widehat{\mathbb{C}}$. Moreover,

$$\mu_v = \frac{\overline{\partial}\widehat{\phi}_v}{\partial\widehat{\phi}_v} = f(v) : V \to M(\mathbb{C})$$

is a continuous map and

$$\widehat{\phi}(v, z) = w^{\mu_v}(z), \quad \forall (v, z) \in V \times \widehat{\mathbb{C}}.$$

This completes the proof. □

REMARK 5. In Theorem 8, the assumption that $\phi : V \times E_n \to \widehat{\mathbb{C}}$ is a quasiconformal motion of $E_n$ over $V$ can be replaced by the assumption that $\phi : V \times E_n \to \widehat{\mathbb{C}}$ is a continuous motion of $E_n$ over $V$. In addition, the extension $\widehat{\phi}$ is unique up to isotopy rel $E_n$. (See [**17**]).

For a quasiconformal motion $\phi(x, z) : X \times \widehat{\mathbb{C}} \to \widehat{\mathbb{C}}$ and each $x \in X$, $\phi_x(\cdot) : \widehat{\mathbb{C}} \to \widehat{\mathbb{C}}$ is a quasiconformal homeomorphism of $\widehat{\mathbb{C}}$. Let

$$\mu_x = \frac{(\phi_x)_{\overline{z}}}{(\phi_x)_z} \in M(\mathbb{C})$$

be the complex dilatation of $\phi_x$, that is, $\phi_x = w^{\mu_x}$. We have the following proposition from [**26**].

PROPOSITION 1. *A motion $\phi(x, z) : X \times \widehat{\mathbb{C}} \to \widehat{\mathbb{C}}$ is a quasiconformal motion if and only if for each $x \in X$, $\phi_x(\cdot) : \widehat{\mathbb{C}} \to \widehat{\mathbb{C}}$ is a quasiconformal homeomorphism of $\widehat{\mathbb{C}}$ and $\mu_x : X \to M(\widehat{\mathbb{C}})$ is continuous.*

An essential problem in quasiconformal motions is to investigate Theorem 8 when $\#E = \infty$. Similar to the previous section, we first found a counter-example in [**19**].

THEOREM 9. *Let $I = [0, 1]$ be the unit interval with basepoint $0$. There exist a closed set $E$ in $\widehat{\mathbb{C}}$, with $\#E = \infty$, and a quasiconformal motion $\phi : I \times E \to \widehat{\mathbb{C}}$, such that for any interval $U = [0, a]$, $0 < a \leq 1$, $\phi$ cannot be extended to a quasiconformal motion of $\widehat{\mathbb{C}}$ over $U$.*

PROOF. The idea behind proving this theorem resembles that of Theorem 3. Here, we use $\infty$ as the accumulation point for the set $E$ instead of using 0 in Theorem 3.

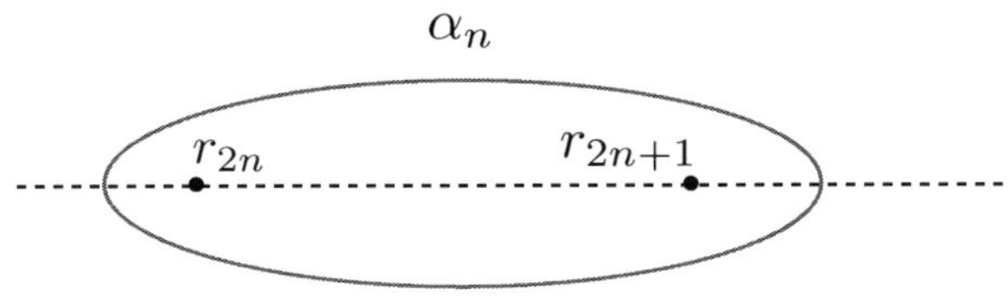


FIGURE 4

We take $1 < r_1 < r_2 < \cdots < r_n < r_{n+1} < \cdots$ so that $r_{n+1}/r_n \to \infty$ as $n \to \infty$. Put $X = \hat{\mathbb{C}} \setminus (\cup_{n=1}^{\infty} r_n \cup \{\infty\})$ and $E = \cup_{n=1}^{\infty} C_n \cup \{\infty\}$, where $C_n = \{|z| = r_n\}$. Let $\alpha_n$ ($n \in \mathbb{N}$) be a simple closed curve in $X$ only surrounding $r_{2n}$ and $r_{2n+1}$ as Figure 4.

Let $A_n := \{r_{2n} \leq |z| \leq r_{2n+1}\}$ and $B_n := \{r_{2n+1} \leq |z| \leq r_{2n+2}\}$. We take $p_n \in \mathbb{N}$ so large that

$$\lim_{n\to\infty} \frac{\ell_X(\tau_n^{p_n}(\alpha_n))}{\ell_X(\alpha_n)} = \infty, \tag{2}$$

where $\tau_n$ is the right Dehn twist in $A_n$ about $D_n := \{|z| = \sqrt{r_{2n}r_{2n+1}}\}$ (see Figure 5) and $\ell_X(c)$ stands for the hyperbolic length of the geodesic on $X$ homotopic to a closed curve $c$ in $X$.

For each $n \in \mathbb{N}$, we define $\Phi : I \times E \to \widehat{\mathbb{C}}$ by

$$\Phi_n(t, z) = \exp\{2\pi i n(n+1)(t - (n+1)^{-1})p_n\}z$$

for $(t, z) \in [(n+1)^{-1}, n^{-1}] \times C_{2n+1}$ and $\Phi_n(t, z) = z$ elsewhere. Note that $n(n+1)(t - (n+1)^{-1})p_n \uparrow p_n$ as $t \uparrow n^{-1}$ and $\Phi_n((n+1)^{-1}, z) = \Phi_n(n^{-1}, z) = z$.

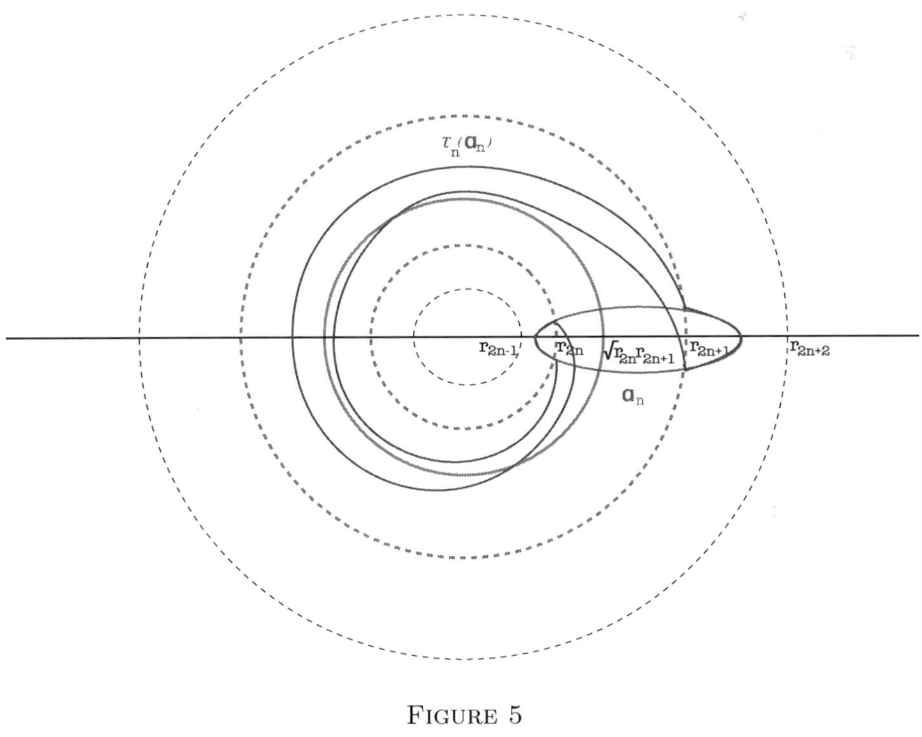


FIGURE 5

Thus $\Phi_n(t,z)$ is continuous at $n^{-1}$ and $(n+1)^{-1}$. Now, we define $\phi : I \times E \to \widehat{\mathbb{C}}$ by

$$\phi(t,z) = \lim_{n\to\infty} \Phi_n \circ \cdots \circ \Phi_1(t,z)$$

for every $(t,z) \in I \times E$. Obviously, $\phi$ is a continuous motion of $E$ over $I$. We claim that $\phi$ is a quasiconformal motion of $E$ over $I$.

Let us give a proof of this claim. We define quasiconformal homeomorphisms $f_{t,n}^+$ and $f_{t,n}^-$ of $\widehat{\mathbb{C}}$ as follows: For any $t \in I$, let $\theta_n(t)$ be in $[0,1)$ with $n(n+1)p_n t - \theta_n(t) \in \mathbb{N}$.

**(i):** For $z = re^{2\pi i\theta} \in A_n$,

$$f_{t,n}^+(z) = r\exp\left\{2\pi i\left(\frac{\log r - \log r_{2n}}{\log r_{2n+1} - \log r_{2n}}\theta_n(t) + \theta\right)\right\}$$

and for $z = re^{2\pi i\theta} \in B_n$,

$$f_{t,n}^+(z) = r\exp\left\{2\pi i\left(\frac{\log r - \log r_{2n+2}}{\log r_{2n+1} - \log r_{2n+2}}\theta_n(t) + \theta\right)\right\},$$

**(ii):** For $z = re^{2\pi i\theta} \in A_n$,

$$f_{t,n}^-(z) = r\exp\left\{2\pi i\left(\frac{\log r - \log r_{2n}}{\log r_{2n+1} - \log r_{2n}}(1-\theta_n(t)) + \theta\right)\right\}$$

and for $z = re^{2\pi i\theta} \in B_n$,

$$f_{t,n}^-(z) = r\exp\left\{2\pi i\left(\frac{\log r - \log r_{2n+2}}{\log r_{2n+1} - \log r_{2n+2}}(1-\theta_n(t)) + \theta\right)\right\};$$

**(iii):** $f_{t,n}^+(z) = f_{t,n}^-(z) = z$ for $z \notin A_n \cup B_n$.

Since $\lim_{n\to\infty}(\log r_{2n} - \log r_{2n-1}) = \lim_{n\to\infty}(\log r_{2n+1} - \log r_{2n}) = \infty$ and $\theta_n(t) \in [0,1)$, we see that

$$\lim_{n\to\infty} \sup\left(K(f_{t,n}^{\pm}) : (n+1)^{-1} \le t \le n^{-1}\right) = 1. \tag{3}$$

and

$$\lim_{t\downarrow(n+1)^{-1}} K(f_{t,n}^+) = \lim_{t\uparrow n^{-1}} K(f_{t,n}^-) = 1, \tag{4}$$

where $K(f)$ denotes the maximal dilatation of a quasiconformal map $f$.

We also see that $f_{t,n}^+(z) = f_{t,n}^-(t,z) = \phi(t,z)$ for $z \in E$ and $t \in [(n+1)^{-1}, n^{-1}]$. Moreover, $f_{n^{-1},n}^+(z) = f_{(n+1)^{-1},n}^+(z) = z$ on $\widehat{\mathbb{C}}$ because $\theta_n((n+1)^{-1}) = \theta_n(n^{-1}) = 0$. However, note that $\lim_{t\uparrow n^{-1}} \theta_n(t) = 1$.

Now, we are ready to show that $\phi : I \times E \to \widehat{\mathbb{C}}$ is a quasiconformal motion. Let $t_0 \in I$ and $\epsilon > 0$.

**Case 1.** $(n+1)^{-1} < t_0 < n^{-1}$**:** Since $K(f_{t,n}^+ \circ (f_{t_0,n}^+)^{-1}) \to 1$ as $t \to t_0$, it follows from the continuity of cross ratios under quasiconformal deformation (see [8]) that there exists a $\delta > 0$ such that if $|t - t_0| < \delta$, then

$$\begin{aligned}\rho(\phi_t(a,b,c,d), \phi_{t_0}(a,b,c,d)) &= \rho(f_{t,n}^+(a,b,c,d), f_{t_0,n}^+(a,b,c,d)) \\ &= \rho(f_{t,n}^+ \circ (f_{t_0,n}^+)^{-1}(a_{t_0}, b_{t_0}, c_{t_0}, d_{t_0}), (a_{t_0}, b_{t_0}, c_{t_0}, d_{t_0})) < \epsilon\end{aligned}$$

for any four distinct points $a,b,c,d$ in $E$, where $a_{t_0} = f_{t_0,n}^+(a)$, $b_{t_0} = f_{t_0,n}^+(b)$, $c_{t_0} = f_{t_0,n}^+(c)$ and $d_{t_0} = f_{t_0,n}^+(d)$.

**Case 2.** $t_0 = (n+1)^{-1}$**:** By using (4) and the continuity of cross ratios as above, we may find a $\delta > 0$ such that for any four distinct points $a, b, c, d$ in $E$,

$$\rho(\phi_t(a,b,c,d), \phi_{t_0}(a,b,c,d)) = \rho(f^+_{t,n}(a,b,c,d), f^+_{t_0,n}(a,b,c,d))$$
$$= \rho(f^+_{t,n}(a,b,c,d), (a,b,c,d)) < \epsilon,$$

if $t_0 < t < t_0 + \delta$ and

$$\rho(\phi_t(a,b,c,d), \phi_{t_0}(a,b,c,d)) = \rho(f^-_{t,n+1}(a,b,c,d), f^-_{t_0,n+1}(a,b,c,d))$$
$$= \rho(f^-_{t,n+1}(a,b,c,d), (a,b,c,d)) < \epsilon,$$

if $t_0 - \delta < t < t_0$.

**Case 3.** $t_0 = 0$**:** Using the continuity of cross ratios again, we see from (3) that

$$\rho(\phi_t(a,b,c,d), \phi_{t_0}(a,b,c,d)) = \rho(f^+_{t,n}(a,b,c,d), f^+_{t_0,n}(a,b,c,d)$$
$$= \rho(f^+_{t,n}(a,b,c,d), (a,b,c,d)) < \epsilon$$

holds for sufficiently small $t > 0$ and large $n \in \mathbb{N}$.

Therefore, we conclude that $\phi : I \times E \to \widehat{\mathbb{C}}$ is a quasiconformal motion.

Our next claim is that $\phi$ cannot be extended to a quasiconformal motion of $\widehat{\mathbb{C}}$ over any interval $U = [0, a]$ for $0 < a \leq 1$.

Let us now prove this claim. Suppose that there exists a quasiconformal motion $\hat{\phi}$ of $\widehat{\mathbb{C}}$ over $U$ which extends $\phi$. It follows from Proposition 2 that $\hat{\phi}_t : \widehat{\mathbb{C}} \to \widehat{\mathbb{C}}$ is a quasiconformal map for $t \in U$ and $U \ni t \mapsto \mu_{\hat{\phi}_t} \in M(\mathbb{C})$ is continuous. Hence, there exists $K \geq 1$ such that $\hat{\phi}_t$ is $K$-quasiconformal for any $t \in U$ since $U$ is a closed interval, thus compact.

Let $N > 0$ such that $1/N \in U$. For any $n > N$, we consider $F_t := \hat{\phi}_t \circ (\hat{\phi}_{(n+1)^{-1}})^{-1}$ for $t \in [(n+1)^{-1}, n^{-1}]$. Since $F_{(n+1)^{-1}} = id$ and $F_{n^{-1}} = \lim_{t \uparrow n^{-1}} F_t$, we verify that $F_{n^{-1}}(\alpha_n) = \lim_{t\uparrow n^{-1}} F_t(\alpha_n)$ is homotopic to $\tau_n^{p_n}(\alpha_n)$ in $X$. (Indeed, $F_{n^{-1}}|_{A_n}$ is a homeomorphism of the annulus $A_n$ which keeps each boundary point fixed. It gives a $p_n$-times rotation on $A_n$. Since $F_t$ is a family of homeomorphisms of $\widehat{\mathbb{C}}$ continuously depending on $t$, so when $t$ changes from $(n+1)^{-1}$ to $n^{-1}$, $\alpha_n$ moves continuously to $\tau_n^{p_n}(\alpha_n)$.) Since $\hat{\phi}_t$ is $K$-quasiconformal,

$$\ell_X(\tau_n^{p_n}(\alpha_n)) = \ell_X(F_{n^{-1}}(\alpha_n)) \leq K^2 \ell_X(\alpha_n).$$

This contradicts (2). Thus, we have shown that $\phi$ cannot be extended to a quasiconformal motion of $\widehat{\mathbb{C}}$ over $U$. □

REMARK 6. Theorem 9 disproves the assertion in [**30**, Theorem 3] when $\#E = \infty$. It is noteworthy that we can still extend the quasiconformal motion $\phi$ of $E$ over $I$ in Theorem 9 to a continuous motion of $\widehat{\mathbb{C}}$ over $I$, similar to the construction of the map $f$ in Theorem 3, which is continuous but fails to adhere the tame condition at some point in $P'_f$.

Therefore, we posed a tameness condition in [**19**], similar to what we did in the previous section. Remember that $X$ is a connected Hausdorff space with basepoint $x_0$ and $E$ is a set in $\widehat{\mathbb{C}}$ (containing the points 0, 1, and $\infty$).

DEFINITION 10. A *tame quasiconformal motion* is a motion $\phi : X \times E \to \widehat{\mathbb{C}}$ of $E$ over $X$, that is, it satisfies (i) and (ii) in Definition 6, with the additional property:

(iii)″ Given any $x$ in $X$, there exists a quasiconformal map $w : \widehat{\mathbb{C}} \to \widehat{\mathbb{C}}$, a neighborhood $N(x)$, with basepoint $x$, and a quasiconformal motion $\psi : N(x) \times \widehat{\mathbb{C}} \to \widehat{\mathbb{C}}$ over $N(x)$ such that $\phi(y, z) = \psi(y, w(z))$ for all $(y, z) \in N(x) \times E$.

REMARK 7. Any tame quasiconformal motion is a quasiconformal motion. Furthermore, every holomorphic motion qualifies as a tame quasiconformal motion in a stronger sense: given a holomorphic motion $\phi : W \times E \to \mathbb{C}$ and any point $w \in W$, there exists a quasiconformal map $h : \mathbb{C} \to \mathbb{C}$, along with a neighborhood $N(w)$ with basepoint $w$, and a holomorphic motion $\psi : N(\tau) \times \widehat{\mathbb{C}} \to \mathbb{C}$ such that $\phi(y, z) = \psi(y, h(z))$ for all $(y, z) \in N(\tau) \times E$. It is important to note that every holomorphic motion $\phi : W \times E \to \mathbb{C}$ falls under the category of tame quasiconformal motion in this stronger sense; it does not imply that it can be extended to a holomorphic motion of $\widehat{\mathbb{C}}$ over $W$. Counterexamples exist, such as those elucidated in [**13**] for the case of non-simply connected hyperbolic Riemann surfaces as $W$ and [**16**, Theorem 1] for the case of simply connected domains in $\mathbb{C}^n$ for $n \geq 2$ as $W$.

Under the tameness condition, we provided a revised statement of the assertion in [**30**, Theorem 3] and gave a detailed proof of the following theorem in [**19**, Theorem II and Corollary 1]. Remember that $V$ is a simply connected Hausdorff space with basepoint $v_0$ and $E$ is a set in $\widehat{\mathbb{C}}$ (containing the points 0, 1, and $\infty$).

THEOREM 10. *Any tame quasiconformal motion $\phi : V \times E \to \widehat{\mathbb{C}}$ can be extended to a quasiconformal motion $\widehat{\phi} : V \times \widehat{\mathbb{C}} \to \widehat{\mathbb{C}}$.*

## Acknowledgment

The author expresses gratitude to the referee for his/her valuable comments and suggestions, which greatly enhanced the presentation. I would also like to thank my colleague, Sudeb Mitra, for inviting me to write this survey and for his insightful suggestions and comments. This manuscript was completed during my visits to IMPA in Brazil, IHES in France, and SCMS in China, and I am thankful to these institutions for their hospitality and support. The material in this survey is based on work supported by the National Science Foundation (grants DMS-1747905 and DMS-1841243), a PSC-CUNY Enhanced Award (66680-54), Simons Foundation awards (523341 and 942077), and an NSFC grant (12271185).

## References

[1] L. Ahlfors and L. Bers, *Riemann's mapping theorem for variable metrics*, Ann. of Math. (2) **72** (1960), 385–404, DOI 10.2307/1970141. MR115006

[2] M. Beck, Y. Jiang, S. Mitra, and H. Shiga, *Extending holomorphic motions and monodromy*, Ann. Acad. Sci. Fenn. Math. **37** (2012), no. 1, 53–67, DOI 10.5186/aasfm.2012.3713. MR2920423

[3] T. Chen and Y. Jiang, *Canonical Thurston obstructions for sub-hyperbolic semi-rational branched coverings*, Conform. Geom. Dyn. **17** (2013), 6–25, DOI 10.1090/S1088-4173-2013-00250-6. MR3010347

[4] G. Cui and Y. Jiang, *Geometrically finite and semi-rational branched coverings of the two-sphere*, Trans. Amer. Math. Soc. **363** (2011), no. 5, 2701–2714, DOI 10.1090/S0002-9947-2010-05211-0. MR2763733

[5] G. Cui, Y. Jiang, and D. Sullivan, *On geometrically finite branched covering maps*, Manuscript, 1996.

[6] G. Cui, Y. Jiang, and D. Sullivan, *On geometrically finite branched coverings. I. Locally combinatorial attracting*, Complex dynamics and related topics: lectures from the Morningside Center of Mathematics, New Stud. Adv. Math., vol. 5, Int. Press, Somerville, MA, [2003], pp. 1–14. MR2504307

[7] G. Cui, Y. Jiang, and D. Sullivan, *On geometrically finite branched coverings. II. Realization of rational maps*, Complex dynamics and related topics: lectures from the Morningside Center of Mathematics, New Stud. Adv. Math., vol. 5, Int. Press, Somerville, MA, [2003], pp. 15–29. MR2504308

[8] G. Cui and L. Tan, *A characterization of hyperbolic rational maps*, Invent. Math. **183** (2011), no. 3, 451–516, DOI 10.1007/s00222-010-0281-8. MR2772086

[9] A. Douady and J. H. Hubbard, *A proof of Thurston's topological characterization of rational functions*, Acta Math. **171** (1993), no. 2, 263–297, DOI 10.1007/BF02392534. MR1251582

[10] C. J. Earle and S. Mitra, *Variation of moduli under holomorphic motions*, In the tradition of Ahlfors and Bers (Stony Brook, NY, 1998), Contemp. Math., vol. 256, Amer. Math. Soc., Providence, RI, 2000, pp. 39–67, DOI 10.1090/conm/256/03996. MR1759669

[11] F. P. Gardiner, Y. Jiang, and Z. Wang, *Holomorphic motions and related topics*, Geometry of Riemann surfaces, London Math. Soc. Lecture Note Ser., vol. 368, Cambridge Univ. Press, Cambridge, 2010, pp. 156–193. MR2665009

[12] J. H. Hubbard, *Teichmüller theory and applications to geometry, topology, and dynamics. Vol. 2*, Matrix Editions, Ithaca, NY, 2016. Surface homeomorphisms and rational functions. MR3675959

[13] Y. Jiang, *Winding numbers and full extendibility in holomorphic motions*, Conform. Geom. Dyn. **24** (2020), 109–117, DOI 10.1090/ecgd/351. MR4102668

[14] Y. Jiang, *A framework toward understanding the characterization of holomorphic dynamics*, Frontiers in complex dynamics, Princeton Math. Ser., vol. 51, Princeton Univ. Press, Princeton, NJ, 2014, pp. 235–257. MR3289913

[15] Y. Jiang, *Quasiconformal Mappings and Riemann Surfaces and Applications to Dynamics*, Lecture Notes in Jiang's webpage.

[16] Y. Jiang and S. Mitra, *Some applications of universal holomorphic motions*, Kodai Math. J. **30** (2007), no. 1, 85–96, DOI 10.2996/kmj/1175287624. MR2319079

[17] Y. Jiang and S. Mitra, *Douady-Earle section, holomorphic motions, and some applications*, Quasiconformal mappings, Riemann surfaces, and Teichmüller spaces, Contemp. Math., vol. 575, Amer. Math. Soc., Providence, RI, 2012, pp. 219–251, DOI 10.1090/conm/575/11400. MR2933902

[18] Y. Jiang and S. Mitra, *Variation of moduli under continuous motions*, In the tradition of Ahlfors-Bers. VII, Contemp. Math., vol. 696, Amer. Math. Soc., Providence, RI, 2017, pp. 205–214, DOI 10.1090/conm/696/14023. MR3715448

[19] Y. Jiang, S. Mitra, H. Shiga, and Z. Wang, *Teichmüller spaces and tame quasiconformal motions*, Tohoku Math. J. (2) **70** (2018), no. 4, 607–631, DOI 10.2748/tmj/1546570827. MR3896139

[20] S. Koch, *Teichmüller theory and critically finite endomorphisms*, Adv. Math. **248** (2013), 573–617, DOI 10.1016/j.aim.2013.08.019. MR3107522

[21] G. S. Lieb, *Holomorphic motions and Teichmuller space*, ProQuest LLC, Ann Arbor, MI, 1989. Thesis (Ph.D.)–Cornell University. MR2638376

[22] M. Yu. Lyubich, *Some typical properties of the dynamics of rational mappings* (Russian), Uspekhi Mat. Nauk **38** (1983), no. 5(233), 197–198. MR718838

[23] R. Mañé, P. Sad, and D. Sullivan, *On the dynamics of rational maps*, Ann. Sci. École Norm. Sup. (4) **16** (1983), no. 2, 193–217. MR732343

[24] C. T. McMullen, *Complex dynamics and renormalization*, Annals of Mathematics Studies, vol. 135, Princeton University Press, Princeton, NJ, 1994. MR1312365

[25] S. Mitra, *Teichmüller spaces and holomorphic motions*, J. Anal. Math. **81** (2000), 1–33, DOI 10.1007/BF02788984. MR1785276

[26] S. Mitra, *Extensions of holomorphic motions to quasiconformal motions*, In the tradition of Ahlfors-Bers. IV, Contemp. Math., vol. 432, Amer. Math. Soc., Providence, RI, 2007, pp. 199–208, DOI 10.1090/conm/432/08309. MR2342817

[27] S. Nag, *The complex analytic theory of Teichmüller spaces*, Canadian Mathematical Society Series of Monographs and Advanced Texts, John Wiley & Sons, Inc., New York, 1988. A Wiley-Interscience Publication. MR927291

[28] K. M. Pilgrim, *Canonical Thurston obstructions*, Adv. Math. **158** (2001), no. 2, 154–168, DOI 10.1006/aima.2000.1971. MR1822682

[29] G. Zhang and Y. Jiang, *Combinatorial characterization of sub-hyperbolic rational maps*, Adv. Math. **221** (2009), no. 6, 1990–2018, DOI 10.1016/j.aim.2009.03.009. MR2522834

[30] D. P. Sullivan and W. P. Thurston, *Extending holomorphic motions*, Acta Math. **157** (1986), no. 3-4, 243–257, DOI 10.1007/BF02392594. MR857674

[31] W. Thurston, *The combinatorics of iterated rational maps*, Preprint, Princeton University, Princeton, N.J., 1983.

Department of Mathematics, Queens College of the City University of New York, Flushing, New York 11367-1597; and The Ph.D. Program in Mathematics, Graduate Center of the City University of New York, New York, New York 10016

*Email address*: `yunping.jiang@qc.cuny.edu`

Contemporary Mathematics
Volume **820**, 2025
https://doi.org/10.1090/conm/820/16411

# Holomorphic motions and related topics – A brief survey

Nishan Chatterjee, Arshiya Farhath. G, and Sudeb Mitra

ABSTRACT. In this survey article, we discuss some results on extensions of holomorphic motions, and some applications in geometric function theory. We also discuss some properties of quasiconformal motions (in the sense of Sullivan and Thurston), tame quasiconformal motions, and continuous motions.

## 1. Introduction

NOTATIONS. We will use the following notation: $\mathbb{C}$ for the complex plane, $\widehat{\mathbb{C}} := \mathbb{C} \cup \{\infty\}$ for the Riemann sphere, and $\Delta$ for the open unit disk $\Delta := \{z \in \mathbb{C} : |z| < 1\}$.

In their study of the dynamics of rational maps Mañe, Sad, and Sullivan (in [**25**]) introduced the concept of holomorphic motions. They defined a holomorphic motion of a set $E$ in $\widehat{\mathbb{C}}$ as a curve $\phi_t(z)$ defined for every $z$ in $E$ and for every $t$ in $\Delta$ with the following properties:

(i) $\phi_0(z) = z$ for all $z \in E$,
(ii) $z \mapsto \phi_t(z)$ is injective as a function from $E$ to $\widehat{\mathbb{C}}$, for each fixed $t$ in $\Delta$, and
(iii) $t \mapsto \phi_t(z)$ is holomorphic for $|t| < 1$, and for each fixed $z$ in $E$.

As $t$ moves in $\Delta$, the set $E_t := \phi_t(E)$ moves in $\widehat{\mathbb{C}}$. We can think of $t$ as the complex time-parameter for the motion. As Gardiner and Keen remark in the introduction in [**15**], although $E$ may start out as smooth as a circle and although the points of $E$ move holomorphically, for every $t \neq 0$, $E_t = \phi_t(E)$ can be an interesting fractal with fractional Hausdorff dimension. Julia sets of rational maps and limit sets of Kleinian groups are almost always fractals. The famous $\lambda$-lemma in [**25**] showed the surprising fact that injectivity and holomorphic dependence automatically lead to quasiconformality. Since then, holomorphic motions have found many interesting applications in Teichmüller theory, dynamical systems, Kleinian groups, and geometric function theory.

A central topic in the study of holomorphic motions is the question of extensions. If $\phi : \Delta \times E \to \widehat{\mathbb{C}}$ is a holomorphic motion, a natural question is whether there exists a holomorphic motion $\widetilde{\phi} : \Delta \times \widehat{\mathbb{C}} \to \widehat{\mathbb{C}}$ that extends $\phi$. Important partial

2020 *Mathematics Subject Classification.* Primary 32G15, 30C62; Secondary 37F31.
*Key words and phrases.* Holomorphic motions, Tame quasiconformal motions, Quasiconformal motions, Teichmüller spaces.

results were obtained in [**7**] and in [**36**]. A complete affirmative answer was given by Slodkowski in [**33**]. He showed that any holomorphic motion of $E$ over $\Delta$ can be extended to $\widehat{\mathbb{C}}$.

The concept of quasiconformal motions was first introduced in the paper by Sullivan and Thurston ([**36**]). Theorem 3 of that paper claimed that every quasiconformal motion of any set over an interval can be extended to the Riemann sphere. The paper [**21**] presented a counterexample to Theorem 3 of [**36**]. In that same paper ([**21**]), a new concept was introduced, called *tame quasiconformal motions*, and it was shown that Theorem 3 of [**36**] holds for tame quasiconformal motions over any simply connected Hausdorff space. It was also shown that this extension can be done in a conformally natural way, for tame quasiconformal motions. The crucial idea was to show that tame quasiconformal motions have a certain "universal property" that quasiconformal motions (in the sense of Sullivan and Thurston) do not have.

In this survey article, we give a broad overview of some fundamental results on holomorphic, quasiconformal, tame quasiconformal, and continuous motions. We also discuss some applications in geometric function theory. These results crucially depend on some generalized Teichmüller spaces, which we briefly discuss in the last section.

For the reader's convenience, we note the following papers that contain several important expository topics and discussions; the chapter "A Supplement to Ahlfors's Lectures" by Earle and Kra in [**1**], and the papers [**19**] and [**34**].

## 2. Quasiconformal mappings

DEFINITION 1. A complex-valued function $w = f(z)$ defined on a region $\Omega$ in $\mathbb{C}$ is called a quasiconformal mapping if it is a sense–preserving homeomorphism of $\Omega$ onto its image and its complex distributional derivatives

$$w_z = \frac{1}{2}\Big(\frac{\partial f}{\partial x} - i\frac{\partial f}{\partial y}\Big) \text{ and } w_{\overline{z}} = \frac{1}{2}\Big(\frac{\partial f}{\partial x} + i\frac{\partial f}{\partial y}\Big)$$

are Lebesgue measurable locally square integrable functions on $\Omega$ that satisfy the inequality $|w_{\overline{z}}| \leq k|w_z|$ almost everywhere in $\Omega$, for some real number $k$ with $0 \leq k < 1$.

If $w = f(z)$ is a quasiconformal mapping defined on the region $\Omega$ then the function $w_z$ is known to be nonzero almost everywhere on $\Omega$. Therefore the function

$$\mu_f = \frac{w_{\overline{z}}}{w_z}$$

is a well–defined $L^\infty$ function on $\Omega$, called the *complex dilatation* or the *Beltrami coefficient* of $f$. The $L^\infty$ norm of every Beltrami coefficient is less than one.

The positive number

$$K(f) = \frac{1 + \|\mu_f\|_\infty}{1 - \|\mu_f\|_\infty}$$

is called the *dilatation* of $f$. We say that $f$ is $K$-quasiconformal if $f$ is a quasiconformal mapping and $K(f) \leq K$.

We call a homeomorphism of $\widehat{\mathbb{C}}$ *normalized* if it fixes the points 0, 1, and $\infty$.

Let $M(\mathbb{C})$ denote the open unit ball of the complex Banach space $L^\infty(\mathbb{C})$. Then, for each $\mu$ in $M(\mathbb{C})$, there exists a unique normalized quasiconformal homeomorphism of $\widehat{\mathbb{C}}$ onto itself that has Beltrami coefficient $\mu$; we denote this quasiconformal map by $w^\mu$. Furthermore, for every fixed $z \in \widehat{\mathbb{C}}$, the map $\mu \mapsto w^\mu(z)$ of $M(\mathbb{C})$ into $\widehat{\mathbb{C}}$ is holomorphic. The basepoint of $M(\mathbb{C})$ is the zero function.

## 3. Holomorphic motions

DEFINITION 2. Let $E \subset \widehat{\mathbb{C}}$, and let $X$ be a connected Hausdorff space with basepoint $x_0$. A *motion of $E$ over $X$* is a map $\phi : X \times E \to \widehat{\mathbb{C}}$ satisfying

(i) $\phi(x_0, z) = z$ for all $z \in E$, and
(ii) for all $x \in X$, the map $\phi(x, \cdot) : E \to \widehat{\mathbb{C}}$ is injective.

We say that $X$ is the *parameter space* of the motion $\phi$.

We will sometimes write $\phi(x, z)$ as $\phi_x(z)$ for $x$ in $X$ and $z$ in $E$.

We will always assume that 0, 1, and $\infty$ belong to $E$ and that the motion $\phi$ is normalized; i.e. 0, 1, and $\infty$ are fixed points of the map $\phi(x, \cdot)$ for every $x$ in $X$. This can always be achieved after a suitable change of coordinates: choose three distinct points $a$, $b$, $c$ in $E$. For each $x \in X$, let $S_x$ be the Möbius transformation that maps $\phi_x(a)$, $\phi_x(b)$, $\phi_x(c)$ to 0, 1, $\infty$ respectively. Let $\widehat{E} = S_{x_0}(E)$. Then the map $\widehat{\phi} : X \times \widehat{E} \to \widehat{\mathbb{C}}$ defined by

$$\widehat{\phi}(x, S_{x_0}(z)) = S_x(\phi(x, z))$$

is a normalized motion.

If $E$ is a proper subset of $\widetilde{E}$ and $\phi : X \times E \to \widehat{\mathbb{C}}$ and $\widetilde{\phi} : X \times \widetilde{E} \to \widehat{\mathbb{C}}$ are two motions, we say that $\widetilde{\phi}$ *extends* $\phi$ if $\widetilde{\phi}(x, z) = \phi(x, z)$ for all $(x, z) \in X \times E$.

DEFINITION 3. Let $V$ be a connected complex manifold with basepoint $x_0$. A *holomorphic motion of $E$ over $V$* is a motion $\phi : V \times E \to \widehat{\mathbb{C}}$ of $E$ over $V$ such that the map $\phi(\cdot, z) : V \to \widehat{\mathbb{C}}$ is holomorphic for each $z$ in $E$.

**An example.** Define $\Phi : M(\mathbb{C}) \times \widehat{\mathbb{C}} \to \widehat{\mathbb{C}}$ as follows:

$$\Phi(\mu, z) = w^\mu(z).$$

By Theorem 11 in [**2**], $\Phi$ is a holomorphic motion of $\widehat{\mathbb{C}}$ over $M(\mathbb{C})$.

As mentioned in the Introduction, holomorphic motions were first studied over the parameter space $\Delta$, motivated by the study of dynamics of rational maps. Let $V = \Delta$. We can think of $\Delta$ as the space of "time-parameters."

EXAMPLE. Fix $\mu \in L^\infty(\mathbb{C})$ such that $\|\mu\|_\infty = 1$. For each $t$ in $\Delta$, let $\phi(t, z) = w^{t\mu}(z)$ for all $z \in \widehat{\mathbb{C}}$, where $w^{t\mu}$ is the (normalized) quasiconformal self-map of $\widehat{\mathbb{C}}$ fixing the points 0, 1, and $\infty$. Then, by [**2**], the map

$$\phi(t, z) = w^{t\mu}(z)$$

is a holomorphic motion of $\widehat{\mathbb{C}}$ over $\Delta$. See also §1.3 of the Chapter "A Supplement to Ahlfors's Lectures" by Earle and Kra in [**1**].

The classical $\lambda$-lemma of Mañé, Sad, and Sullivan in [**25**] states three surprising properties of holomorphic motions.

**The $\lambda$-lemma.** Let $\phi : \Delta \times E \to \widehat{\mathbb{C}}$ be a holomorphic motion. Then

(i) $\phi$ is a continuous map from $\Delta \times E$ to $\widehat{\mathbb{C}}$,
(ii) the map $z \mapsto \phi(\lambda, z)$ is the restriction to $E$ of a quasiconformal mapping of the sphere.
(iii) $\phi$ extends to a holomorphic motion $\widetilde{\phi}$ of the closure $\overline{E}$ of $E$.

**Slodkowski's extension theorem.** Let $\phi : \Delta \times E \to \widehat{\mathbb{C}}$ be a holomorphic motion. Then, there exists a holomorphic motion $\widehat{\phi} : \Delta \times \widehat{\mathbb{C}} \to \widehat{\mathbb{C}}$, such that $\widehat{\phi}$ extends $\phi$.

See the papers [**3**], [**9**], [**16**], [**33**], and the book [**17**] for complete proofs. The paper [**22**] gives an alternative proof, using Teichmüller spaces.

Slodkowski's theorem does not, in general, hold for higher-dimensional parameter spaces. By definition, a holomorphic motion $\phi : B \times E \to \widehat{\mathbb{C}}$ is called *maximal* if there is no holomorphic motion $\widehat{\phi} : B \times \widehat{E} \to \widehat{\mathbb{C}}$ such that $E$ is a proper subset of $\widehat{E}$ and $\widehat{\phi}$ extends $\phi$. The following example was given in [**14**].

EXAMPLE. Give the domain

$$B = \{(\alpha, \beta) \in \mathbb{C}^2 : |e^{i\alpha}| + |\beta| < 1\}$$

the basepoint $(i, 0)$, and let

$$E = \{z \in \widehat{\mathbb{C}} : |z| \geq 1\} \cup \{z \in \mathbb{C} : |z| \leq e^{-1}\}.$$

Then, $\phi((\alpha, \beta), z) = z$ if $(\alpha, \beta) \in B, z \in E$, and $|z| \geq 1$, and $\phi((\alpha, \beta), z) = e^{i\alpha}ez + \beta$, if $(\alpha, \beta) \in B, z \in E$, and $|z| \leq e^{-1}$ defines a maximal holomorphic motion $\phi : B \times E \to \widehat{\mathbb{C}}$.

See Theorem 8.1 in [**14**].

DEFINITION 4. Let $\phi : \Delta \times E \to \widehat{\mathbb{C}}$ be a holomorphic motion. Let $G$ be a group of Möbius transformations, and suppose that $E$ is invariant under $G$ (which means, $g(E) = E$ for all $g$ in $G$). We say that $\phi$ is *G-equivariant* if and only if for each $g$ in $G$, and $t$ in $\Delta$, there is a Möbius transformation $\theta_t(g)$ such that

$$\phi(t, g(z)) = \theta_t(g)(\phi(t, z)) \tag{3.1}$$

for all $z \in E$.

The following result was proved in the foundational paper [**13**]. It has important applications in Teichmüller theory.

**Group equivariant form of Slodkowski's theorem.** Suppose that $\phi : \Delta \times E \to \widehat{\mathbb{C}}$ is a $G$-equivariant holomorphic motion. Then, there exists a $G$-equivariant holomorphic motion $\widehat{\phi} : \Delta \times \widehat{\mathbb{C}} \to \widehat{\mathbb{C}}$ such that $\widehat{\phi}$ extends $\phi$; which means $\widehat{\phi}(t, z) = \phi(t, z)$ for all $(t, z) \in \Delta \times E$ and

$$\widehat{\phi}(t, g(z)) = \theta_t(g)(\widehat{\phi}(t, z))$$

holds for all $g \in G$, $t \in \Delta$, and $z \in \widehat{\mathbb{C}}$.

## 4. Some results on extending holomorphic motions

THEOREM 1. *Let $E_0$ be any subset of $\widehat{\mathbb{C}}$, not necessarily closed; as usual, we assume that $0$, $1$, and $\infty$ belong to $E_0$. Let $E$ be the closure of $E_0$. Let $\phi : V \times E_0 \to \widehat{\mathbb{C}}$ be a holomorphic motion where $V$ is a connected complex Banach manifold with a basepoint. There exists a holomorphic motion $\widehat{\phi} : V \times E \to \widehat{\mathbb{C}}$ that extends $\phi$.*

See Theorem 1 in [**18**].

REMARK 1. Henceforth, in this section, we will always assume that $E$ is a closed set in $\widehat{\mathbb{C}}$, and as usual, $0$, $1$, and $\infty$ are in $E$, and are fixed points of $\phi(x, \cdot)$ for each $x \in V$.

THEOREM 2. *Let $\phi : V \times E \to \widehat{\mathbb{C}}$ be a holomorphic motion, where $V$ is a simply connected complex Banach manifold with basepoint $x_0$. Then, for every $x \in V$, $\phi(x, \cdot)$ is the restriction to $E$ of a quasiconformal self map of $\widehat{\mathbb{C}}$ with dilatation not exceeding $exp(2\rho_V(x, x_0))$ where $\rho_V$ denotes the Kobayashi pseudometric on $V$.*

We are assuming that the hyperbolic metric has constant curvature $-4$.

See Theorem D in [**27**]. This generalizes Theorem 1 of the paper [**7**]. It was used in [**18**] to prove the following result. As above, $V$ is a simply connected complex Banach manifold with a basepoint $x_0$, and $\rho_V$ denotes the Kobayashi pseudometric on $V$.

THEOREM 3. *Suppose $R(x, z)$ is a hyperbolic family of holomorphic dynamical systems over $V$. Then, for any $x$ in $V$, the Julia set $J_x$ is quasiconformally equivalent to $J_{x_0}$ by a quasiconformal map of $\widehat{\mathbb{C}}$ whose dilatation does not exceed $exp(2\rho_V(x, x_0))$. Moreover, $J_x$ depends holomorphically on $x$ over $V$.*

**Generalized harmonic $\lambda$-lemma.** Let $\Omega$ be a plane region whose complement in $\mathbb{C}$ contains at least two points. We call a Beltrami coefficient $\mu$ on $\Omega$ *harmonic* if the function $\overline{\mu}\lambda^2$ on $\Omega$ is holomorphic, where $\lambda$ is the Poincaré density on $\Omega$.

In the next theorem, $V$ is a simply connected complex Banach manifold with a basepoint $x_0$, $\rho_V$ is the Kobayashi metric on $V$ and $\rho_\Delta$ is the Poincaré metric on $\Delta$. (We assume that $V$ is hyperbolic.)

THEOREM 4. *Let $\phi : V \times E \to \widehat{\mathbb{C}}$ be a holomorphic motion. There exists a **unique** holomorphic motion $\widehat{\phi} : U \times \widehat{\mathbb{C}} \to \widehat{\mathbb{C}}$, where*

$$U = \{x \in V : \rho_V(x, x_0) < \rho_\Delta(0, 1/3)\}$$

*such that $\widehat{\phi}(x, z) = \phi(x, z)$ for all $(x, z) \in U \times E$ and the Beltrami coefficient of the map $z \mapsto \widehat{\phi}(x, z)$ is harmonic on each component of $\widehat{\mathbb{C}} \setminus E$ for each $x \in U$.*

See Theorem B in [**27**].

REMARKS.

1. For $V = \Delta$ this is Theorem 3 in [**7**]. We call it the *Bers-Royden extenstion theorem.*

2. In [**11**], Douady observed that the Bers-Royden extension theorem can be generalized to holomorphic motions over open unit balls in complex Banach spaces. A proof was sketched by Sugawa in his important paper [**35**].

3. The Bers-Royden extension theorem was called the "harmonic $\lambda$-lemma" by McMullen and Sullivan in [**26**]. The uniqueness of the extension has special advantages. That was used in [**26**] to study quasiconformal conjugacies.

4. For some other applications of the Bers-Royden extension theorem see the papers [**5**], [**6**], and [**32**].

5. Theorem 4 proves the harmonic $\lambda$-lemma in its fullest generality.

## 5. Quasiconformal motions

Let $X$ be a connected Hausdorff space with a basepoint $x_0$, and let $E$ be any subset of $\widehat{\mathbb{C}}$, not necessarily closed; as usual, we assume that 0, 1, and $\infty$ belong to $E$. For any motion $\phi : X \times E \to \widehat{\mathbb{C}}$, $x$ in $X$, and any quadruplet of distinct points $a, b, c, d$ of points in $E$, let $\phi_x(a, b, c, d)$ denote the cross-ratio of the values $\phi(x, a), \phi(x, b), \phi(x, c)$ and $\phi(x, d)$. As mentioned before, we will often write $\phi(x, z)$ as $\phi_x(z)$ for $x$ in $X$ and $z$ in $E$. So we have;

$$\phi_x(a, b, c, d) = \frac{(\phi_x(a) - \phi_x(c))(\phi_x(b) - \phi_x(d))}{(\phi_x(a) - \phi_x(d))(\phi_x(b) - \phi_x(c))}$$

for each $x$ in $X$.

It is obvious that condition (ii) in Definition 2 holds if and only if $\phi_x(a, b, c, d)$ is a well-defined point in the thrice-punctured sphere $\widehat{\mathbb{C}} \setminus \{0, 1, \infty\}$ for all $x$ in $X$ and all quadruplets $a, b, c, d$ of distinct points in $E$.

Let $\rho$ be the Poincaré distance on $\widehat{\mathbb{C}} \setminus \{0, 1, \infty\}$. In their paper [**36**], Sullivan and Thurston introduced the following definition.

DEFINITION 5. A *quasiconformal motion* is a motion $\phi : X \times E \to \widehat{\mathbb{C}}$ of $E$ over $X$ with the following additional property:

(iii) given any $x$ in $X$ and any $\epsilon > 0$, there exists a neighborhood $U_x$ of $x$ such that for any quadruplet of distinct points $a, b, c, d$ in $E$, we have

$$\rho\Big(\phi_y(a, b, c, d), \phi_{y'}(a, b, c, d)\Big) < \epsilon$$

for all $y$ and $y'$ in $U_x$.

REMARK 2. It was shown in Proposition 14 of [**21**] that every quasiconformal motion of $E$ can be extended to its closure. Therefore, we will assume that $E$ is a closed set in $\widehat{\mathbb{C}}$, containing the points 0, 1, and $\infty$.

REMARK 3. Suppose that $\phi : V \times E \to \widehat{\mathbb{C}}$ is a holomorphic motion, where $V$ is a connected complex manifold with a basepoint $x_0$, and let $E$ be a closed set in $\widehat{\mathbb{C}}$, containing 0, 1, and $\infty$. For any quadruplet of points $a$, $b$, $c$, $d$ in $E$, the map $x \mapsto \phi_x(a, b, c, d)$ from $V$ into $\widehat{\mathbb{C}} \setminus \{0, 1, \infty\}$ is holomorphic. Therefore, it is distance-decreasing with respect to the Kobayashi metrics on $V$ and $\widehat{\mathbb{C}} \setminus \{0, 1, \infty\}$. It easily follows that $\phi$ is also a quasiconformal motion.

The following characterization of quasiconformal motions of $\widehat{\mathbb{C}}$ was proved in [**29**].

PROPOSITION 1. *A motion $\phi : X \times \widehat{\mathbb{C}} \to \widehat{\mathbb{C}}$ is quasiconformal if and only if it satisfies:*

(i) *the map $\phi_x : \widehat{\mathbb{C}} \to \widehat{\mathbb{C}}$ is quasiconformal for each $x$ in $X$, and*

(ii) *the map from $X$ to $M(\mathbb{C})$ that sends $x$ to the Beltrami coefficient of $\phi_x$ for each $x$ in $X$ is continuous.*

Part (ii) means that the map $x \mapsto \mu_x = \frac{(\phi_x)_{\bar{z}}}{(\phi_x)_z}$, $x \in X$, is continuous.

We now introduce the concept of *continuous motions.*

DEFINITION 6. Let $X$ be a connected Hausdorff space with a basepoint $x_0$. A *continuous motion* of $\widehat{\mathbb{C}}$ over $X$ is a motion $\phi : X \times \widehat{\mathbb{C}} \to \widehat{\mathbb{C}}$ such that the map $\phi$ is continuous.

REMARK 4. Recall that all motions are normalized. If $\phi$ is a continuous motion of $\widehat{\mathbb{C}}$, then each $\phi_x$, $x$ in $X$, is a map from $\widehat{\mathbb{C}}$ to itself that fixes 0, 1, and $\infty$. Since $\phi_x$ is injective and continuous, it is a homeomorphism of $\widehat{\mathbb{C}}$ onto itself, by invariance of domain. Alternatively, note that $\phi_x$ maps compact sets to compact set. Since $\widehat{\mathbb{C}}$ is compact, any closed subset of $\widehat{\mathbb{C}}$ is compact. Therefore, $\phi_x$ is a closed map, which in turn means that $\phi_x^{-1}$ is continuous.

REMARK 5. Let $X$ be a connected Hausdorff space with a basepoint $x_0$, and let $E$ be a closed set in $\widehat{\mathbb{C}}$ such that $E$ contains the points 0, 1, and $\infty$. We note that a *normalized continuous motion* of $E$ over $X$ is a continuous map $\phi : X \times E \to \widehat{\mathbb{C}}$ such that:

(i) $\phi(x_0, z) = z$ for all $z$ in $E$, and

(ii) for each $x$ in $X$, the map $\phi(x, \cdot)$ is a homeomorphism onto its image, that fixes the points 0, 1, and $\infty$.

The following was proved in [**29**].

PROPOSITION 2. *Every quasiconformal motion of $\widehat{\mathbb{C}}$ is a continuous motion.*

The following stronger result was proved in [**21**]; see Proposition 14 in that paper.

PROPOSITION 3. *Let $\phi : X \times E \to \widehat{\mathbb{C}}$ be a quasiconformal motion of $E$, where $X$ is a connected Hausdorff space with a basepoint $x_0$. Then, $\phi : X \times E \to \widehat{\mathbb{C}}$ is also a continuous motion.*

## 6. Extensions of holomorphic motions to quasiconformal motions

We noted in §3 that Slodkowski's theorem does not, in general, hold for higher-dimensional parameter spaces. However, we have the following extension theorem of holomorphic motions to quasiconformal motions of the sphere.

THEOREM 5. *Let $V$ be a simply connected complex Banach manifold with a basepoint $x_0$, and let $\phi : V \times E \to \widehat{\mathbb{C}}$ be a holomorphic motion. Then, there exists a quasiconformal motion $\widetilde{\phi} : V \times \widehat{\mathbb{C}} \to \widehat{\mathbb{C}}$ such that:*

(i) *For each $x$ in $V$, the map $\widetilde{\phi}_x : \widehat{\mathbb{C}} \to \widehat{\mathbb{C}}$ is a quasiconformal map; let its Beltrami coefficient be $\mu_x$.*
(ii) *The map $x \mapsto \mu_x$ is continuous for $x$ in $V$.*
(iii) *$\widetilde{\phi}(x, z) = \phi(x, z)$ for all $(x, z) \in V \times E$.*
(iv) *The $L^\infty$ norm of $\mu_x$ is bounded above by a number less than 1, that depends only on the Kobayashi distance from $x$ to $x_0$, denoted by $\rho_V(x, x_0)$.*

*Furthermore, if $\phi$ is a $G$-equivariant holomorphic motion (see Definition 3), then $\widetilde{\phi}$ can be chosen to be a $G$-equivariant quasiconformal motion.*

For a complete proof, see the papers [**19**], and [**30**].

## 7. Tame quasiconformal motions

As mentioned in §5, the concept of quasiconformal motions was first introduced in the paper by Sullivan and Thurston [**36**]. Theorem 3 of that paper claimed that every quasiconformal motion of any set over an interval can be extended to the Riemann sphere.

A natural question is:

If $\phi : X \times E \to \widehat{\mathbb{C}}$ is a quasiconformal motion, where $X$ is any simply connected Hausdorff space with a basepoint $x_0$, does there exist a quasiconformal motion $\widetilde{\phi} : X \times \widehat{\mathbb{C}} \to \widehat{\mathbb{C}}$ such that $\widetilde{\phi}$ extends $\phi$?

The answer is affirmative if $E$ is finite. See Theorem 14 in §9.3 of this paper (also proved in [**21**]).

However, Theorem I of [**21**] showed that the answer is negative for an infinite closed set, where $X$ is an interval $I := [0, 1]$. That gives a counterexample to Theorem 3 of [**36**]. Here is the precise statement of Theorem 1 in [**21**].

THEOREM 6. *There exists a closed, infinite set $E$ in $\widehat{\mathbb{C}}$ and a quasiconformal motion $\phi : I \times E \to \widehat{\mathbb{C}}$, such that $\phi$ can be extended to a continuous motion of $\widehat{\mathbb{C}}$ over $I$. However, for any neighborhood $U$ about 0, $\phi$ cannot be extended to a quasiconformal motion of $\widehat{\mathbb{C}}$ over $U$.*

In that same paper [**21**], the authors introduced a new concept, called *tame quasiconformal motions*, and showed that Theorem 3 of Sullivan–Thurston holds for tame quasiconformal motions over any simply connected Hausdorff space. It was also shown that this extension can be done in a conformally natural way, for tame quasiconformal motions.

DEFINITION 7. Let $X$ be a connected Hausdorff space with a basepoint $x_0$, and $E$ be any set in $\widehat{\mathbb{C}}$ (containing the points 0, 1, and $\infty$). A *tame quasiconformal motion* is a motion $\phi : X \times E \to \widehat{\mathbb{C}}$ of $E$ over $X$ with the additional property:

(iii) Given any $x$ in $X$, there exists a quasiconformal map $w : \widehat{\mathbb{C}} \to \widehat{\mathbb{C}}$, a neighborhood $N(x)$, with basepoint $x$, and a quasiconformal motion $\psi : N(x) \times \widehat{\mathbb{C}} \to \widehat{\mathbb{C}}$ over $N(x)$ such that $\phi(y, z) = \psi(y, w(z))$ for all $(y, z) \in N(x) \times E$.

The following equivalent formulation was proved in [**21**]; for the reader's convenience, we include a proof.

LEMMA 1. *A motion $\phi : X \times E \to \widehat{\mathbb{C}}$ is a tame quasiconformal motion if and only if given any $x$ in $X$, there exists a neighborhood $N(x)$, and a continuous map $g_x : N(x) \to M(\mathbb{C})$ such that $\phi(y, z) = w^{g_x(y)}(z)$ for all $(y, z) \in N(x) \times E$.*

PROOF. Let $\phi : X \times E \to \widehat{\mathbb{C}}$ be a motion. Suppose, for each $x$ in $X$, there exists a neighborhood $N(x)$, and a continuous map $g_x : N(x) \to M(\mathbb{C})$ such that $\phi(y, z) = w^{g_x(y)}(z)$ for all $(y, z) \in N(x) \times E$. Set $w = w^{g_x(x)}$ and $\psi(y, z) = w^{g_x(y)}(w^{-1}(z))$ in $N(x) \times \widehat{\mathbb{C}}$. It now follows that $\phi$ is a tame quasiconformal motion of $E$ over $X$.

Conversely, if $\phi : X \times E \to \widehat{\mathbb{C}}$ is a tame quasiconformal motion, then by Proposition 1, the condition of our lemma immediately follows. □

PROPOSITION 4. *Let $\phi : X \times E \to \widehat{\mathbb{C}}$ be a tame quasiconformal motion. Then, $\phi$ is a quasiconformal motion.*

PROOF. The proof follows immediately from Lemma 1 and the quasi-invariance of cross ratios (see Theorem 1 in [**23**]). □

Here is the extension theorem for tame quasiconformal motions. In what follows, $V$ is a simply connected Hausdorff space with basepoint $x_0$, and $E$ is a closed set in $\widehat{\mathbb{C}}$; as usual, 0, 1, and $\infty$ belong to $E$. Also, let $G$ be a group of Möbius transformations, such that $E$ is invariant under $G$ (which means $g(E) = E$ for all $g$ in $G$). As in Definition 1, a tame quasiconformal motion $\phi : X \times E \to \widehat{\mathbb{C}}$ is called *$G$-equivariant* if and only if for each $g$ in $G$, and $x$ in $X$, there exists a Möbious transformation $\theta_x(g)$ such that

$$\phi(x, g(z)) = \theta_x(g)(\phi(x, z)) \tag{7.1}$$

for all $z \in E$.

THEOREM 7. *Let $\phi : V \times E \to \widehat{\mathbb{C}}$ be a $G$-equivariant tame quasiconformal motion. Then, there exists a $G$-equivariant quasiconformal motion $\widetilde{\phi} : V \times \widehat{\mathbb{C}} \to \widehat{\mathbb{C}}$ such that $\widetilde{\phi}$ extends $\phi$.*

*This means the following:*

(i) *For each $x$ in $V$, the map $\widetilde{\phi}_x : \widehat{\mathbb{C}} \to \widehat{\mathbb{C}}$ is a quasiconformal map; let its Beltrami coefficient be $\mu_x$.*
(ii) *The map $x \mapsto \mu_x$ is continuous for $x$ in $V$.*
(iii) *$\widetilde{\phi}(x, z) = \phi(x, z)$ for all $(x, z) \in V \times E$.*
(iv) *$\widetilde{\phi}_x \circ g \circ \widetilde{\phi}_x^{-1} = \theta_x(g)$ for each $g$ in $G$.*

## 8. Some geometrical applications

In March 1996, at the Oberwolfach conference, Dieter Gaier posed four problems about the dependence of conformal invariants on parameters, that were hard to solve using classical methods. Three of them were on the modules of ring domains or quadrilaterals. The general question can be stated as follows:

*Suppose a quadrilateral or ring domain $D$ in $\widehat{\mathbb{C}}$ is varied by letting its boundary points depend analytically on some parameters. Is the module of $D$ a real analytic function of the parameters?*

It was shown in [**14**] that the answer is positive if the parameter space is a connected complex Banach manifold $V$ and the boundary of $D$ undergoes a holomorphic motion over $V$. In this section we state the basic results. For all details, the reader is referred to the paper [**14**].

Let $V$ be a connected complex Banach manifold with a basepoint $x_0$, and let $\phi : V \times E \to \widehat{\mathbb{C}}$ be a holomorphic motion of a closed set $E$ over $V$. For each $x$ in $V$ there exists a quasiconformal map $\widetilde{\phi}_x$ of $\widehat{\mathbb{C}}$ onto itself such that $\widetilde{\phi}_x(z) = \phi(x, z)$ for all $z$ in $E$. One can show that, if $D$ is a component of $\widehat{\mathbb{C}} \setminus E$, the region $\widetilde{\phi}_x(D)$ depends only on $x$, $\phi$, and $D$, not on the choice of $\widetilde{\phi}_x$. For $x$ in $V$, we set $\phi_x(E) := \{\phi(x, z) : z \in E\}$ and we define $\phi_x(D)$ to be the component of $\widehat{\mathbb{C}} \setminus \phi_x(E)$ such that $\phi_x(D) = \widetilde{\phi}_x(D)$ for any quasiconformal map $\widetilde{\phi}_x$ of $\widehat{\mathbb{C}}$ onto itself satisfying $\widetilde{\phi}_x(z) = \phi(x, z)$ for all $z$ in $E$. So, if $D$ is a region in $\widehat{\mathbb{C}}$ and $E$ is its boundary, we get a well-defined region $\phi_x(D)$ for each $x$ in $V$.

**Variable ring domains.** We call the region $D$ in $\widehat{\mathbb{C}}$ a *nondegenerate ring domain* if and only if it is conformally equivalent to the annulus

$$A(R) = \{z \in \mathbb{C} : 1 < |z| < R\}$$

for some finite number $R > 1$. The number $R$ is uniquely determined by $D$, and the number

$$mod(D) = \frac{1}{2\pi} \log(R)$$

is called the *module* of $D$.

THEOREM 8. *Let $E$ be the boundary of the nondegenerate ring domain $D$. If $\phi$ is a holomorphic motion of $E$ over the parameter space $V$, (where $V$ is a connected complex Banach manifold with a basepoint), then the set $\phi_x(D)$ is a nondegenerate ring domain for each $x$ in $V$, and the function $x \mapsto mod(\phi_x(D))$ is real analytic in $V$.*

See Theorem 1.3 in [**14**].

**Variable quadrilaterals.** By definition, a quadrilateral is a Jordan region $D$ in $\widehat{\mathbb{C}}$ with four distinguished points $a$, $b$, $c$, and $d$ on its boundary. We list them in the order of their occurrence as the boundary is traversed in the positive direction (i.e. with $D$ on the left) starting at the point $a$. Let $f$ be a conformal mapping of $D$ onto a rectangle in $\mathbb{C}$ whose vertices are the points $f(a)$, $f(b)$, $f(c)$, and $f(d)$. By definition, the *module* of the quadrilateral $(D; a, b, c, d)$ is the ratio

$$mod(D; a, b, c, d) = \frac{|f(b) - f(a)|}{|f(c) - f(b)|}.$$

THEOREM 9. *Let $E$ be the boundary of the quadrilateral $(D; a, b, c, d)$, and let $\phi$ be a holomorphic motion of $E$ over $V$, where $V$ is a connected complex Banach manifold with a basepoint. The function*

$$x \mapsto mod\Big(\phi_x(D); \phi(x, a), \phi(x, b), \phi(x, c), \phi(x, d)\Big)$$

*is real analytic in $V$.*

See Theorem 1.6 in [**14**].

**Poincaré geodesics in variable regions.** Let $\phi : V \times E \to \widehat{\mathbb{C}}$ be a holomorphic motion of a closed set $E$, and let $\gamma$ be a closed curve in the component $D$ of $\widehat{\mathbb{C}} \setminus E$. Let $\widetilde{\phi}$ and $\widetilde{\psi}$ be two quasiconformal motions that extend $\phi$. For each $x$ in $V$, the regions $\widetilde{\phi}_x(D)$ and $\widetilde{\psi}_x(D)$ are equal, and the closed curves $\widetilde{\phi}_x(\gamma)$ and $\widetilde{\psi}_x(\gamma)$ are homotopic in $\widetilde{\phi}_x(D)$. If $\gamma$ is a Poincaré geodesic in $D$, then the homotopy class of $\widetilde{\phi}_x(\gamma)$ in $\widetilde{\phi}_x(D)$ contains a unique Poincaré geodesic, which we denote by $\phi_x(\gamma)$.

THEOREM 10. *Let $D$ be a region in $\widehat{\mathbb{C}}$ whose boundary $E$ contains at least three points, and let $\phi$ be a holomorphic motion of $E$ over the simply connected complex Banach manifold $V$. For any closed Poincaré geodesic $\gamma$ in $D$, the length of the closed Poincaré geodesic $\phi_x(\gamma)$ in $\phi_x(D)$ is a real analytic function of $x$ in $V$.*

See Theorem 1.7 in [**14**]

## 9. Teichmüller space of a closed set in $\widehat{\mathbb{C}}$

All theorems discussed in the previous sections depend on some analytic properties of the Teichmüller space of a closed set $E$ in $\widehat{\mathbb{C}}$. These Teichmüller spaces were first studied by G. S. Lieb in his 1990 Cornell University dissertation [**24**]. Subsequently, they were extensively studied in the papers [**8**], [**10**], [**14**], [**19**], [**27**], and [**30**].

Let $E$ be a closed set in $\widehat{\mathbb{C}}$, containing 0, 1, and $\infty$.

DEFINITION 8. Two normalized quasiconformal self-mappings $f$ and $g$ of $\widehat{\mathbb{C}}$ are said to be $E$-equivalent if and only if $f^{-1} \circ g$ is isotopic to the identity rel $E$. The *Teichmüller space* $T(E)$ is the set of all $E$-equivalence classes of normalized quasiconformal self-mappings of $\widehat{\mathbb{C}}$. The basepoint of $T(E)$ is the $E$-equivalence class of the identity map.

The reader should note that by "$f^{-1} \circ g$ is isotopic to the identity rel $E$" we mean that, at each stage of the isotopy, $f^{-1} \circ g$ keeps the set $E$ pointwise fixed.

DEFINITION 9. Let $X$ and $Y$ be connected Hausdorff spaces with basepoints, and $f$ be a continuous basepoint preserving map of $X$ into $Y$. If $\phi$ is a motion of $E$ over $Y$ its *pullback* by $f$ is the motion

$$f^*(\phi)(x,z) = \phi(f(x),z) \qquad \forall (x,z) \in X \times E$$

of $E$ over $X$.

REMARK 6. If the motion $\phi$ is continuous, or tame quasiconformal, $f^*(\phi)$ has the same property. If $X$ and $Y$ are complex manifolds, $f$ holomorphic, and $\phi$ is a holomorphic motion, then so is $f^*(\phi)$.

Recall that $M(\mathbb{C})$ denotes the open unit ball of the complex Banach space $L^\infty(\mathbb{C})$. Each $\mu$ in $M(\mathbb{C})$ is the Beltrami coefficient of a unique normalized quasiconformal homeomorphism $w^\mu$ of $\widehat{\mathbb{C}}$ onto itself. The basepoint of $M(\mathbb{C})$ is the zero function.

We can define the quotient map

$$P_E : M(\mathbb{C}) \to T(E)$$

by setting $P_E(\mu)$ equal to the $E$-equivalence class of $w^\mu$, written as $[w^\mu]_E$. Clearly, $P_E$ maps the basepoint of $M(\mathbb{C})$ to the basepoint of $T(E)$.

In his doctoral dissertation (see [**24**]), G. Lieb proved that $T(E)$ is a complex Banach manifold such that the projection map $P_E$ from $M(\mathbb{C})$ to $T(E)$ is a holomorphic split submersion. The result was also proved in [**14**].

### 9.1. Universal holomorphic motion of $E$.

The *universal holomorphic motion* $\Psi_E$ *of* $E$ *over* $T(E)$ is defined as follows:

$$\Psi_E(P_E(\mu), z) = w^\mu(z) \text{ for } \mu \in M(\mathbb{C}) \text{ and } z \in E.$$

The definition of $P_E$ above guarantees that $\Psi_E$ is well-defined. It is a holomorphic motion since $P_E$ is a holomorphic split submersion and $\mu \mapsto w^\mu(z)$ is a holomorphic map from $M(\mathbb{C})$ to $\widehat{\mathbb{C}}$ for every fixed $z$ in $\widehat{\mathbb{C}}$ (by Theorem 11 in [**2**]). This holomorphic motion is "universal" in the following sense:

THEOREM 11. *Let $\phi : W \times E \to \widehat{\mathbb{C}}$ be a holomorphic motion where $W$ is a simply connected complex Banach manifold with a basepoint $x_0$. Then, there exists a unique basepoint preserving holomorphic map $f : W \to T(E)$ such that $f^*(\Psi_E) = \phi$; which means, $\phi(x, z) = \Psi_E(f(x), z)$ for all $(x, z) \in W \times E$.*

For a proof, see §14 in [**27**].

Here is a generalization of Theorem 11; here the parameter space is not necessarily simply connected (see the papers [**28**] and [**29**]).

THEOREM 12. *Let $V$ be a connected complex Banach manifold with a basepoint $x_0$, and let $\phi : V \times E \to \widehat{\mathbb{C}}$ be a holomorphic motion. The following are equivalent:*

(i) *There exists a continuous motion $\widehat{\phi} : V \times \widehat{\mathbb{C}} \to \widehat{\mathbb{C}}$ such that $\widehat{\phi}$ extends $\phi$.*
(ii) *There exists a quasiconformal motion $\widetilde{\phi} : V \times \widehat{\mathbb{C}} \to \widehat{\mathbb{C}}$ such that $\widetilde{\phi}$ extends $\phi$.*
(iii) *There exists a unique basepoint preserving holomorphic map $F : V \to T(E)$ such that $F^*(\Psi_E) = \phi$.*

Every holomorphic motion is a tame quasiconformal motion. This was shown in [**21**]; for the reader's convenience we include a sketch here.

PROPOSITION 5. *Let $\phi : X \times E \to \widehat{\mathbb{C}}$ be a holomorphic motion where $X$ is a connected complex Banach manifold with a basepoint $x_0$. Then, $\phi$ is a tame quasiconformal motion.*

PROOF. It is sufficient to consider a simply connected neighborhood $N(x_0)$ of the basepoint $x_0$. By Theorem 5, there exists a quasiconformal motion $\widetilde{\phi}$ : $N(x_0) \times \widehat{\mathbb{C}} \to \widehat{\mathbb{C}}$ such that $\widetilde{\phi}(x, z) = \phi(x, z)$ for all $(x, z) \in N(x_0) \times E$. Our conclusion follows from Proposition 1 and Lemma 1. □

By Proposition 5, every holomorphic motion is also a tame quasiconformal motion. Hence, $\Psi_E : T(E) \times E \to \widehat{\mathbb{C}}$ is also a tame quasiconformal motion. In [**21**], it was proved that this is the universal tame quasiconformal motion of the closed set $E$ over a simply connected Hausdorff space. Here is the precise statement:

THEOREM 13. *Let $\phi : X \times E \to \widehat{\mathbb{C}}$ be a tame quasiconformal motion where $X$ is a simply connected Hausdorff space with a basepoint $x_0$. There exists a unique basepoint preserving continuous map $f : X \to T(E)$ such that $f^*(\Psi_E) = \phi$.*

See Theorem II in [**21**].

**9.2. Douady-Earle section.** The following fact was crucial in the paper [**21**].

PROPOSITION 6. *There is a continuous basepoint preserving map $s$ from $T(E)$ to $M(\mathbb{C})$ such that $P_E \circ s$ is the identity map on $T(E)$.*

See [**14**] or [**18**] for a proof. It immediately implies that

COROLLARY 1. *The Teichmüller space $T(E)$ is contractible.*

DEFINITION 10. The map $s$ from $T(E)$ to $M(\mathbb{C})$ is called the *Douady-Earle section of $P_E$* for the Teichmüller space $T(E)$.

We give an application of the importance of Douady-Earle section. In what follows, $X$ is a connected Hausdorff space with a basepoint $x_0$, and $E$ is a closed set in $\widehat{\mathbb{C}}$; as usual, 0, 1, and $\infty$ belong to $E$. This is Proposition 5 in [**20**]. For the reader's convenience, we include the proof.

PROPOSITION 7. *A continuous motion $\phi : X \times E \to \widehat{\mathbb{C}}$ extends to a quasiconformal motion $\widetilde{\phi} : X \times \widehat{\mathbb{C}} \to \widehat{\mathbb{C}}$ if and only if there exists a basepoint-preserving continuous map $f : X \to T(E)$ such that $\phi(x, z) = \Psi_E(f(x), z)$ for all $(x, z) \in X \times E$.*

PROOF. Suppose there exists a basepoint-preserving continuous map $f : X \to T(E)$ such that $\phi(x, z) = \Psi_E(f(x), z)$ for all $(x, z) \in X \times E$. Define $\widetilde{f} = s \circ f$, where $s$ is the Douady-Earle section given in Proposition 5; then, $\widetilde{f} : X \to M(\mathbb{C})$ is a basepoint preserving continuous map. Let

$$\widetilde{\phi}(x, z) = w^{\widetilde{f}(x)}(z) \qquad (x, z) \in X \times \widehat{\mathbb{C}}.$$

By Proposition 1, $\widetilde{\phi}$ is a quasiconformal motion of $\widehat{\mathbb{C}}$ over $X$. Also, for all $(x, z) \in X \times E$, we have

$$\phi(x, z) = \Psi_E(f(x), z) = \Psi_E(P_E(s(f(x))), z) = \Psi_E(P_E(\widetilde{f}(x)), z) = w^{\widetilde{f}(x)}(z).$$

This shows that $\widetilde{\phi}$ extends $\phi$.

Next, suppose that there exists a quasiconformal motion $\widetilde{\phi} : X \times \widehat{\mathbb{C}} \to \widehat{\mathbb{C}}$ extending $\phi$. Then, by Proposition 1, the map $\widetilde{\phi}_x : \widehat{\mathbb{C}} \to \widehat{\mathbb{C}}$ is quasiconformal, for each $x$ in $X$. Let $\widetilde{f}(x)$ be the Beltrami coefficient of $\widetilde{\phi}_x$; by Proposition 1, $\widetilde{f} : X \to M(\mathbb{C})$ is a basepoint-preserving continuous map. Now, set $f = P_E \circ \widetilde{f}$. Clearly, $f$ is a basepoint-preserving continuous map. Note that

$$\Psi_E(f(x), z) = \Psi_E(P_E(\widetilde{f}(x)), z) = w^{\widetilde{f}(x)}(z) = \widetilde{\phi}(x, z) = \phi(x, z)$$

for all $(x, z) \in X \times E$. □

**9.3. When $E$ is finite.** The case when $E$ is finite has some interesting applications. Let $E$ be a finite set in $\widehat{\mathbb{C}}$ containing the points 0, 1, and $\infty$. In this case, its complement $\Omega = \widehat{\mathbb{C}} \setminus E$ is the Riemann sphere with punctures at the points of $E$. Since $T(E)$ and the classical Teichmüller space are quotients of $M(\mathbb{C})$ by the same equivalence relation, $T(E)$ can be naturally identified with $Teich(\Omega)$. See Example 3.1 in [**27**] for the details.

The above identification was used in the paper [**20**] to prove the following result. It shows the remarkable connection between continuous motions and quasiconformality. In what follows, $V$ is a simply connected Hausdorff space with a basepoint $x_0$.

THEOREM 14. *Suppose $E$ consists of $n + 3$ distinct points $\{0, 1, \infty, \zeta_1, \cdots \zeta_n\}$. Let $\phi : V \times E \to \widehat{\mathbb{C}}$ be a continuous motion. Then, there exists a quasiconformal motion $\widetilde{\phi} : V \times \widehat{\mathbb{C}} \to \widehat{\mathbb{C}}$ that extends $\phi$. Moreover, $\widetilde{\phi}$ is unique up to isotopy rel $E$.*

We give a sketch of the proof of Theorem 14. We first recall the following important result; see [**31**].

PROPOSITION 8. *Given $n > 0$, choose a point $(\zeta_1, \cdots, \zeta_n)$ in the domain*

$$Y_n = \{(z_1, \cdots, z_n) \in \mathbb{C}^n : z_i \neq z_j \text{ for } i \neq j \text{ and } z_i \neq 0, 1 \text{ for all } i = 1, \cdots, n\}$$

*and let $E = \{0, 1, \infty, \zeta_1, \cdots, \zeta_n\}$. Then, the map $p : T(E) \to Y_n$ defined by*

$$p([w^\mu]_E) = (w^\mu(\zeta_1), \cdots, w^\mu(\zeta_n)) \text{ for all } \mu \in M(\mathbb{C})$$

*is a holomorphic universal covering.*

We now show how to construct the quasiconformal motion $\widetilde{\phi} : V \times \widehat{\mathbb{C}} \to \widehat{\mathbb{C}}$ in Theorem 14.

We define a map $F : V \to Y_n$ by

$$F(x) = (\phi(x, \zeta_1), \cdots, \phi(x, \zeta_n)).$$

Clearly, the map $F$ is a basepoint preserving continuous map. Since $V$ is simply connected, by Proposition 7, this map $F$ has a unique basepoint preserving lift

$$f : V \to T(E)$$

under $p$ that is also a continuous map. The definition of the universal covering map $p$ then shows that $f^*(\Psi_E) = \phi$.

It now follows by Proposition 6, that there exists a quasiconformal motion $\widetilde{\phi} : V \times \widehat{\mathbb{C}} \to \widehat{\mathbb{C}}$ that extends $\phi$.

For the uniqueness of $\widetilde{\phi}$ upto isotopy rel $E$, we refer to Proposition 3 in [**20**].

The following are immediate applications of Theorem 14. See Theorems 1 and 2 in [**20**]. In what follows, $V$ is a simply connected Hausdorff space with a basepoint $x_0$.

THEOREM 15. *Let $D$ be a region in $\widehat{\mathbb{C}}$ and $E$ be a finite subset of the boundary $\partial D$, such that $E$ contains at least four points. Suppose $\Gamma$ is a family of curves in $D$. Let $\phi : V \times E \to \widehat{\mathbb{C}}$ be a continuous motion. Then, for each $x$ in $V$, there exists a region $D_x$ and a family $\Gamma_x$ of curves in $D_x$ with $D_{x_0} = D$, $\Gamma_{x_0} = \Gamma$, such that $E_x := \phi_x(E) \subset \partial D_x$, and the extremal length $\Lambda(\Gamma_x)$ varies continuously with respect to $x$.*

THEOREM 16. *Under the assumptions of Theorem 15, let $\gamma$ be a closed Poincaré geodesic in $D$. For each $x$ in $V$, there is a region $D_x$ and a closed Poincaré geodesic $\gamma_x$ in $D_x$ with $D_{x_0} = D$, $\gamma_{x_0} = \gamma$, such that $E_x := \phi_x(E) \subset \partial D_x$, and the hyperbolic length of $\gamma_x$ varies continuously with respect to $x$.*

The following is an example of a continuous motion of a four-point set over the punctured unit disk, that cannot be extended to a quasiconformal motion of the sphere.

EXAMPLE. Let $\Delta^* := \{z \in \mathbb{C} : 0 < |z| < 1\}$ and choose some basepoint $a$ in $\Delta^*$. Let $E := \{0, 1, \infty, a\}$. Set $\phi(t, z) = z$ for all $(t, z) \in \Delta^* \times \{0, 1, \infty\}$, and $\phi(t, a) = t$ for all $t$ in $\Delta^*$. Then, clearly, $\phi$ is a continuous motion of $E$ over $\Delta^*$. Suppose there exists a quasiconformal motion $\widetilde{\phi} : \Delta^* \times \widehat{\mathbb{C}} \to \widehat{\mathbb{C}}$ that extends $\phi$. By Proposition 2, $\widetilde{\phi}$ is a continuous motion. For each $\zeta$ in $\mathbb{C} \setminus \{0\}$, let $\gamma_\zeta : [0, 2\pi] \to \mathbb{C} \setminus \{0\}$ be the closed curve

$$\gamma_\zeta(\theta) = \widetilde{\phi}(ae^{i\theta}, \zeta)$$

for $\theta$ in $[0, 2\pi]$. Since $\widetilde{\phi}$ is a continuous motion, the winding number of $\gamma_\zeta$ about zero is a continuous function of $\zeta$. But that winding number is zero when $\zeta = 1$ and one when $\zeta = a$.

REMARK. This appeared in [**28**] for holomorphic motions. We thank Clifford Earle for showing us this example. For other such examples, the reader is referred to the paper [**4**], where we introduce the concepts of *trace monodromy* and *monodromy* of a holomorphic motion.

The universal holomorphic motion of a finite set gives an interesting example of a maximal holomorphic motion. The following was proved in [**19**]; see Proposition 5.4 of that paper.

PROPOSITION 9. *Let* $E = \{0, 1, \infty, \zeta_1, \zeta_2, \cdots, \zeta_n\}$ *where* $\zeta_i \neq \zeta_j$ *if* $i \neq j$ *and* $n \geq 2$. *Consider the universal holomorphic motion* $\Psi_E : T(E) \times E \to \widehat{\mathbb{C}}$.

(i) *There exists a quasiconformal motion* $\widetilde{\Psi}_E : T(E) \times \widehat{\mathbb{C}} \to \widehat{\mathbb{C}}$ *such that* $\widetilde{\Psi}_E$ *extends* $\Psi_E$.
(ii) *If* $a \in \widehat{\mathbb{C}} \setminus E$, *there does not exist any holomorphic motion* $\widehat{\phi} : T(E) \times (E \cup \{a\}) \to \widehat{\mathbb{C}}$ *such that* $\widehat{\phi}$ *extends* $\Psi_E$.

REMARK. The relationship between holomorphic, quasiconformal, tame quasiconformal, and continuous motions can be summarized as follows:

**Holomorphic motions** $\Longrightarrow$ **Tame quasiconformal motions**
$\Longrightarrow$ **Quasiconformal motions** $\Longrightarrow$ **Continuous motions.**

## Acknowledgment

We want to thank the referee for his meticulous reading of our paper, and for his very valuable and careful suggestions. We are very grateful to the referee for bringing to our attention the reference [**34**], and also for showing an alternative argument in Remark 4.

## References

[1] L. V. Ahlfors, *Lectures on quasiconformal mappings*, 2nd ed., University Lecture Series, vol. 38, American Mathematical Society, Providence, RI, 2006. With supplemental chapters by C. J. Earle, I. Kra, M. Shishikura and J. H. Hubbard, DOI 10.1090/ulect/038. MR2241787

[2] L. Ahlfors and L. Bers, *Riemann's mapping theorem for variable metrics*, Ann. of Math. (2) **72** (1960), 385–404, DOI 10.2307/1970141. MR115006

[3] K. Astala and G. J. Martin, *Holomorphic motions*, Papers on analysis, Rep. Univ. Jyväskylä Dep. Math. Stat., vol. 83, Univ. Jyväskylä, Jyväskylä, 2001, pp. 27–40. MR1886611

[4] M. Beck, Y. Jiang, S. Mitra, and H. Shiga, *Extending holomorphic motions and monodromy*, Ann. Acad. Sci. Fenn. Math. **37** (2012), no. 1, 53–67, DOI 10.5186/aasfm.2012.3713. MR2920423

[5] L. Bers, *On a theorem of Abikoff*, Ann. Acad. Sci. Fenn. Ser. A I Math. **10** (1985), 83–87, DOI 10.5186/aasfm.1985.1010. MR802469

[6] L. Bers, *Holomorphic families of isomorphisms of Möbius groups*, J. Math. Kyoto Univ. **26** (1986), no. 1, 73–76, DOI 10.1215/kjm/1250520965. MR827159

[7] L. Bers and H. L. Royden, *Holomorphic families of injections*, Acta Math. **157** (1986), no. 3-4, 259–286, DOI 10.1007/BF02392595. MR857675

[8] N. Chatterjee, *Some metric properties of the Teichmüller space of a closed set in the Riemann sphere*, Ann. Acad. Sci. Fenn. Math. **42** (2017), no. 2, 755–770, DOI 10.5186/aasfm.2017.4247. MR3701646

[9] E. M. Chirka, *On the extension of holomorphic motions*, Doklady Mathematics **70** (1) (2004), 516–519.

[10] X. Dong and S. Mitra, *Carathéodory metric on some generalized Teichmüller spaces* (English, with English and Finnish summaries), Ann. Fenn. Math. **48** (2023), no. 2, 797–807. MR4673802

[11] A. Douady, *Prolongement de mouvements holomorphes (d'après Słodkowski et autres)* (French, with French summary), Astérisque **227** (1995), Exp. No. 775, 3, 7–20. Séminaire Bourbaki, Vol. 1993/94. MR1321641

[12] A. Douady and C. J. Earle, *Conformally natural extension of homeomorphisms of the circle*, Acta Math. **157** (1986), no. 1-2, 23–48, DOI 10.1007/BF02392590. MR857678

[13] C. J. Earle, I. Kra, and S. L. Krushkal′, *Holomorphic motions and Teichmüller spaces*, Trans. Amer. Math. Soc. **343** (1994), no. 2, 927–948, DOI 10.2307/2154750. MR1214783
[14] C. J. Earle and S. Mitra, *Variation of moduli under holomorphic motions*, In the tradition of Ahlfors and Bers (Stony Brook, NY, 1998), Contemp. Math., vol. 256, Amer. Math. Soc., Providence, RI, 2000, pp. 39–67, DOI 10.1090/conm/256/03996. MR1759669
[15] F. P. Gardiner and L. Keen, *Conformal dynamical systems and Lipman Bers*, in Selected Works of Lipman Bers, Papers on Complex Analysis, Part 2, AMS, 1998.
[16] F. P. Gardiner, Y. Jiang, and Z. Wang, *Holomorphic motions and related topics*, Geometry of Riemann surfaces, London Math. Soc. Lecture Note Ser., vol. 368, Cambridge Univ. Press, Cambridge, 2010, pp. 156–193. MR2665009
[17] J. H. Hubbard, *Teichmüller theory and applications to geometry, topology, and dynamics; Volume 1*, Matrix Editions, Ithaca, NY, 2006.
[18] Y. Jiang and S. Mitra, *Some applications of universal holomorphic motions*, Kodai Math. J. **30** (2007), no. 1, 85–96, DOI 10.2996/kmj/1175287624. MR2319079
[19] Y. Jiang and S. Mitra, *Douady-Earle section, holomorphic motions, and some applications*, Quasiconformal mappings, Riemann surfaces, and Teichmüller spaces, Contemp. Math., vol. 575, Amer. Math. Soc., Providence, RI, 2012, pp. 219–251, DOI 10.1090/conm/575/11400. MR2933902
[20] Y. Jiang and S. Mitra, *Variation of moduli under continuous motions*, In the tradition of Ahlfors-Bers. VII, Contemp. Math., vol. 696, Amer. Math. Soc., Providence, RI, 2017, pp. 205–214, DOI 10.1090/conm/696/14023. MR3715448
[21] Y. Jiang, S. Mitra, H. Shiga, and Z. Wang, *Teichmüller spaces and tame quasiconformal motions*, Tohoku Math. J. (2) **70** (2018), no. 4, 607–631, DOI 10.2748/tmj/1546570827. MR3896139
[22] Y. Jiang, S. Mitra, and Z. Wang, *Liftings of holomorphic maps into Teichmüller spaces*, Kodai Math. J. **32** (2009), no. 3, 547–563, DOI 10.2996/kmj/1257948895. MR2582017
[23] I. Kra, *On Teichmüller's theorem on the quasi-invariance of cross ratios*, Israel J. Math. **30** (1978), no. 1-2, 152–158, DOI 10.1007/BF02760836. MR508260
[24] G. S. Lieb, *Holomorphic motions and Teichmuller space*, ProQuest LLC, Ann Arbor, MI, 1989. Thesis (Ph.D.)–Cornell University. MR2638376
[25] R. Mañé, P. Sad, and D. Sullivan, *On the dynamics of rational maps*, Ann. Sci. École Norm. Sup. (4) **16** (1983), no. 2, 193–217. MR732343
[26] C. T. McMullen and D. P. Sullivan, *Quasiconformal homeomorphisms and dynamics. III. The Teichmüller space of a holomorphic dynamical system*, Adv. Math. **135** (1998), no. 2, 351–395, DOI 10.1006/aima.1998.1726. MR1620850
[27] S. Mitra, *Teichmüller spaces and holomorphic motions*, J. Anal. Math. **81** (2000), 1–33, DOI 10.1007/BF02788984. MR1785276
[28] S. Mitra, *Extensions of holomorphic motions*, Israel J. Math. **159** (2007), 277–288, DOI 10.1007/s11856-007-0047-1. MR2342482
[29] S. Mitra, *Extensions of holomorphic motions to quasiconformal motions*, In the tradition of Ahlfors-Bers. IV, Contemp. Math., vol. 432, Amer. Math. Soc., Providence, RI, 2007, pp. 199–208, DOI 10.1090/conm/432/08309. MR2342817
[30] S. Mitra and H. Shiga, *Extensions of holomorphic motions and holomorphic families of Möbius groups*, Osaka J. Math. **47** (2010), no. 4, 1167–1187. MR2791561
[31] S. Nag, *The Torelli spaces of punctured tori and spheres*, Duke Math. J. **48** (1981), no. 2, 359–388. MR620255
[32] B. Rodin, *Behavior of the Riemann mapping function under complex analytic deformations of the domain*, Complex Variables Theory Appl. **5** (1986), no. 2-4, 189–195, DOI 10.1080/17476938608814139. MR846487
[33] Z. Slodkowski, *Holomorphic motions and polynomial hulls*, Proc. Amer. Math. Soc. **111** (1991), no. 2, 347–355, DOI 10.2307/2048323. MR1037218
[34] H. Shiga and T. Sugawa, *Kleinian groups and geometric function theory*, In the tradition of Thurston III. Geometry and dynamics, Springer, Cham, 2024, pp. 249–274, DOI 10.1007/978-3-031-43502-7_7. MR4759045
[35] T. Sugawa, *The Bers projection and the $\lambda$-lemma*, J. Math. Kyoto Univ. **32** (1992), no. 4, 701–713, DOI 10.1215/kjm/1250519403. MR1194110
[36] D. P. Sullivan and W. P. Thurston, *Extending holomorphic motions*, Acta Math. **157** (1986), no. 3-4, 243–257, DOI 10.1007/BF02392594. MR857674

Department of Mathematics, Naba Barrackpur Prafulla Chandra Mahavidyalaya, P.O. New Barrackpur, Dist.–North 24 Parganas, Kolkata 700131, India
*Email address*: nishan6392@gmail.com

Department of Mathematics, Pondicherry University, Kalapet, Puducherry 605014, India
*Email address*: arshiyafarhathg@gmail.com

Department of Mathematics, Queens College of the City University of New York, Flushing, New York 11367-1597; and Department of Mathematics, Graduate Center of the City University of New York, New York, New York 10016
*Email address*: sudeb.mitra@qc.cuny.edu

Contemporary Mathematics
Volume **820**, 2025
https://doi.org/10.1090/conm/820/16412

# On generalized extended Voigt-type function and associated bounds

Rakesh K. Parmar and Uthara S

ABSTRACT. In this present paper, we establish functional bounding inequalities, monotonicity properties, log-convexity properties and Turán-type inequality results for the generalized extended Voigt-type function. The computable series expansions for the extended Voigt-type function are also obtained in terms of Kampé de Fériet function of two variables.

## 1. Introduction and preliminaries

In 1899, Woldemar Voigt (1850-1919) introduced "*Zur Theorie der Beugung ebener inhomogener Wellen an einem geradlinig begrentzen unendlichen und absolut schwarzen Schirm*" the so-called Voigt functions $K(x,y)$ and $L(x,y)$ [**28**]. The Voigt functions $K(x,y)$ and $L(x,y)$ which play an essential role in astrophysical spectroscopy and neutron physics are investigated and generalized from the viewpoint of integral operators. Also, the function $K(x,y) + i\,L(x,y)$ is identical to the so-called plasma dispersion function, which has been tabulated by Fried and Conte[**8**]. On the other hand, the Doppler broadened Breit-Wigner resonances in neutron reactions are essentially the same as the Voigt functions. A broad spectrum of physics problems, including astrophysical spectroscopy, radiation emission, absorption, and transfer of radiation in heated atmospheres, plasma dispersion, and even the theory of neutron reaction, commonly involve the Voigt functions $K(x,y)$ and $L(x,y)$, see for example [**1**,**2**]. The Voigt functions $K(x,y)$ and $L(x,y)$ can be represented by the convolution integrals [**9**]:

$$K(x,y) = \frac{y}{\pi}\int_{-\infty}^{\infty} \frac{\mathrm{e}^{-t^2}}{(x-t)^2+y^2}\,\mathrm{d}t \quad (x,\, y \in \mathbb{R}) \tag{1.1}$$

and

$$L(x,y) = \frac{1}{\pi}\int_{-\infty}^{\infty} \frac{(x-t)\,\mathrm{e}^{-t^2}}{(x-t)^2+y^2}\,\mathrm{d}t \quad (x,\, y \in \mathbb{R}). \tag{1.2}$$

---

2020 *Mathematics Subject Classification.* Primary33C05, 33C20; Secondary 33C65.

*Key words and phrases.* Voigt functions $K(x,y)$ and $L(x,y)$; Generalized hypergeometric function; Fox–Wright function; Kampé de Fériet function; Bessel function of the first kind; functional bounding inequalities; log–convexity; Turán inequality.

The first author is the corresponding author.

A further representation of the Voigt functions $K(x,y)$ and $L(x,y)$ are obtained by introducing the cosine Fourier integral transform of $\frac{y}{(x-t)^2+y^2}$ and the sine Fourier integral transform of $\frac{(x-t)}{(x-t)^2+y^2}$ in equations (1.1) and (1.2), respectively, and inverting the order of integration, which are due essentially to Reiche[**22**]

$$K(x,y)=\frac{1}{\sqrt{\pi}}\int_0^\infty \mathrm{e}^{-yt-\frac{t^2}{4}}\cos(xt)\,\mathrm{d}t \quad (x\in\mathbb{R},\ y\in\mathbb{R}^+) \tag{1.3}$$

and

$$L(x,y)=\frac{1}{\sqrt{\pi}}\int_0^\infty \mathrm{e}^{-yt-\frac{t^2}{4}}\sin(xt)\,\mathrm{d}t \quad (x\in\mathbb{R},\ y\in\mathbb{R}^+). \tag{1.4}$$

Srivastava and Miller[**24**] introduced an unification of the Voigt function $K(x,y)$ and $L(x,y)$ in the form

$$V_{\mu,\nu}(x,y)=\sqrt{\frac{x}{2}}\int_0^\infty t^\mu \mathrm{e}^{-yt-\frac{t^2}{4}}J_\nu(xt)\,\mathrm{d}t \quad (\mu,x,y\in\mathbb{R}^+;\Re(\mu+\nu)>-1) \tag{1.5}$$

where $J_\nu$ is the Bessel function of first kind of the order $\nu$ and has the power series[**29**],

$$J_\nu(z)=\sum_{k\geq 0}\frac{(-1)^k(\frac{z}{2})^{\nu+2k}}{k!\,\Gamma(\nu+k+1)} \quad (z\in\mathbb{C}). \tag{1.6}$$

Using the fact[**17**]

$$J_{-\frac{1}{2}}(z)=\sqrt{\frac{2}{\pi z}}\cos(z) \qquad \text{and} \qquad J_{\frac{1}{2}}(z)=\sqrt{\frac{2}{\pi z}}\sin(z),$$

in (1.6), we obtain

$$K(x,y)=V_{\frac{1}{2},-\frac{1}{2}}(x,y) \qquad L(x,y)=V_{\frac{1}{2},\frac{1}{2}}(x,y).$$

Klush [**10**] proposed a unification of the Voigt functions $K(x,y)$ and $L(x,y)$ in the form :

$$\Omega_{\mu,\nu}[x,y,z]=\sqrt{\frac{x}{2}}\int_0^\infty t^\mu e^{-yt-zt^2}J_\nu(xt)dt \quad (\mu,x,y\in\mathbb{R}^+;\Re(\mu+\nu)>-1). \tag{1.7}$$

In fact, it is easily verified by comparing (1.5) and (1.7) that

$$V_{\mu,\nu}(x,y)=(2\sqrt{z})^{\mu+\frac{1}{2}}\Omega_{\mu,\nu}(2x\sqrt{z},2y\sqrt{z},z)$$

or, equivalently, that

$$\Omega_{\mu,\nu}[x,y,z]=(2\sqrt{z})^{-\mu-\frac{1}{2}}V_{\mu,\nu}\left(\frac{x}{2\sqrt{z}},\frac{y}{2\sqrt{z}}\right).$$

Here, we consider the extended generalized Voigt-type function in the following form:

$$\mathfrak{V}_{\mu,\nu}^{\boldsymbol{\alpha},\boldsymbol{\beta}}(x,y,z)=\sqrt{\frac{x}{2}}\int_0^\infty t^\mu \mathrm{e}^{-yt-zt^2}{}_pF_{p+1}\left[\begin{matrix}\alpha_1,\cdots,\alpha_p\\ \beta_1,\cdots,\beta_p,\nu+1\end{matrix}\,\Big|\,\frac{-x^2t^2}{4}\right]\mathrm{d}t, \tag{1.8}$$

where ${}_rF_s[z]$ is the generalized hypergeometric function[**17**]

$$ {}_rF_s\left[\begin{array}{c} a_1,\cdots,a_r \\ b_1,\cdots,b_s \end{array}\Big| z\right] = \sum_{k\geq 0} \frac{\prod\limits_{l=1}^{r}(a_l)_k}{\prod\limits_{l=1}^{s}(b_l)_k} \frac{z^k}{k!}, $$

where, as usual, the Pochhammer symbol (or raising factorial) is defined by

$$ (\tau)_0 = 1; \quad (\tau)_k = \tau(\tau+1)\cdots(\tau+k-1) = \frac{\Gamma(\tau+k)}{\Gamma(\tau)}, \qquad k \in \mathbb{N}, $$

and $\boldsymbol{\alpha} = \{\alpha_1, \alpha_2, \cdots, \alpha_p\}$, $\boldsymbol{\beta} = \{\beta_1, \beta_2, \cdots, \beta_p\}$ and $\mu, x, y, z \in \mathbb{R}^+$.

In the present article, we investigate the generalized extended Voigt function $\mathfrak{V}^{\boldsymbol{\alpha},\boldsymbol{\beta}}_{\mu,\nu}(x,y,z)$ by replacing Bessel function $J_\nu$ by generalized hypergeometric function ${}_pF_{p+1}$ in the integrand of (1.7). In section 2, representations of generalized extended Voigt function $\mathfrak{V}^{\boldsymbol{\alpha},\boldsymbol{\beta}}_{\mu,\nu}(x,y,z)$ is expressed in terms of Kampé de Fériet function of two variables. Moreover, functional bounding inequalities are also derived. In section 3, monotonicity properties, log-convexity properties and Turán-type inequality results are presented for the investigated generalized extended Voigt-type function $\mathfrak{V}^{\boldsymbol{\alpha},\boldsymbol{\beta}}_{\mu,\nu}(x,y,z)$.

## 2. Computable series representation and bounds for $\mathfrak{V}^{\boldsymbol{\alpha},\boldsymbol{\beta}}_{\mu,\nu}(x,y,z)$

The definition of Fox-Wright function, which is a generalization of hypergeometric function ${}_rF_s[z]$, is defined as follows [**7**], [**31**, p. 4, Eq. (2.4)]:

$$ {}_r\Psi_s\left[\begin{array}{c} (a_1,A_1),\cdots,(a_r,A_r) \\ (b_1,B_1),\cdots,(b_s,B_s) \end{array}\Bigg| z\right] = \sum_{k\geq 0} \frac{\prod\limits_{j=1}^{r}\Gamma(a_j+kA_j)}{\prod\limits_{j=1}^{s}\Gamma(b_j+kB_j)} \frac{z^k}{k!}, \tag{2.1} $$

where $A_j > 0,\ j = 1,\cdots,r; B_j > 0,\ j = 1,\cdots,s$. The convergence conditions for the series at the right-hand side of (2.1) follow from the known asymptotic of the Euler Gamma–function. The defining series in (2.1) converges in the whole complex $z$-plane when

$$ \Delta = 1 + \sum_{j=1}^{q} B_j - \sum_{i=1}^{p} A_i > 0. $$

If $\Delta = 0$, then the series in (2.1) converges for $|z| < \rho$, and $|z| = \rho$ under the condition $\Re(\theta) > \frac{1}{2}$ where

$$ \rho = \prod_{j=1}^{r} A_j^{-A_j} \cdot \prod_{j=1}^{s} B_j^{B_j}, \qquad \theta = \sum_{j=1}^{s} b_j - \sum_{k=1}^{r} a_k + \frac{r-s}{2}. \tag{2.2} $$

In the special case $A_i = B_j = 1; i = 1,\cdots,r; j = 1,\cdots,s$, the Fox–Wright function ${}_r\Psi_s[z]$ reduces (up to the multiplicative constant) to the generalized hypergeometric function

$$ {}_r\Psi_s\left[\begin{array}{c} (a_1,1),\cdots,(a_r,1) \\ (b_1,1),\cdots,(b_s,1) \end{array}\Big| z\right] = \frac{\Gamma(a_1)\cdots\Gamma(a_r)}{\Gamma(b_1)\cdots\Gamma(b_s)} {}_rF_s\left[\begin{array}{c} a_1,\cdots,a_r \\ b_1,\cdots,b_s \end{array}\Big| z\right]. $$

Next, the defintion of the Kampé de Fériet function of two variables is defined by [**26**, p. 27, Eq. (28)](see also, [**6**]):

$$F^{p:q;k}_{l:m;n}\left[\begin{array}{c}(a_p):(b_q);(c_k);\\(\alpha_l):(\beta_m);(\gamma_n);\end{array} x,\, y\right] = \sum_{r,s=0}^{\infty} \frac{\prod_{j=1}^{p}(a_j)_{r+s}\prod_{j=1}^{q}(b_j)_r\prod_{j=1}^{k}(c_j)_s}{\prod_{j=1}^{l}(\alpha_j)_{r+s}\prod_{j=1}^{m}(\beta_j)_r\prod_{j=1}^{n}(\gamma_j)_s}\frac{x^r}{r!}\frac{y^s}{s!},$$

with the convergence,

$(i)$ $p+q<l+m+1$ and $p+k<l+n+1$, $|x|<\infty$ and $|y|<\infty$,

or

$(ii)$ $p+q=l+m+1$ and $p+k=l+n+1$,

and $\begin{cases}|x|^{\frac{1}{(p-1)}}+|y|^{\frac{1}{(p-1)}}<1, & \text{if} p>1\\ \max\left(|x|,|y|<1\right), & \text{if} p\leq 1.\end{cases}$

**2.1. Computable series representation.** First, we establish the computable series expansions for the extended Voigt-type function in terms of Kampé de Fériet function of two variables.

THEOREM 1. *Let $\mu$, $x$, $y\in\mathbb{R}^+$ and $\Re(\mu)>-1$. Then the following representation holds true:*

$$\mathfrak{V}^{\boldsymbol{\alpha},\boldsymbol{\beta}}_{\mu,\nu}(x,y,z) = \frac{1}{2z^{\frac{\mu+1}{2}}}\sqrt{\frac{x}{2}}\left\{\Gamma\left(\frac{\mu+1}{2}\right)F^{1:0,p}_{0:1,p+1}\left[\begin{array}{c}\frac{\mu+1}{2}:-,(\alpha_p);\\ -:\frac{1}{2},(\beta_p),1+\nu;\end{array}\frac{-x^2}{4z},\frac{y^2}{4z}\right]\right.$$
$$\left.-\frac{y}{\sqrt{z}}\Gamma\left(\frac{\mu+2}{2}\right)F^{1:0,p}_{0:1,p+1}\left[\begin{array}{c}\frac{\mu+2}{2}:-,(\alpha_p);\\ -:\frac{3}{2},(\beta_p),1+\nu;\end{array}\frac{-x^2}{4z},\frac{y^2}{4z}\right]\right\},$$

*where $F^{p:q;k}_{l:m;n}[x,y]$ denotes the Kampé de Fériet function of two variables.*

PROOF. Using the series representation of generalized hypergeometric function ${}_pF_{p+1}$ and the exponential function $\exp(-yt)$ and then integrating the resulting (absolutely convergent) double series term by term in the definition of extended generalized Voigt-type function $\mathfrak{V}^{\boldsymbol{\alpha},\boldsymbol{\beta}}_{\mu,\nu}(x,y,z)$ (1.8), we get

$$\mathfrak{V}^{\boldsymbol{\alpha},\boldsymbol{\beta}}_{\mu,\nu}(x,y,z) = \frac{1}{2z^{\frac{\mu+1}{2}}}\sqrt{\frac{x}{2}}\sum_{m,n=0}^{\infty}\frac{\prod_{j=1}^{p}\{(\alpha_j)_n\}}{\prod_{j=1}^{p}\{(\beta_j)_n\}(1+\nu)_n}\frac{\left(\frac{-x^2}{4z}\right)^n}{n!}\frac{\left(\frac{-y}{\sqrt{z}}\right)^m}{m!}$$
$$\times\Gamma\left(\frac{\mu+2n+m+1}{2}\right).$$

Now, finally separating the $m$-series into its even and odd terms and which after simple calculations, we obtain the representation in terms of Kampé de Fériet function of two variables as

$$\mathfrak{V}^{\boldsymbol{\alpha},\boldsymbol{\beta}}_{\mu,\nu}(x,y,z) = \frac{1}{2z^{\frac{\mu+1}{2}}}\sqrt{\frac{x}{2}}\left\{\Gamma\left(\frac{\mu+1}{2}\right)F^{1:0,p}_{0:1,p+1}\left[\begin{array}{c}\frac{\mu+1}{2}:-,(\alpha_p);\\ -:\frac{1}{2},(\beta_p),1+\nu;\end{array}\frac{-x^2}{4z},\frac{y^2}{4z}\right]\right.$$
$$\left.-\frac{y}{\sqrt{z}}\Gamma\left(\frac{\mu+2}{2}\right)F^{1:0,p}_{0:1,p+1}\left[\begin{array}{c}\frac{\mu+2}{2}:-,(\alpha_p);\\ -:\frac{3}{2},(\beta_p),1+\nu;\end{array}\frac{-x^2}{4z},\frac{y^2}{4z}\right]\right\}.$$

□

**2.2. Bounding inequalities.** Next, we recall the bounds for $J_\nu$, which ones we will employ on the positive real half–axis. Firstly, we mention von Lommel's results [**12**], [**13**, pp. 548–549] (see also [**29**, p. 406])

$$|J_\nu(t)| \le 1, \qquad |J_{\nu+1}(t)| \le \frac{1}{\sqrt{2}}, \qquad \nu \in \mathbb{R}^+,\, t \in \mathbb{R}, \tag{2.3}$$

and the bound by Minakshisundaram and Szász [**14**, p. 37]

$$|J_\nu(t)| \le \frac{1}{\Gamma(\nu+1)} \left(\frac{|t|}{2}\right)^\nu, \qquad t \in \mathbb{R}. \tag{2.4}$$

Another bounds were derived by Landau [**11**], who gave in a sense best possible bounds for $J_\nu$ with respect to $\nu$ and $t$:

$$|J_\nu(t)| \le b_L\, \nu^{-1/3}, \qquad b_L = \sqrt[3]{2} \sup_{t \ge 0} \mathrm{Ai}(t), \tag{2.5}$$

$$|J_\nu(t)| \le c_L\, |t|^{-1/3}, \qquad c_L = \sup_{t \ge 0} t^{1/3} J_0(t)\,, \tag{2.6}$$

where $\mathrm{Ai}(\cdot)$ stands for the Airy function

$$\mathrm{Ai}(x) = \frac{\pi}{2}\sqrt{\frac{x}{3}}\left(J_{-\frac{1}{3}}\left(2(x/3)^{3/2}\right) + J_{\frac{1}{3}}\left(2(x/3)^{3/2}\right)\right).$$

Olenko established the upper bound [**16**, Theorem 1]

$$\sup_{t \ge 0} \sqrt{t}|J_\nu(t)| \le b_L \sqrt{\nu^{1/3} + \frac{\alpha_1}{\nu^{1/3}} + \frac{3\alpha_1^2}{10\nu}} = d_O, \qquad \nu \in \mathbb{R}^+, \tag{2.7}$$

where $\alpha_1$ is the smallest positive zero of the Airy–function Ai and $b_L$ is the Landau's constant from above. Further considerable upper bounds are listed, for example, in [**3, 4, 21, 27**]. In recent papers[**18, 19**], certain sharp bounding inequalities for the generalized Voigt-type function were derived by making use of the known integral representation [**5**, p. 200, Eq. 13.1(95)]

$$\begin{aligned} &{}_{r+1}F_{s+1}[\rho, a_1, \cdots, a_r; \rho+\sigma, b_1, \cdots, b_s; z] \\ &\quad = \frac{\Gamma(\rho+\sigma)}{\Gamma(\rho)\Gamma(\sigma)} \int_0^1 u^{\rho-1}(1-u)^{\sigma-1}\, {}_rF_s[a_1, \cdots, a_r; b_1, \cdots, b_s; zu]\, du, \end{aligned} \tag{2.8}$$

$$(r \le s+1;\ \min\{\Re(\rho), \Re(\sigma)\} > 0;\ |z| < 1 \text{ when } p = q+1),$$

by specializing the parameters and applying several known upper bounds for the first-kind of the Bessel function $J_\nu(x)$ for the generalized hypergeometric function ${}_1F_2$ and ${}_2F_3$ with appealing the bounds of Lommel's, Minakshisundaram and Szász, Landau and Olenko, respectively given in (2.3), (2.4), (2.5), (2.6) and (2.7). Now by using the integral representation (2.8), if we specify $r = s-1 = p-1$, $\rho = \alpha_1$, $\rho+\sigma = \beta_1$, $a_1 = \alpha_2, \cdots, a_{p-1} = \alpha_p$, $b_1 = \beta_2, \cdots, b_{p-1} = \beta_p$, $b_{p+1} = 1+\nu$, and $z = -\frac{x^2t^2}{4}$, we conclude for $\beta_1 > \alpha_1 > 0$

$$\begin{aligned} {}_pF_{p+1}\left[\begin{matrix} \alpha_1, \cdots, \alpha_p \\ \beta_1, \cdots, \beta_p, \nu+1 \end{matrix} \,\Big|\, \frac{-x^2t^2}{4}\right] &= \frac{\Gamma(\beta_1)}{\Gamma(\alpha_1)\Gamma(\beta_1-\alpha_1)} \int_0^1 u^{\alpha_1-1}(1-u)^{\beta_1-\alpha_1-1} \\ &\quad \times {}_{p-1}F_p\left[\begin{matrix} \alpha_2, \cdots, \alpha_p \\ \beta_2, \cdots, \beta_p, \nu+1 \end{matrix} \,\Big|\, \frac{-ux^2t^2}{4}\right] du. \end{aligned} \tag{2.9}$$

Now by using the integral representation (2.9) and same methodology used to obtain the upper bounds for ${}_1F_2$ and ${}_2F_3$ in [**18**, pp. 194, Theorem 1] and [**19**, pp. 50, Theorem 1], respectively, we can obtain bounds for ${}_pF_{p+1}$. Here we are giving the bounds for ${}_pF_{p+1}$ in the Lemma 1 without proof.

LEMMA 1. *Let* $\boldsymbol{\alpha},\ \boldsymbol{\beta},\ x,\ t,\ \nu \in \mathbb{R}^+$. *When* $2\min\{\boldsymbol{\alpha},\ \boldsymbol{\beta}\} > \nu$ *we have*

$$\left|{}_pF_{p+1}\left[\begin{matrix}\alpha_1,\cdots,\alpha_p\\ \beta_1,\cdots,\beta_p,\nu+1\end{matrix}\,\Big|\,\frac{-x^2t^2}{4}\right]\right| \le \frac{2^{\nu-\frac{1}{2}}\,\Gamma(\nu+1)\prod\limits_{j=1}^{p}\{\Gamma(\beta_j)\,\Gamma\left(\alpha_j-\frac{\nu}{2}\right)\}}{|xt|^{\nu}\prod\limits_{j=1}^{p}\{\Gamma(\alpha_j)\Gamma\left(\beta_j-\frac{\nu}{2}\right)\}}. \tag{2.10}$$

*Moreover,*

$$\left|{}_pF_{p+1}\left[\begin{matrix}\alpha_1,\cdots,\alpha_p\\ \beta_1,\cdots,\beta_p,\nu+1\end{matrix}\,\Big|\,\frac{-x^2t^2}{4}\right]\right| \le \frac{b_L\,2^{\nu}\,\Gamma(\nu+1)\prod\limits_{j=1}^{p}\{\Gamma(\beta_j)\,\Gamma\left(\alpha_j-\frac{\nu}{2}\right)\}}{\sqrt[3]{\nu}\,|xt|^{\nu}\prod\limits_{j=1}^{p}\{\Gamma(\alpha_j)\Gamma\left(\beta_j-\frac{\nu}{2}\right)\}} \tag{2.11}$$

*and*

$$\left|{}_pF_{p+1}\left[\begin{matrix}\alpha_1,\cdots,\alpha_p\\ \beta_1,\cdots,\beta_p,\nu+1\end{matrix}\,\Big|\,\frac{-x^2t^2}{4}\right]\right| \le \begin{cases}\dfrac{c_L\,2^{\nu}\,\Gamma(\nu+1)\prod\limits_{j=1}^{p}\{\Gamma(\beta_j)\,\Gamma\left(\alpha_j-\frac{\nu}{2}-\frac{1}{6}\right)\}}{|xt|^{\nu+\frac{1}{3}}\prod\limits_{j=1}^{p}\{\Gamma(\alpha_j)\Gamma\left(\beta_j-\frac{\nu}{2}-\frac{1}{6}\right)\}}, \\[2ex] \dfrac{d_O\,2^{\nu}\,\Gamma(\nu+1)\prod\limits_{j=1}^{p}\{\Gamma(\beta_j)\,\Gamma\left(\alpha_j-\frac{\nu}{2}-\frac{1}{4}\right)\}}{|xt|^{\nu+\frac{1}{2}}\prod\limits_{j=1}^{p}\{\Gamma(\alpha_j)\Gamma\left(\beta_j-\frac{\nu}{2}-\frac{1}{4}\right)\}}.\end{cases} \tag{2.12}$$

*Here the bound above holds if* $6\min\{\boldsymbol{\alpha},\ \boldsymbol{\beta}\} > 3\nu+1$, *while the expression below appears for* $4\min\{\boldsymbol{\alpha},\ \boldsymbol{\beta}\} > 2\nu+1$.

Our next goal is to give a functional bound for the function $\mathfrak{V}_{\mu,\nu}^{\boldsymbol{\alpha},\boldsymbol{\beta}}(x,y,z)$ by exploring its integral representation (1.8) with the help of Lemma 1.

THEOREM 2. *Let* $\boldsymbol{\alpha},\ \boldsymbol{\beta}, x, y, z, \mu, \nu > 0$. *For all* $\min\{\boldsymbol{\alpha},\ \boldsymbol{\beta}\} > \nu$ *it is*

$$\left|\mathfrak{V}_{\mu,\nu}^{\boldsymbol{\alpha},\boldsymbol{\beta}}(x,y,z)\right| \le \frac{2^{\nu-2}\,\Gamma(\nu+1)\prod\limits_{j=1}^{p}\{\Gamma(\beta_j)\,\Gamma\left(\alpha_j-\frac{\nu}{2}\right)\}}{|x|^{\nu-\frac{1}{2}}z^{\frac{\mu-\nu+1}{2}}\prod\limits_{j=1}^{p}\{\Gamma(\alpha_j)\Gamma\left(\beta_j-\frac{\nu}{2}\right)\}}\,{}_1\Psi_0\left[\begin{matrix}\left(\frac{\mu-\nu+1}{2},\frac{1}{2}\right)\\ -\end{matrix}\,\Big|\,-\frac{y}{\sqrt{z}}\right]. \tag{2.13}$$

*In the same parameter range holds*

$$\left|\mathfrak{V}_{\mu,\nu}^{\boldsymbol{\alpha},\boldsymbol{\beta}}(x,y,z)\right| \le \frac{b_L\,2^{\nu-\frac{3}{2}}\,\Gamma(\nu+1)\prod\limits_{j=1}^{p}\{\Gamma(\beta_j)\,\Gamma\left(\alpha_j-\frac{\nu-\frac{3}{2}}{2}\right)\}}{\sqrt[3]{\nu}\,|x|^{\nu-\frac{1}{2}}z^{\frac{\mu-\nu+1}{2}}\prod\limits_{j=1}^{p}\{\Gamma(\alpha_j)\Gamma\left(\beta_j-\frac{\nu}{2}\right)\}}\,{}_1\Psi_0\left[\begin{matrix}\left(\frac{\mu-\nu+1}{2},\frac{1}{2}\right)\\ -\end{matrix}\,\Big|\,-\frac{y}{\sqrt{z}}\right] \tag{2.14}$$

*and*

(2.15)

$$\left|\mathfrak{V}_{\mu,\nu}^{\boldsymbol{\alpha},\boldsymbol{\beta}}(x,y,z)\right| \leq \begin{cases} \dfrac{c_L\, 2^{\nu-\frac{3}{2}}\,\Gamma(\nu+1)\prod\limits_{j=1}^{p}\{\Gamma(\beta_j)\,\Gamma\big(\alpha_j-\frac{\nu}{2}-\frac{1}{6}\big)\}}{|x|^{\nu-\frac{1}{6}}\, z^{\frac{\mu}{2}-\frac{\nu}{2}+\frac{1}{3}}\prod\limits_{j=1}^{p}\{\Gamma(\alpha_j)\Gamma\big(\beta_j-\frac{\nu}{2}-\frac{1}{6}\big)\}}\; {}_1\Psi_0\left[\begin{matrix}(\frac{\mu}{2}-\frac{\nu}{2}+\frac{1}{3},\frac{1}{2})\\ -\end{matrix}\,\middle|\, -\frac{y}{\sqrt{z}}\right], \\[4ex] \dfrac{d_O\, 2^{\nu-\frac{3}{2}}\,\Gamma(\nu+1)\prod\limits_{j=1}^{p}\{\Gamma(\beta_j)\,\Gamma\big(\alpha_j-\frac{\nu}{2}-\frac{1}{4}\big)\}}{|x|^{\nu}\, z^{\frac{\mu}{2}-\frac{\nu}{2}+\frac{1}{4}}\prod\limits_{j=1}^{p}\{\Gamma(\alpha_j)\Gamma\big(\beta_j-\frac{\nu}{2}-\frac{1}{4}\big)\}}\; {}_1\Psi_0\left[\begin{matrix}(\frac{\mu}{2}-\frac{\nu}{2}+\frac{1}{4},\frac{1}{2})\\ -\end{matrix}\,\middle|\, -\frac{y}{\sqrt{z}}\right], \end{cases}$$

*where the bound above holds if* $6\min\{\boldsymbol{\alpha},\ \boldsymbol{\beta}\} > 3\nu+1$, *while the expression below appears for* $4\min\{\boldsymbol{\alpha},\ \boldsymbol{\beta}\} > 2\nu+1$. *Here* ${}_1\Psi_0$ *is the confluent Fox–Wright function for the case* $r=1$ *and* $s=0$ *in* (2.1).

PROOF. Let us evaluate the integral similar to given in [**27**, p. 198, Eq.(15)]

$$\mathcal{I}_\mu(y,z) = \int_0^\infty t^{\mu-1}\, e^{-yt-zt^2}\, dt,$$

by the series expansion of the exponential $e^{-yt}$ in the kernel into its Taylor-Maclaurin series and legitimate integral–sum interchange and then using the familiar Gamma formula [**17**]

$$\Gamma(\eta)\xi^{-\eta} = \int_0^\infty e^{-\xi t} t^{\eta-1}\,\mathrm{d}t, \qquad \Re(\xi) > 0,\ \Re(\eta) > 0,$$

we obtain

$$\mathcal{I}_\mu(y,z) = \frac{1}{2\, z^{\frac{\mu}{2}}} \sum_{m=0}^{\infty} \frac{\Gamma\left(\dfrac{\mu+m}{2}\right)}{m!} \left(-\frac{y}{\sqrt{z}}\right)^m = \frac{1}{2\, z^{\frac{\mu}{2}}}\, {}_1\Psi_0\left[\begin{matrix}(\frac{\mu}{2},\frac{1}{2})\\ -\end{matrix}\,\middle|\, -\frac{y}{\sqrt{z}}\right],$$

where we have used the defintion of Fox–Wright hypergeometric function for the case $r=1$ and $s=0$ in (2.1), say confluent Fox–Wright function ${}_1\Psi_0$.

Applying the first inequality (2.10) to the modulus of the generalized hypergeometric function ${}_pF_{p+1}$ occurring in the integrand of (1.8), we easily conclude that

(2.16)

$$\left|\mathfrak{V}_{\mu,\nu}^{\boldsymbol{\alpha},\boldsymbol{\beta}}(x,y,z)\right| \leq \frac{2^{\nu-1}\,\Gamma(\nu+1)\prod\limits_{j=1}^{p}\{\Gamma(\beta_j)\,\Gamma\left(\alpha_j-\frac{\nu}{2}\right)\}}{|x|^{\nu-\frac{1}{2}}\prod\limits_{j=1}^{p}\{\Gamma(\alpha_j)\Gamma\left(\beta_j-\frac{\nu}{2}\right)\}} \int_0^\infty t^{\mu-\nu}\, e^{-yt-zt^2}\, dt.$$

Now applying integral representation $\mathcal{I}_\mu(y,z)$ given above in (2.16), we arrive at (2.13). Similarly using the inequalities (2.11) and (2.12) as above and simplifying, we get the results (2.14) and (2.15), respectively. □

## 3. Complete monotonicity, log–convexity and Turán type inequalities

In this section, we prove the complete monotonicity and log–convexity properties, and certain Turán–type inequalities for the extended generalized Voigt–type function $\mathfrak{V}_{\mu,\nu}^{\boldsymbol{\alpha},\boldsymbol{\beta}}(x,y,z)$ in (1.8).

The function $f : (0,\infty) \to \mathbb{R}$ is completely monotonic (or totally monotonic), if $f$ has derivatives of all (positive integer) orders and satisfies

$$(-1)^m f^{(m)}(x) \geq 0, \qquad m \in \mathbb{N}_0;\ x > 0.$$

We say that a function $f : [a,b] \subseteq \mathbb{R} \to \mathbb{R}$ is said to be logrithimically convex, or simply log-convex, if its natural logarithm $\log h$ is convex, that is, for all $x, y \in [a,b]$ and $\lambda \in [0,1]$, we have

$$f(\lambda x + (1-\lambda)y) \leq [f(x)]^{\lambda}\,[f(y)]^{1-\lambda}.$$

Every completely monotonic function is log–convex, see [**30**, p. 167].

We recall the Hölder–Rogers inequality for integrals [**15**, p. 54]

$$\int_a^b |f(t)\,g(t)|\,\mathrm{d}t = \left(\int_0^1 |f(t)|^p\ \mathrm{d}t\right)^{1/p} \left(\int_0^1 |g(t)|^q\ \mathrm{d}t\right)^{1/q}, \tag{3.1}$$

where $p > 1$, $1/p + 1/q = 1$, and the real functions $f \in \mathrm{L}^p[a,b]$ and $g \in \mathrm{L}^p[a,b]$.

THEOREM 3. *Let $\nu > -1$ and $x > 0$ . Then*

- *the function $y \mapsto \mathfrak{V}^{\boldsymbol{\alpha},\boldsymbol{\beta}}_{\mu,\nu}(x,y,z)$ is completely monotonic and log–convex on $(0,\infty)$,*
- *the function $z \mapsto \mathfrak{V}^{\boldsymbol{\alpha},\boldsymbol{\beta}}_{\mu,\nu}(x,y,z)$ is completely monotonic and log–convex on $(0,\infty)$,*
- *the function $\mu \mapsto \mathfrak{V}^{\boldsymbol{\alpha},\boldsymbol{\beta}}_{\mu,\nu}(x,y,z)$ is log–convex on $(0,\infty)$ for all $y,\ z > 0$,*
- *the function $(y,z) \mapsto \mathfrak{V}^{\boldsymbol{\alpha},\boldsymbol{\beta}}_{\mu,\nu}(x,y,z)$ is log–convex on $(0,\infty)$.*

PROOF. Differentiating $\mathfrak{V}^{\boldsymbol{\alpha},\boldsymbol{\beta}}_{\mu,\nu}(x,y,z)$ with respect to $y$ we arrive at

$$\left(\frac{\mathrm{d}}{\mathrm{d}y}\right)^m \mathfrak{V}^{\boldsymbol{\alpha},\boldsymbol{\beta}}_{\mu,\nu}(x,y,z) = \sqrt{\frac{x}{2}} \int_0^1 t^{\mu+m}\ \mathrm{e}^{-yt-zt^2}\ {}_pF_{p+1}\left[\begin{matrix}\alpha_1,\cdots,\alpha_p \\ \beta_1,\cdots,\beta_p,\nu+1\end{matrix}\ \Bigg|\ \frac{-x^2t^2}{4}\right]\mathrm{d}t,$$

where $m \in \mathbb{N}_0$. Consequently

$$(-1)^m \left(\frac{\mathrm{d}}{\mathrm{d}y}\right)^m \mathfrak{V}^{\boldsymbol{\alpha},\boldsymbol{\beta}}_{\mu,\nu}(x,y,z) \geq 0,$$

for all $t \in (0,1)$. So the function $y \mapsto \mathfrak{V}^{\boldsymbol{\alpha},\boldsymbol{\beta}}_{\mu,\nu}(x,y,z)$ is completely monotone in $y$.

By the complete monotone behavior of $y \mapsto \mathfrak{V}^{\boldsymbol{\alpha},\boldsymbol{\beta}}_{\mu,\nu}(x,y,z)$ it is also log–convex for $y > 0$ for all $\mu > 0$, $z > 0$. Hence, for a suitably used $y_1, y_2 > 0$ and $\lambda \in [0,1]$, we can write

$$\mathfrak{V}^{\boldsymbol{\alpha},\boldsymbol{\beta}}_{\mu,\nu}(x,\lambda y_1 + (1-\lambda)y_2, z) \leq \left[\mathfrak{V}^{\boldsymbol{\alpha},\boldsymbol{\beta}}_{\mu,\nu}(x,y_1,z)\right]^{\lambda} \left[\mathfrak{V}^{\boldsymbol{\alpha},\boldsymbol{\beta}}_{\mu,\nu}(x,y_2,z)\right]^{1-\lambda}. \tag{3.2}$$

In a similar manner, we can prove the next assertions concerning the complete monotonicity and log–convexity of $\mathfrak{V}^{\boldsymbol{\alpha},\boldsymbol{\beta}}_{\mu,\nu}(x,y,z)$ in $z$.

Applying the integral representation (1.8), with the help of (3.1) we conclude

$$\mathfrak{V}^{\boldsymbol{\alpha},\boldsymbol{\beta}}_{\lambda\mu_1+(1-\lambda)\mu_2,\nu}(x,y,z)=\sqrt{\frac{x}{2}}\int_0^1 \frac{t^{\lambda\mu_1+(1-\lambda)\mu_2}}{e^{yt+zt^2}}\,{}_pF_{p+1}\left[\begin{array}{c}\alpha_1,\cdots,\alpha_p\\ \beta_1,\cdots,\beta_p,\nu+1\end{array}\Big|\frac{-x^2t^2}{4}\right]dt$$

$$=\sqrt{\frac{x}{2}}\int_0^1\left\{t^{\mu_1}e^{-yt-zt^2}\,{}_pF_{p+1}\left[\begin{array}{c}\alpha_1,\cdots,\alpha_p\\ \beta_1,\cdots,\beta_p,\nu+1\end{array}\Big|\frac{-x^2t^2}{4}\right]\right\}^{\lambda}$$

$$\cdot\left\{t^{\mu_2}e^{-yt-zt^2}\,{}_pF_{p+1}\left[\begin{array}{c}\alpha_1,\cdots,\alpha_p\\ \beta_1,\cdots,\beta_p,\nu+1\end{array}\Big|\frac{-x^2t^2}{4}\right]\right\}^{1-\lambda}dt$$

$$\le\left\{\sqrt{\frac{x}{2}}\int_0^1 t^{\mu_1}\,e^{-yt-zt^2}\,{}_pF_{p+1}\left[\begin{array}{c}\alpha_1,\cdots,\alpha_p\\ \beta_1,\cdots,\beta_p,\nu+1\end{array}\Big|\frac{-x^2t^2}{4}\right]dt\right\}^{\lambda}$$

$$\cdot\left\{\sqrt{\frac{x}{2}}\int_0^1 t^{\mu_2}\,e^{-yt-zt^2}\,{}_pF_{p+1}\left[\begin{array}{c}\alpha_1,\cdots,\alpha_p\\ \beta_1,\cdots,\beta_p,\nu+1\end{array}\Big|\frac{-x^2t^2}{4}\right]dt\right\}^{1-\lambda}.$$

This is equivalent to

$$\mathfrak{V}^{\boldsymbol{\alpha},\boldsymbol{\beta}}_{\lambda\mu_1+(1-\lambda)\mu_2,\alpha,\beta,\nu}(x,y,z)\le\left[\mathfrak{V}^{\boldsymbol{\alpha},\boldsymbol{\beta}}_{\mu_1,\nu}(x,y,z)\right]^{\lambda}\left[\mathfrak{V}^{\boldsymbol{\alpha},\boldsymbol{\beta}}_{\mu_2,\nu}(x,y,z)\right]^{1-\lambda},$$

which proves the assertion that the function $\mu\mapsto\mathfrak{V}^{\boldsymbol{\alpha},\boldsymbol{\beta}}_{\mu,\nu}(x,y,z)$ is log–convex for all $\mu_1,\mu_2>0,\ \lambda\in[0,1]$ and $y,z>0$.

Moreover, the function $(y,z)\mapsto\mathfrak{V}^{\boldsymbol{\alpha},\boldsymbol{\beta}}_{\mu,\nu}(x,y,z)$ is log–convex for all $y_1,y_2,z_1,z_2>0,\mu>0,\ \lambda\in[0,1]$ and $x>0$ and hence equivalent to

$$\mathfrak{V}^{\boldsymbol{\alpha},\boldsymbol{\beta}}_{\mu,\nu}(x,\lambda y_1+(1-\lambda)y_2,z_1+(1-\lambda)z_2)\le\left[\mathfrak{V}^{\boldsymbol{\alpha},\boldsymbol{\beta}}_{\mu,\nu}(x,y_1,z_1)\right]^{\lambda}\left[\mathfrak{V}^{\boldsymbol{\alpha},\boldsymbol{\beta}}_{\mu,\nu}(x,y_2,z_2)\right]^{1-\lambda},$$

which finishes the proof of the theorem. □

THEOREM 4. *For the same parameter range as in Theorem 3 there holds the Turán inequality*

$$\{\mathfrak{V}^{\boldsymbol{\alpha},\boldsymbol{\beta}}_{\mu,\nu}(x,y,z)\}^2-\mathfrak{V}^{\boldsymbol{\alpha},\boldsymbol{\beta}}_{\mu,\nu}(x,y-1,z)\,\mathfrak{V}^{\boldsymbol{\alpha},\boldsymbol{\beta}}_{\mu,\nu}(x,y+1,z)\le 0.$$

*Moreover, for the same parameter space we have the Turán inequalities*

$$\{\mathfrak{V}^{\boldsymbol{\alpha},\boldsymbol{\beta}}_{\mu,\nu}(x,y,z)\}^2-\mathfrak{V}^{\boldsymbol{\alpha},\boldsymbol{\beta}}_{\mu-1,\nu}(x,y,z)\mathfrak{V}^{\boldsymbol{\alpha},\boldsymbol{\beta}}_{\mu,\nu}(x,y,z)\le 0,$$

$$\{\mathfrak{V}^{\boldsymbol{\alpha},\boldsymbol{\beta}}_{\mu,\nu}(x,y,z)\}^2-\mathfrak{V}^{\boldsymbol{\alpha},\boldsymbol{\beta}}_{\mu,\nu}(x,y,z-1)\,\mathfrak{V}^{\boldsymbol{\alpha},\boldsymbol{\beta}}_{\mu,\nu}(x,y,z+1)\le 0.$$

*Finally, we have*

$$\{\mathfrak{V}^{\boldsymbol{\alpha},\boldsymbol{\beta}}_{\mu,\nu}(x,y,z)\}^2-\mathfrak{V}^{\boldsymbol{\alpha},\boldsymbol{\beta}}_{\mu,\nu}(x,y-1,z-1)\,\mathfrak{V}^{\boldsymbol{\alpha},\boldsymbol{\beta}}_{\mu,\nu}(x,y+1,z+1)\le 0.$$

PROOF. Choosing $y_1=y-1,\ y_2=y+1$ and $\lambda=\frac{1}{2}$ in (3.2), we conclude the first Turán inequality in Theorem 4. Finally, for other assertion, specifying $z_1=z-1,\ \mu_1=\mu-1,\ z_2=z+1,\ \mu_2=\mu+1,\ y_1=y-1,\ z_1=z-1,\ y_2=y+1,\ z_2=z+1$ and $\lambda=\frac{1}{2}$, respectively, we deduce the remaining Turán inequalities. □

## 4. Discussion and remarks

In this paper, we have presented functional bounding inequalities, monotonicity properties, log-convexity properties and Turán-type inequality results for the generalized extended Voigt-type function $\mathfrak{V}^{\boldsymbol{\alpha},\boldsymbol{\beta}}_{\mu,\nu}(x,y,z)$. Moreover, it is worth to mention that the generalized extended Voigt-type function $\mathfrak{V}^{\boldsymbol{\alpha},\boldsymbol{\beta}}_{\mu,\nu}(x,y,z)$ can be explicitly expressed in terms of Kampé de Fériet function $F^{p:q;k}_{l:m;n}[x,y]$[**6, 26**] of two variables.

The results established in this paper may be useful in some areas of mathematical physics as well as astrophysical spectroscopy, neutron physics, plasma physics and statistical communication theory, (see, *e.g.*,[**1, 2, 8–10, 20, 23, 25, 27**]).

## Acknowledgment

The authors appreciate the corrections and constructive suggestions made by the referee, which strongly improved the article.

## References

[1] B. H. Armstrong, *Spectrum line profiles: the Voigt function*, J. Quant. Spectrosc. Radiat. Transfer **7** (1967), 61–88.

[2] B. H. Armstrong and R. W. Nicholls, *Emission, Absorption and Transfer of Radiation in Heated Atmosphere*, Pergamon Press, New York, 1972.

[3] Á. Baricz, P. L. Butzer, and T. K. Pogány, *Alternating Mathieu series, Hilbert-Eisenstein series and their generalized Omega functions*, Analytic number theory, approximation theory, and special functions, Springer, New York, 2014, pp. 775–808. MR3329260

[4] Á. Baricz, D. Jankov Maširević, and T. K. Pogány, *Series of Bessel and Kummer-type functions*, Lecture Notes in Mathematics, vol. 2207, Springer, Cham, 2017, DOI 10.1007/978-3-319-74350-9. MR3753667

[5] A. Erdélyi, W. Magnus, F. Oberhettinger, and F. G. Tricomi, *Tables of Integral Transforms*, Vol. **II**, McGraw-Hill Book Company, New York, Toronto and London, 1954.

[6] H. Exton, *Multiple hypergeometric functions and applications*, Mathematics & its Applications, Ellis Horwood Ltd., Chichester; Halsted Press [John Wiley & Sons, Inc.], New York-London-Sydney, 1976. Foreword by L. J. Slater. MR422713

[7] C. Fox, *The Asymptotic Expansion of Generalized Hypergeometric Functions*, Proc. London Math. Soc. (2) **27** (1928), no. 5, 389–400, DOI 10.1112/plms/s2-27.1.389. MR1575398

[8] B. D. Fried and S. D. Conte, *The Plasma Dispersion Function*, Academic Press, New York, 1961.

[9] H. J. Haubold and R. W. John, *Spectral line profiles and neutron cross sections: new results concerning the analysis of Voigt functions*, Astrophys. and Space Sci. **65** (1979), no. 2, 477–491, DOI 10.1007/BF00648512. MR549917

[10] D. Klusch, *Astrophysical spectroscopy and neutron reactions: integral transforms and Voigt functions*, Astrophys. Space Sci. **175** (1991), no. 2, 229–240, DOI 10.1007/BF00644285. MR1098132

[11] L. Landau, *Monotonicity and bounds on Bessel functions*, Proceedings of the Symposium on Mathematical Physics and Quantum Field Theory (Berkeley, CA, 1999), Electron. J. Differ. Equ. Conf., vol. 4, Southwest Texas State Univ., San Marcos, TX, 2000, pp. 147–154. MR1785371

[12] E. C. J. von Lommel, *Die Beugungserscheinungen einer kreisrunden Öffnung und eines kreisrunden Schirmchens theoretisch und experimentell bearbeitet*, Abh. Math.-Phys. Kl., König. Bayer. Akad. Wiss. **15** (1884-1886), 229–328.

[13] E. C. J. von Lommel, *Die Beugungserscheinungen geradlinig begrenzter Schirme*, Abh. Math.-Phys. Kl., König. Bayer. Akad. Wiss. **15** (1884–1886), 529–664.

[14] S. Minakshisundaram and O. Szász, *On absolute convergence of multiple Fourier series*, Trans. Amer. Math. Soc. **61** (1947), 36–53, DOI 10.2307/1990288. MR19148

[15] D. S. Mitrinović, *Analytic inequalities*, Die Grundlehren der mathematischen Wissenschaften, Band 165, Springer-Verlag, New York-Berlin, 1970. In cooperation with P. M. Vasić. MR274686

[16] A. Ya. Olenko, *Upper bound on $\sqrt{x}J_v(x)$ and its applications*, Integral Transforms Spec. Funct. **17** (2006), no. 6, 455–467, DOI 10.1080/10652460600643445. MR2238583

[17] F. W. J. Olver, D. W. Lozier, R. F. Boisvert, and C. W. Clark (eds.), *NIST handbook of mathematical functions*, U.S. Department of Commerce, National Institute of Standards and Technology, Washington, DC; Cambridge University Press, Cambridge, 2010. With 1 CD-ROM (Windows, Macintosh and UNIX). MR2723248

[18] R. K. Parmar, *Bounding inequalities for the generalized Voigt function*, J. Anal. **28** (2020), no. 1, 191–197, DOI 10.1007/s41478-017-0054-5. MR4077311

[19] R. Kumar Parmar and S. Saravanan, *Extended generalized Voigt-type functions and related bounds*, J. Class. Anal. **21** (2023), no. 1, 45–56, DOI 10.7153/jca-2023-21-05. MR4567974

[20] M. A. Pathan and M. J. S. Shahwan, *New representations of the Voigt functions*, Demonstratio Math. **39** (2006), no. 1, 75–80. MR2223876

[21] T. K. Pogány, *Bounds for Incomplete Confluent Fox-Wright Generalized Hypergeometric Functions*, Mathematics **10(17)** (2022), 3106.

[22] F. Reiche, *Über die Emission, Absorption und Intesitätsverteilung von Spektrallinien*, Ber. Deutsch. Phys. Ges. **15** (1913), 3–21.

[23] H. M. Srivastava and M. P. Ch'ên, *Some unified presentations of the Voigt functions*, Astrophys. Space Sci. **192** (1992), no. 1, 63–74, DOI 10.1007/BF00653260. MR1172183

[24] H. M. Srivastava and E. A. Miller, *A unified presentation of the Voigt functions*, Astrophys. and Space Sci. **135** (1987), no. 1, 111–118, DOI 10.1007/BF00644466. MR899371

[25] H. M. Srivastava, M. A. Pathan, and M. Kamarujjama, *Some unified presentations of the generalized Voigt functions*, Commun. Appl. Anal. **2** (1998), no. 1, 49–64. MR1612909

[26] H. M. Srivastava and P. W. Karlsson, *Multiple Gaussian hypergeometric series*, Ellis Horwood Series: Mathematics and its Applications, Ellis Horwood Ltd., Chichester; Halsted Press [John Wiley & Sons, Inc.], New York, 1985. MR834385

[27] H. M. Srivastava and T. K. Pogány, *Inequalities for a unified family of Voigt functions in several variables*, Russ. J. Math. Phys. **14** (2007), no. 2, 194–200, DOI 10.1134/S1061920807020082. MR2318831

[28] W. Voigt, *Zur Theorie der Beugung ebener inhomogener Wellen an einem geradlinig begrentzen unendlichen und absolut schwarzen Schirm*, Gött. Nachr. **1** (1889), 1–33.

[29] G. N. Watson, *A Treatise on the Theory of Bessel Functions*, Cambridge University Press, Cambridge; The Macmillan Company, New York, 1944. MR10746

[30] D. V. Widder, *The Laplace Transform*, Princeton Mathematical Series, vol. 6, Princeton University Press, Princeton, NJ, 1941. MR5923

[31] E. M. Wright, *The asymptotic expansion of the generalized hypergeometric function*, J. London Math. Soc. **10** (1935), 287–293.

[RKP] Department of Mathematics, Ramanujan School of Mathematical Sciences, Pondicherry University, Puducherry-605014, India

*Email address*: `rakeshparmar27@gmail.com,rakeshparmar@pondiuni.ac.in`

[US] Department of Mathematics, Ramanujan School of Mathematical Sciences, Pondicherry University, Puducherry-605014, India

*Email address*: `utharasabu@gmail.com,utharasabu@pondiuni.ac.in`

Contemporary Mathematics
Volume **820**, 2025
https://doi.org/10.1090/conm/820/16413

# Generalized multiple Mordell–Tornheim zeta function, related integral forms, bounds and inequalities

Rakesh K. Parmar, Tibor K. Pogány, and Agnihotri Roy

ABSTRACT. The purpose of this paper is to present integral representations of the generalized multiple Mordell–Tornheim zeta function $\zeta_{MT,m}(\boldsymbol{r}, s, \boldsymbol{x})$, to obtain different kind bounding inequalities and establish Turán type inequalities for $\zeta_{MT,m}$ with respect to the multiparameter $\boldsymbol{r}$ and parameter $s$.

## 1. Introduction and preliminaries

In 1950, Leonard Tornheim [**14**] introduced the following harmonic double series

$$T(p,q,r) = \sum_{n_1,n_2 \geq 1} \frac{1}{n_1^p n_2^q (n_1+n_2)^r}, \qquad p,q,r>0, r+\min\{p,q\}>1, \tag{1.1}$$

and made a systematic and extended study of this series. In honor to the author it is called Tornheim double series. However, there are also the nomenclature Witten zeta, Tornheim–Witten or Mordell–Tornheim–Witten zeta function, [**4**, **10**]. We will take the Mordell–Tornheim (MT) notation throughout. Matsumoto introduced the related, so-called Mordell–Tornheim $m$–tuple zeta function in the form [**9**, p. 1, Eq. (1.4)]

$$\zeta_{MT,m}(\boldsymbol{r}, s) = \sum_{n_1,n_2,\cdots n_m \geq 1} \frac{1}{n_1^{r_1} n_2^{r_2} \cdots n_m^{r_m} (n_1+n_2+\cdots+n_m)^s}, \tag{1.2}$$

where $\boldsymbol{r} = (r_1, r_2, \cdots, r_m)$; $r_j$, $s \in \mathbb{C}$ and showed that the function $\zeta_{MT,m}(\boldsymbol{r}, s)$ can be continued meromorphically to the whole $\mathbb{C}^{m+1}$, the $m$–dimensional complex space and the multiple series in (1.2) is absolutely convergent when $\Re(r_j) > 1, j = 1, \cdots, m$ and $\Re(s) > 0$, see [**9**]. Obviously, setting $m = 2$ with real parameters in (1.2) reduces to Tornheim double series (1.1). Alternating analogues of Tornheim's

2020 *Mathematics Subject Classification.* Primary 11M32, 11M99; Secondary 33B15, 33B99, 40B05.

*Key words and phrases.* Mordell–Tornheim double zeta function; Generalized multiparameter Mordell–Tornheim zeta; Integral representations; Bounds for Mordell–Tornheim zeta functions; Turán inequality.

The research of the second author was partially supported by the University of Rijeka under the project uniri-iskusni-prirod-23-98.

double series (1.1) was considered by Subbarao and Sitaramachandra Rao [**12**]:

$$R(p,q,r) = \sum_{n_1,n_2\geq 1} \frac{(-1)^{n_1}}{n_1^p n_2^q (n_1+n_2)^r} \tag{1.3}$$

$$S(p,q,r) = \sum_{n_1,n_2\geq 1} \frac{(-1)^{n_1+n_2}}{n_1^p n_2^q (n_1+n_2)^r}, \tag{1.4}$$

and further studied by Tsumura [**15, 16, 19**], [**17**, p. 43, Eq. (3)] and [**18**, p. 398, Eqs. (3.4–5)] for another kind two–variable double MT functions; also consult [**1, 7, 21, 22**] for double MT series.

Motivated by the functional series introduced by Tsumura [**17**, p. 45, Eq. (19)]

$$S(p,q,r;x,y) = \sum_{n_1,n_2\geq 1} \frac{x^{n_1} y^{n_2}}{n_1^p n_2^q (n_1+n_2)^r},$$

where the absolute convergence occurs for all three parameters $p,q,r \in \mathbb{C}$ with positive real parts for which $\Re(r)+\min\{\Re(p),\Re(q)\} > 1$ and $3\min\{\Re(p),\Re(q),\Re(r)\} > 2$, when $\max\{|x|,|y|\} < 1$ (compare [**17**, p. 45]), we introduce the generalized multiparameter and multivariable Mordell–Tornheim zeta function

$$\zeta_{MT,m}(\boldsymbol{r},s,\boldsymbol{x}) = \sum_{n_1,\cdots,n_m\geq 1} \frac{x_1^{n_1}\cdots x_m^{n_m}}{n_1^{r_1}\cdots n_m^{r_m}(n_1+\cdots+n_m)^s}.$$

It is fortunate coincidence that the newly introduced multiple MT zeta function's notation can be used also as Matsumoto–Tsumura generalized multiple zeta function. So, we leave it to the interested reader to interpret the abbreviation MT.

Our goals are here: I. to obtain integral representations for the real parameter generalized MT zeta function; II. to derive several different upper and lower bounds, and finally III. to establish Turán type inequalities for this multiple series. Related corollaries and special cases are presented either for the inferred integral expression of $\zeta_{MT,m}(\boldsymbol{r},s,\boldsymbol{x})$ or for the associated MT double series as well.

## 2. Integral representation of multiple generalized MT zeta function

In the present paper we consider generalized *Mordell-Tornheim m-tuple zeta function*

$$\zeta_{MT,m}(\boldsymbol{r},s,\boldsymbol{x}) = \sum_{\boldsymbol{n}\geq \boldsymbol{1}} \frac{x_1^{n_1}\cdots x_m^{n_m}}{n_1^{r_1}\cdots n_m^{r_m}(n_1+\cdots+n_m)^s}, \qquad |x_j|\leq 1, \tag{2.1}$$

and its alternating variant

$$\widetilde{\zeta}_{MT,m}(\boldsymbol{r},s,\boldsymbol{x}) = \sum_{\boldsymbol{n}\geq \boldsymbol{1}} \frac{(-1)^{n_1+\cdots+n_m}\, x_1^{n_1}\cdots x_m^{n_m}}{n_1^{r_1}\cdots n_m^{r_m}(n_1+\cdots+n_m)^s}, \qquad |x_j|\leq 1, \tag{2.2}$$

respectively. Here $\boldsymbol{a} = (a_1,\cdots,a_m) \in \mathbb{C}^m$, $a \in \{r,x\}$; $\boldsymbol{n} = (n_1,\cdots,n_m) \in \mathbb{N}$ and $s \in \mathbb{C}$ provided that for all considered parameters the multiple series (2.1) converges. Obviously, setting $\boldsymbol{x} = \boldsymbol{1}$ and $\boldsymbol{x} = -\boldsymbol{1}$, we obtain *Mordell-Tornheim m–tuple zeta function* in (2.1) and its alternating variant (2.2) which turn out to be $\zeta_{MT,m}(\boldsymbol{r},s,\boldsymbol{1}) \equiv \zeta_{MT,m}(\boldsymbol{r},s)$ and $\widetilde{\zeta}_{MT,m}(\boldsymbol{r},s,-\boldsymbol{1}) \equiv \zeta_{MT,m}(\boldsymbol{r},s)$.

It is worth to mention that for $s=0$, we have

$$\zeta_{MT,m}(\boldsymbol{r},0,\boldsymbol{x}) = \mathrm{Li}_{r_1}(x_1)\cdots\mathrm{Li}_{r_m}(x_m), \tag{2.3}$$

where $\mathrm{Li}_s(z)$ stands for the polylogarithmic function (or *de Jonquière's function* [**8**]) is defined as [**11**, p. 610, Eq. 25.12.10]

$$\mathrm{Li}_s(z) = \sum_{n\ge 1}\frac{z^n}{n^s},$$

provided $s\in\mathbb{C}$ when $|z|<1$ and $\Re(s)>1$ on the unit circle $|z|=1$. We notice that the multiple polylogarithm reads

$$\mathrm{Li}_{\boldsymbol{n}}(\boldsymbol{x}) = \sum_{1\le k_1<\cdots<k_m}\frac{x_1^{k_1}\cdots x_m^{k_m}}{k_1^{n_1}\cdots k_m^{n_m}},$$

with the weight $w=n_1+\cdots+n_m$ and depth $m$, see for instance [**6**, p. 617, Eq. (1)]. However, we will not apply here this generalization.

Particularly, putting $m=2$ in (2.1), we obtain generalized *Mordell-Tornheim double zeta function* and its alternating variants as

(2.4)

$$\begin{aligned}
T(p,q,r;x,y) &= \sum_{n_1,n_2\ge 1}\frac{x^{n_1}y^{n_2}}{n_1^p n_2^q(n_1+n_2)^r}, && p,q>0;\ r\ge 0;\ r+\min\{p,q\}>1,\\
R(p,q,r;x,y) &= \sum_{n_1,n_2\ge 1}\frac{(-1)^{n_1}\,x^{n_1}y^{n_2}}{n_1^p n_2^q(n_1+n_2)^r}, && p,q>0;\ r\ge 0;\ r+p>0\\
S(p,q,r;x,y) &= \sum_{n_1,n_2\ge 1}\frac{(-1)^{n_1+n_2}x^{n_1}y^{n_2}}{n_1^p n_2^q(n_1+n_2)^r}, && p,q>0;\ r\ge 0;\ r+\min\{p,q\}>0,
\end{aligned}$$

with the domain $\max\{|x|,|y|\}<1$, covering the multiplicative representation

$$X(p,q,0;\pm x,\pm y) = \mathrm{Li}_p(\pm x)\,\mathrm{Li}_q(\pm y), \qquad X\in\{T,R,S\}. \tag{2.5}$$

Moreover, for $x=y=1; -x=y=1$ and $x=y=-1$ in (2.4), we obtain generalized the *Mordell-Tornheim double zeta function* and its alternating variants (1.1), (1.3) and (1.4), respectively. We point out that Tsumura devoted his article [**17**] - among others - for expressing the double MT function $\mathcal{T}(2p,2p,2p;x)$ in terms of polylogarithmic functions, compare [**19**, p. 48, Proposition 3 *et seq.*].

THEOREM 1. *Let* $\min(r_j)>1$, $j=1,\cdots,m$ *and* $\Re(s)>0$. *Then the series* $\zeta_{MT,m}(\boldsymbol{r},s,\boldsymbol{x})$ *possesses the integral representation*

$$\zeta_{MT,m}(\boldsymbol{r},s,\boldsymbol{x}) = \frac{1}{\Gamma(s)}\int_0^\infty w^{s-1}\zeta_{MT,m}(\boldsymbol{r},0,\boldsymbol{x}\cdot \mathrm{e}^{-w})\,\mathrm{d}w, \qquad |x_j|<1. \tag{2.6}$$

*Moreover, there holds*

$$\zeta_{MT,m}(\boldsymbol{r},s,\boldsymbol{x}) = \frac{\prod_{j=1}^m x_j}{\Gamma(s)\prod_{j=1}^m\Gamma(r_j)}\int_{\mathbb{R}_+^{m+1}} w^{s-1}\prod_{j=1}^m\frac{t_j^{r_j-1}}{\mathrm{e}^{w+t_j}-x_j}\,\mathrm{d}w\,\mathbf{dt}, \tag{2.7}$$

*where* $\mathbf{dt} := \mathrm{d}t_1\cdots\mathrm{d}t_m$.

PROOF. Transforming (2.1) by the Gamma function formula

$$\frac{\Gamma(s)}{t^s} = \int_0^\infty x^{s-1} \mathrm{e}^{-tx}\, \mathrm{d}x, \qquad \min\{\Re(s), \Re(A)\} > 0,$$

setting $t = n_1 + \cdots + n_m$, we derive the expression

$$\begin{aligned}
\zeta_{MT,m}(\boldsymbol{r}, s, \boldsymbol{x}) &= \frac{1}{\Gamma(s)} \sum_{\boldsymbol{n}\geq 1} \Big[\prod_{j=1}^m \frac{x_j^{n_j}}{n_j^{r_j}}\Big] \int_0^\infty w^{s-1}\, \mathrm{e}^{-(n_1+\cdots+n_m)w}\, \mathrm{d}w \\
&= \frac{1}{\Gamma(s)} \int_0^\infty w^{s-1} \prod_{j=1}^m \sum_{n_j\geq 1} \frac{(x_j \mathrm{e}^{-w})^{n_j}}{n_j^{r_j}}\, \mathrm{d}w \\
&= \frac{1}{\Gamma(s)} \int_0^\infty w^{s-1} \prod_{j=1}^m \mathrm{Li}_{r_j}\left(x_j\, \mathrm{e}^{-w}\right) \mathrm{d}w, \qquad (2.8)
\end{aligned}$$

which proves (2.6) by virtue of the representation formula (2.3). Next, rewriting the inner series in (2.8) again with the aid of the Gamma function formula using now $\sigma = n_j$ and $s = r_j$, we get

(2.9)

$$\begin{aligned}
\zeta_{MT,m}(\boldsymbol{r}, s, \boldsymbol{x}) &= \frac{1}{\Gamma(s) \prod_{j=1}^m \Gamma(r_j)} \int_0^\infty w^{s-1} \left[\prod_{j=1}^m \int_0^\infty t_j^{r_j-1} \sum_{n_j\geq 1} \left(\frac{x_j}{\mathrm{e}^{w+t_j}}\right)^{n_j} \mathrm{d}\, t_j\right] \mathrm{d}w \\
&= \frac{\prod_{j=1}^m x_j}{\Gamma(s) \prod_{j=1}^m \Gamma(r_j)} \int_0^\infty w^{s-1} \int_{\mathbb{R}_+^m} \prod_{j=1}^m t_j^{r_j-1} \frac{\mathrm{d}t_j}{\mathrm{e}^{w+t_j} - x_j}\, \mathrm{d}w.
\end{aligned}$$

Here the binomial series $\sum_{n_j\geq 1} \left(x_j\, \mathrm{e}^{-(w+t_j)}\right)^{n_j}$ is summable since

$$|x_j\, \mathrm{e}^{-(w+t_j)}| \leq \mathrm{e}^{-(w+t_j)} < 1, \qquad \boldsymbol{x} \in (-1,1)^m;\ (w, \boldsymbol{t}) \in \mathbb{R}_+^{m+1}.$$

This ensures the convergence of the binomial series in (2.9). □

REMARK 1. Alternative proof of the second integral form (2.7) of $\zeta_{MT,m}(\boldsymbol{r}, s, \boldsymbol{x})$ can be deduced in the following way. The integrand of (2.6) is nothing else than the product of $m$ polylogarithm functions, compare (2.3). In turn, the integral representation formula [**11**, p. 611, Eq. 25.12.11]

$$\mathrm{Li}_s(z) = \frac{z}{\Gamma(s)} \int_0^\infty \frac{x^{s-1}\, \mathrm{d}x}{\mathrm{e}^x - z},$$

holds for $\Re(s) > 0$, $|\arg(1-z)| < \pi$ or $\Re(s) > 1$ when $|z| = 1$. The rest is obvious. However, above in the proof of Theorem 1 we confirm the integral form of polylogarithm function and the representation of the generalized MT zeta function.

For the sake of completeness we list the following special cases generated by Theorem 1 as by–products. The choice $\boldsymbol{x} = \pm\boldsymbol{1}$ in (2.1) has several implications which lead to the results, some of which have already presented in [**5**, pp. 541–542, Theorem 2, Corollary 1]. Also, we point out that $\widetilde{\zeta}_{MT,m}(\boldsymbol{r}, s, \boldsymbol{x}) = \zeta_{MT,m}(\boldsymbol{r}, s, -\boldsymbol{x})$, hence the results for $\zeta_{MT,m}$ generate the ones for $\widetilde{\zeta}_{MT,m}$.

COROLLARY 1.1. *Let* $\min(r_j) > 1$, $j = 1, \cdots, m$ *and* $s > 0$. *Then there hold the integral expressions*

$$\zeta_{MT,m}(\boldsymbol{r}, s) = \frac{1}{\Gamma(s)\prod_{j=1}^{m}\Gamma(r_j)} \int_{\mathbb{R}_+^{m+1}} w^{s-1} \prod_{j=1}^{m} \frac{t_j^{r_j-1}}{\mathrm{e}^{w+t_j}-1} \,\mathrm{d}w\,\mathbf{dt}$$
$$= \frac{1}{2^m\Gamma(s)\prod\limits_{j=1}^{m}\Gamma(r_j)} \int_{\mathbb{R}_+^{m+1}} \frac{w^{s-1}}{\mathrm{e}^{\frac{mw+t_1+\cdots+t_m}{2}}} \prod_{j=1}^{m} \frac{t_j^{r_j-1}}{\sinh\left(\frac{w+t_j}{2}\right)} \,\mathrm{d}w\,\mathbf{dt}.$$

Moreover, for $m = 2$ the formula (2.7) reduces to the generalized MT function $T$, mentioning that $S(p, q, r; x, y) = T(p, q, r; -x, -y)$.

COROLLARY 1.2. *For all* $p, q > 1, r > 0$ *we have*

$$T(p, q, r; x, y) = \frac{x\,y}{\Gamma(p)\Gamma(q)\Gamma(r)} \int_{\mathbb{R}_+^3} \frac{w^{s-1}\,t_1^{p-1}t_2^{q-1}}{(\mathrm{e}^{w+t_1}-x)(\mathrm{e}^{w+t_2}-y)} \,\mathrm{d}w\,\mathrm{d}t_1\,\mathrm{d}t_2.$$

For $x = y = 1$ we reproduce the following two integral expressions for the MT functions $T$. Their alternating variant $S$ can be readily obtained by the reader.

$$T(p, q, r) = \frac{1}{\Gamma(p)\Gamma(q)\Gamma(r)} \int_{\mathbb{R}_+^3} \frac{w^{s-1}\,t_1^{p-1}t_2^{q-1}}{(\mathrm{e}^{w+t_1}-1)(\mathrm{e}^{w+t_2}-1)} \,\mathrm{d}w\,\mathrm{d}t_1\,\mathrm{d}t_2$$
$$= \frac{1}{4\,\Gamma(p)\Gamma(q)\Gamma(r)} \int_{\mathbb{R}_+^3} \frac{w^{s-1}\,t_1^{p-1}t_2^{q-1}\mathrm{e}^{-w-\frac{t_1+t_2}{2}}}{\sinh\left(\frac{w+t_1}{2}\right)\,\sinh\left(\frac{w+t_2}{2}\right)} \,\mathrm{d}w\,\mathrm{d}t_1\,\mathrm{d}t_2.$$

In both formulae we assume that $p, q > 1, r > 0$ for convergence reasons.

The immediate consequence of the formula (2.5) is:

COROLLARY 1.3. *Suppose that* $p, q > 1, r > 0$. *Then we have*

$$T(p, q, r; x, y) = \frac{1}{\Gamma(r)} \int_0^\infty w^{r-1}\ \mathrm{Li}_p(x\ \mathrm{e}^{-w})\ \mathrm{Li}_q(y\ \mathrm{e}^{-w})\,\mathrm{d}w.$$

Next, by making use of the known results

$$\mathrm{Li}_0(z) = -\log(1-z) \quad \text{and} \quad \mathrm{Li}_1(z) = \frac{z}{1-z},$$

in (2.8), we obtain certain further results below.

COROLLARY 1.4. *Let* $s > m$. *Then it follows that*

$$\zeta_{MT,m}(\mathbf{1}, s; \boldsymbol{x}) = \frac{1}{\Gamma(s)} \int_0^\infty w^{s-1} \prod_{j=1}^{m} \frac{x_j}{\mathrm{e}^w - x_j} \,\mathrm{d}w,$$

*and*

$$\zeta_{MT,m}(\mathbf{0}, s; \boldsymbol{x}) = \frac{(-1)^m}{\Gamma(s)} \int_0^\infty w^{s-1} \prod_{j=1}^{m} \log(1 - x_j\mathrm{e}^{-w})\,\mathrm{d}w.$$

COROLLARY 1.5. *Assuming* $r > 0$. *Then the following two integral representations are valid:*

$$T(1, 1, r; x, y) = \frac{x\ y}{\Gamma(r)} \int_0^\infty \frac{w^{r-1}}{(\mathrm{e}^w - x)(\mathrm{e}^w - y)} \,\mathrm{d}w$$
$$T(0, 0, r; x, y) = \frac{1}{\Gamma(r)} \int_0^\infty w^{r-1} \log(1 - x\mathrm{e}^{-w})\,\log(1 - y\mathrm{e}^{-w})\,\mathrm{d}w.$$

## 3. Bounds upon $\zeta_{MT,m}(\boldsymbol{r}, s, \boldsymbol{x})$

The defining multiple series of $\zeta_{MT,m}(\boldsymbol{r}, s, \boldsymbol{x})$ contains a product and the sum of summing indices which challenges us to apply so–called A–G (Arithmetic mean–Geometric mean) inequality in estimating these expressions, which immediate consequences are refinements of the obvious upper bounds

$$\left|\zeta_{MT,m}(\boldsymbol{r}, s, \boldsymbol{x})\right| \le \zeta_{MT,m}(\boldsymbol{r}, s, |\boldsymbol{x}|) \le \zeta_{MT,m}(\boldsymbol{r}, s),$$

since $|\boldsymbol{x}| = (|x_1|, \cdots, |x_m|) \le \mathbf{1}$.

THEOREM 2. *Let* $\min(r_j) > 1, j = 1, \cdots, m$; $s > 0$ *and* $|x_j| < 1, j = 1, \cdots, m$. *Then we have the bilateral bounding inequality*

$$m^{mR}\zeta_{MT,m}(\mathbf{0}, s + mR, |\boldsymbol{x}|) \le \zeta_{MT,m}(\boldsymbol{r}, s, |\boldsymbol{x}|) \le m^{-s} \prod_{j=1}^{m} \mathrm{Li}_{r_j+\frac{s}{m}}\left(|x_j|\right), \tag{3.1}$$

*where* $\boldsymbol{r} + \alpha := (r_1 + \alpha, \cdots, r_m + \alpha)$ *and* $R = \max\limits_{1\le j\le m}(r_j)$.

*Moreover, for the alternating generalized multiple MT zeta functions there holds*

$$\left|\widetilde{\zeta}_{MT,m}(\boldsymbol{r}, s, \boldsymbol{x})\right| \le m^{-s} \prod_{j=1}^{m} \mathrm{Li}_{r_j+\frac{s}{m}}\left(|x_j|\right),$$

PROOF. Apply the A–G inequality in the following way

$$\begin{aligned} n_1^{r_1} \cdots n_m^{r_m}(n_1 + \cdots + n_m)^s &= m^s\, n_1^{r_1} \cdots n_m^{r_m}\left(\frac{n_1 + \cdots + n_m}{m}\right)^s \\ &\ge m^s\, n_1^{r_1+\frac{s}{m}} \cdots n_m^{r_m+\frac{s}{m}}. \end{aligned}$$

Using this lower bound for the denominator in (2.1) we conclude

$$\zeta_{MT,m}(\boldsymbol{r}, s, |\boldsymbol{x}|) \le m^{-s}\, \zeta_{MT,m}\left(\boldsymbol{r} + \frac{s}{m}, 0, |\boldsymbol{x}|\right).$$

In turn, the right–hand–side expression can be presented as a product of $m$ polylogarithm functions by virtue of the relation (2.3), *viz.*

$$\zeta_{MT,m}(\boldsymbol{r}, s, |\boldsymbol{x}|) \le m^{-s}\, \mathrm{Li}_{r_1+\frac{s}{m}}\left(|x_1|\right) \cdots \mathrm{Li}_{r_m+\frac{s}{m}}\left(|x_m|\right),$$

which is the asserted upper bound (3.1) follows.

The lower bound we infer when we apply the G–A inequality for the product $n_1^{r_1} \cdots n_m^{r_m}$ in (2.1). Indeed,

$$\begin{aligned} n_1^{r_1} \cdots n_m^{r_m}(n_1 + \cdots + n_m)^s &\le (n_1 \cdots n_m)^{\frac{mR}{m}}(n_1 + \cdots + n_m)^s \\ &\le m^{-mR}(n_1 + \cdots + n_m)^{s+mR}, \end{aligned}$$

so we conclude

$$\zeta_{MT,m}(\boldsymbol{r}, s, |\boldsymbol{x}|) \ge m^{mR} \sum_{\boldsymbol{n}\ge 1} \frac{|x_1|^{n_1} \cdots |x_m|^{n_m}}{(n_1 + \cdots + n_m)^{s+mR}} = m^{mR}\zeta_{MT,m}(\mathbf{0}, s + mR, |\boldsymbol{x}|). \tag{3.2}$$

The proof is completed. □

REMARK 2. Regarding the lower bound, let $H$ be the RHS term in (3.2). Thus,

$$\begin{aligned} H &= \frac{m^{mR}}{\Gamma(s+mR)} \sum_{\boldsymbol{n}\geq 1} |x_1|^{n_1}\cdots|x_m|^{n_m} \int_0^\infty t^{s+mR-1}\,\mathrm{e}^{-(n_1+\cdots+n_m)t}\,\mathrm{d}t \\ &= \frac{m^{mR}}{\Gamma(s+mR)} \int_0^\infty t^{s+mR-1} \prod_{j=1}^m \sum_{n_j\geq 1} \left(|x_j|\mathrm{e}^{-t}\right)^{n_j}\,\mathrm{d}t \\ &= \frac{m^{mR}\prod_{j=1}^m |x_j|}{\Gamma(s+mR)} \int_0^\infty \frac{t^{s+mR-1}\,\mathrm{d}t}{\prod_{j=1}^m (\mathrm{e}^t - |x_j|)}. \end{aligned}$$

We see that no another closed formula exists for the lower bound in Theorem 2.

COROLLARY 2.1. *Let $p, q > 1$; $r > 0$. Then for all $\max\{|x|, |y|\} < 1$ we have*

$$2^{2\max\{p,q\}} T(0,0,r+2\max\{p,q\}; |x|, |y|) \leq T(p,q,r;x,y) \leq 2^{-r}\ \mathrm{Li}_{p+\frac{r}{2}}(|x|)\ \mathrm{Li}_{q+\frac{r}{2}}(|y|).$$

## 4. Inequalities for $\zeta_{MT,m}$ with respect to parameters $s$ and $\boldsymbol{r}$

In this section we present inequalities which exploit the behavior of the generalized multiparameter MT zeta function with respect to the parameter $s$. Similar fashion results were established for the so–called Bickley function in [**2**].

THEOREM 3. *Let $\min_{1\leq j\leq m} \Re(r_j) > 1$, $s > \max\{0, \alpha\}$ and $|\boldsymbol{x}| < 1$. Then we have*

$$\zeta_{MT,m}(\boldsymbol{r}, s-\alpha, \boldsymbol{x}) + \zeta_{MT,m}(\boldsymbol{r}, s+\alpha, \boldsymbol{x}) \geq \frac{2\,\Gamma(s)}{\sqrt{\Gamma(s-\alpha)\Gamma(s+\alpha)}}\,\zeta_{MT,m}(\boldsymbol{r}, s, \boldsymbol{x}).$$

PROOF. Firstly, let $\boldsymbol{r}, \boldsymbol{x}$ be fixed. By virtue of the initial series representation (2.6) we get the integral expression

$$\begin{aligned} X_{MT,m} &:= \zeta_{MT,m}(\boldsymbol{r}, s-\alpha, \boldsymbol{x}) + \zeta_{MT,m}(\boldsymbol{r}, s+\alpha, \boldsymbol{x}) \\ &= \frac{1}{\Gamma(s+\alpha)} \int_0^\infty w^{s+\alpha-1}\ \zeta_{MT,m}(\boldsymbol{r}, 0, \boldsymbol{x}\ \mathrm{e}^{-w})\,\mathrm{d}w \\ &\quad + \frac{1}{\Gamma(s-\alpha)} \int_0^\infty w^{s-\alpha-1}\ \zeta_{MT,m}(\boldsymbol{r}, 0, \boldsymbol{x}\ \mathrm{e}^{-w})\,\mathrm{d}w \\ &= \int_0^\infty w^{s-1} \left(\frac{w^\alpha}{\Gamma(s+\alpha)} + \frac{w^{-\alpha}}{\Gamma(s-\alpha)}\right) \zeta_{MT,m}(\boldsymbol{r}, 0, \boldsymbol{x}\ \mathrm{e}^{-w})\,\mathrm{d}w. \end{aligned}$$

Since for all $a, b, x > 0$ it is

$$\left(\sqrt{\frac{x}{a}} - \frac{1}{\sqrt{b\,x}}\right)^2 = \frac{x}{a} + \frac{1}{b\,x} - \frac{2}{\sqrt{a\,b}} \geq 0, \tag{4.1}$$

we conclude for $x = w^\alpha$, $a = \Gamma(s+\alpha)$ and $b = \Gamma(s-\alpha)$ that

$$X_{MT,m} \geq \frac{2}{\sqrt{\Gamma(s-\alpha)\Gamma(s+\alpha)}} \int_0^\infty w^{s-1} \zeta_{MT,m}(\boldsymbol{r}, 0, \boldsymbol{x}\mathrm{e}^{-w})\,\mathrm{d}w,$$

which is equivalent to the claim of theorem. □

THEOREM 4. *Let* $\min_{1\le j\le m}\Re(r_j) > 1$, $s > \max\{0,\mu,\nu\}$ *and* $|\boldsymbol{x}| < 1$. *Then there holds*

$$\begin{aligned}\zeta_{MT,m}(\boldsymbol{r}, s+\nu, \boldsymbol{x})\,\zeta_{MT,m}(\boldsymbol{r}, s-\mu, \boldsymbol{x}) + \zeta_{MT,m}(\boldsymbol{r}, s-\nu, \boldsymbol{x})\,\zeta_{MT,m}(\boldsymbol{r}, s+\mu, \boldsymbol{x})\\ \ge \frac{2\,\Gamma^2(s)}{\sqrt{\Gamma(s+\nu)\Gamma(s-\nu)\Gamma(s+\mu)\Gamma(s-\mu)}}\,\zeta^2_{MT,m}(\boldsymbol{r}, s, \boldsymbol{x}).\end{aligned}$$

PROOF. In the similar way as in the previous theorem's proof, fixing the values $\boldsymbol{r}$ and $\boldsymbol{x}$, applying the integral representation (2.6), we write

$$\begin{aligned}\zeta_{MT,m}(\boldsymbol{r}, s+\nu, \boldsymbol{x})\,\zeta_{MT,m}(\boldsymbol{r}, s-\mu, \boldsymbol{x}) + \zeta_{MT,m}(\boldsymbol{r}, s-\nu, \boldsymbol{x})\,\zeta_{MT,m}(\boldsymbol{r}, s+\mu, \boldsymbol{x})\\ = \int_{\mathbb{R}^2_+} (uv)^{s-1}\,\zeta_{MT,m}(\boldsymbol{r}, 0, \boldsymbol{x}\,\mathrm{e}^{-u})\,\zeta_{MT,m}(\boldsymbol{r}, 0, \boldsymbol{x}\,\mathrm{e}^{-u})\\ \cdot\Big(\frac{u^\nu v^{-\mu}}{\Gamma(s+\nu)\,\Gamma(s-\mu)} + \frac{v^\mu u^{-\nu}}{\Gamma(s-\nu)\,\Gamma(s+\mu)}\Big)\,\mathrm{d}u\,\mathrm{d}v\\ \ge \frac{2\,\Gamma^2(s)\left[\dfrac{1}{\Gamma(s)}\displaystyle\int_0^\infty w^{s-1}\zeta_{MT,m}(\boldsymbol{r}, 0, \boldsymbol{x}\,\mathrm{e}^{-w})\,\mathrm{d}w\right]^2}{\sqrt{\Gamma(s+\nu)\Gamma(s-\nu)\Gamma(s+\mu)\Gamma(s-\mu)}},\end{aligned} \tag{4.2}$$

where we get the lower bound by the estimate (4.1) specifying $x = u^\nu v^{-\mu}$, $a = \Gamma(s+\nu)\,\Gamma(s-\mu)$ and $b = \Gamma(s-\nu)\,\Gamma(s+\mu)$. Obviously, (4.2) implies the claim of the theorem. □

Now, we are focused to a variant of the celebrated Turán inequality [**20**] pioneered in 1950, but firstly published two years earlier by Gábor Szegő in 1948, *viz.* [**13**], also see [**3**] and the appropriate references therein. Here we expose the Turán–type inequality for the generalized MT multiparameter function $\zeta_{MT,m}$ with respect to the multiparameter $\boldsymbol{r}$.

We start with defining the Turánian difference for $\zeta_{MT,m}$ with respect for $\boldsymbol{r}$:

$$\Delta_{\boldsymbol{r}}\big(\zeta_{MT,m}\big) = \zeta^2_{MT,m}(\boldsymbol{r}, s, \boldsymbol{x}) - \zeta_{MT,m}(\boldsymbol{\rho}+\eta, \boldsymbol{r}-\boldsymbol{\rho}, s, \boldsymbol{x})\,\zeta_{MT,m}(\boldsymbol{\rho}-\eta, \boldsymbol{r}-\boldsymbol{\rho}, s, \boldsymbol{x}),$$

where $\boldsymbol{r}-\boldsymbol{\rho}, \dim(\boldsymbol{\rho}) = \ell \le m$, stands by convention for the unchanged sub–vector of $r_j$'s; in turn, the case $\ell = m$, $\boldsymbol{r}$ and $\boldsymbol{\rho}$ coincide.

THEOREM 5. *Let* $1 \le j_1 < j_2 < \cdots < j_\ell \le m$ *and* $\boldsymbol{\rho} = (r_{j_1}, \cdots, r_{j_\ell})$ *a choice of some elements of* $\boldsymbol{r}$. *For all* $\min\limits_{1\le k\le \ell}\Re(\rho_{j_k}) > \max\{1,\eta\}$, $s > 0$ *and* $0 \le \boldsymbol{x} \le 1$ *we have*

$$\zeta^2_{MT,m}(\boldsymbol{r}, s, \boldsymbol{x}) \le \zeta_{MT,m}(\boldsymbol{\rho}+\eta, \boldsymbol{r}-\boldsymbol{\rho}, s, \boldsymbol{x})\,\zeta_{MT,m}(\boldsymbol{\rho}-\eta, \boldsymbol{r}-\boldsymbol{\rho}, s, \boldsymbol{x}). \tag{4.3}$$

PROOF. Firstly fix $\boldsymbol{r}, s$ and $\boldsymbol{x} \in [0,1]^m$. Obviously, it is enough to prove our assertion when only one power exponent $r_j$ out of $\boldsymbol{r}$ is translated by $\eta$, that is $\rho = (r_1)$. Consider in this case the Turánian difference built by the initial multiple series definition (2.1):

$$\Delta_{\boldsymbol{r}}(\zeta_{MT,m}) = \sum_{\boldsymbol{k},\boldsymbol{n}\geq \boldsymbol{1}} \frac{x_1^{k_1+n_1}\cdots x_m^{k_m+n_m}}{(k_1n_1)^{r_1}\cdots(k_mn_m)^{r_m}(k_1+\cdots+k_m)^s(n_1+\cdots+n_m)^s}$$
$$- \sum_{\boldsymbol{k},\boldsymbol{n}\geq \boldsymbol{1}} \frac{x_1^{k_1+n_1}\cdots x_m^{k_m+n_m}}{k_1^{r_1+\eta}n_1^{r_1-\eta}\cdots(k_mn_m)^{r_m}(k_1+\cdots+k_m)^s(n_1+\cdots+n_m)^s}$$
$$= \sum_{\boldsymbol{k},\boldsymbol{n}\geq \boldsymbol{1}} \frac{x_1^{k_1+n_1}\cdots x_m^{k_m+n_m}\left[1-\left(\frac{n_1}{k_1}\right)^{\eta}\right]}{(k_1n_1)^{r_1}\cdots(k_mn_m)^{r_m}(k_1+\cdots+k_m)^s(n_1+\cdots+n_m)^s},$$

by which we deduce the final outlook reached by the invariance of the difference with respect the permutation of summation indices as:

$$\Delta_{\boldsymbol{r}}(\zeta_{MT,m}) = \frac{1}{2}\sum_{\boldsymbol{k},\boldsymbol{n}\geq \boldsymbol{1}} \frac{x_1^{k_1+n_1}\cdots x_m^{k_m+n_m}\left[2-\left(\frac{n_1}{k_1}\right)^{\eta}-\left(\frac{k_1}{n_1}\right)^{\eta}\right]}{(k_1n_1)^{r_1}\cdots(k_mn_m)^{r_m}(k_1+\cdots+k_m)^s(n_1+\cdots+n_m)^s}.$$

The application of (4.1) to the expression in the squared brackets shows that $\Delta_{\boldsymbol{r}}(\zeta_{MT,m}) \leq 0$, therefore the Turán inequality (4.3) is proved. □

THEOREM 6. *For all* $\min\limits_{1\leq k\leq \ell} \Re(\rho_{j_k}) > 1$, $s > \max\{0,\tau\}$ *and* $0 \leq \boldsymbol{x} \leq 1$ *we have*

$$\zeta_{MT,m}^2(\boldsymbol{r},s,\boldsymbol{x}) \leq \zeta_{MT,m}(\boldsymbol{r},s+\tau,\boldsymbol{x})\ \zeta_{MT,m}(\boldsymbol{r},s-\tau,\boldsymbol{x}).$$

PROOF. The Turánian difference with respect to the parameter $s$ reads

$$\Delta_s(\zeta_{MT,m}) = \frac{1}{2}\sum_{\boldsymbol{k},\boldsymbol{n}\geq \boldsymbol{1}} \frac{x_1^{k_1+n_1}\cdots x_m^{k_m+n_m}\left[2-\left(\frac{\sum_{j=1}^m n_j}{\sum_{j=1}^m k_j}\right)^{\tau}-\left(\frac{\sum_{j=1}^m k_j}{\sum_{j=1}^m n_j}\right)^{\tau}\right]}{(k_1n_1)^{r_1}\cdots(k_mn_m)^{r_m}(k_1+\cdots+k_m)^s(n_1+\cdots+n_m)^s}.$$

The use of the estimate (4.1) proves the assertion of the theorem. □

## Acknowledgment

The authors would like to thank the editor and the unknown referee for careful reading and comments which greatly improved the article.

## References

[1] D. H. Bailey, D. Borwein, and J. M. Borwein, *On Eulerian log-gamma integrals and Tornheim-Witten zeta functions*, Ramanujan J. **36** (2015), no. 1-2, 43–68, DOI 10.1007/s11139-012-9427-1. MR3296709

[2] Á. Baricz and T. K. Pogány, *Functional inequalities for the Bickley function*, Math. Inequal. Appl. **17** (2014), no. 3, 989–1003, DOI 10.7153/mia-17-72. MR3224849

[3] Á. Baricz and T. K. Pogány, *Turán determinants of Bessel functions*, Forum Math. **26** (2014), no. 1, 295–322, DOI 10.1515/form.2011.160. MR3176632

[4] J. M. Borwein, *Hilbert's inequality and Witten's zeta-function*, Amer. Math. Monthly **115** (2008), no. 2, 125–137, DOI 10.1080/00029890.2008.11920505. MR2384265

[5] B. Draščić Ban, J. Pečarić, I. Perić, and T. Pogány, *Discrete multiple Hilbert type inequality with non-homogeneous kernel*, J. Korean Math. Soc. **47** (2010), no. 3, 537–546, DOI 10.4134/JKMS.2010.47.3.537. MR2662436

[6] A. B. Goncharov, *The double logarithm and Manin's complex for modular curves*, Math. Res. Lett. **4** (1997), no. 5, 617–636, DOI 10.4310/MRL.1997.v4.n5.a1. MR1484694

[7] J. G. Huard, K. S. Williams, and N.-Y. Zhang, *On Tornheim's double series*, Acta Arith. **75** (1996), no. 2, 105–117, DOI 10.4064/aa-75-2-105-117. MR1379394

[8] A. Jonquière, *Note sur la série* $\sum_{n=1}^{\infty} \frac{x^n}{n^s}$ (French), Bull. Soc. Math. France **17** (1889), 142–152. MR1504064

[9] K. Matsumoto, *On Mordell-Tornheim and other multiple zeta-functions*, Proceedings of the Session in Analytic Number Theory and Diophantine Equations, Bonner Math. Schriften, vol. 360, Univ. Bonn, Bonn, 2003, pp. 17. MR2075634

[10] L. J. Mordell, *On the evaluation of some multiple series*, J. London Math. Soc. **33** (1958), 368–371, DOI 10.1112/jlms/s1-33.3.368. MR100181

[11] F. W. J. Olver, D. W. Lozier, R. F. Boisvert, and C. W. Clark (eds.), *NIST handbook of mathematical functions*, U.S. Department of Commerce, National Institute of Standards and Technology, Washington, DC; Cambridge University Press, Cambridge, 2010. With 1 CD-ROM (Windows, Macintosh and UNIX). MR2723248

[12] M. V. Subbarao and R. Sitaramachandra Rao, *On some infinite series of L. J. Mordell and their analogues*, Pacific J. Math. **119** (1985), no. 1, 245–255. MR797027

[13] G. Szegő, *On an inequality of P. Turán concerning Legendre polynomials*, Bull. Amer. Math. Soc. **54** (1948), 401–405, DOI 10.1090/S0002-9904-1948-09017-6. MR23954

[14] L. Tornheim, *Harmonic double series*, Amer. J. Math. **72** (1950), 303–314, DOI 10.2307/2372034. MR34860

[15] H. Tsumura, *On alternating analogues of Tornheim's double series*, Proc. Amer. Math. Soc. **131** (2003), no. 12, 3633–3641, DOI 10.1090/S0002-9939-03-07186-7. MR1998168

[16] H. Tsumura, *On Mordell-Tornheim zeta values*, Proc. Amer. Math. Soc. **133** (2005), no. 8, 2387–2393, DOI 10.1090/S0002-9939-05-08132-3. MR2138881

[17] H. Tsumura, *On certain polylogarithmic double series*, Arch. Math. (Basel) **88** (2007), no. 1, 42–51, DOI 10.1007/s00013-006-1809-4. MR2289599

[18] H. Tsumura, *On functional relations between the Mordell-Tornheim double zeta functions and the Riemann zeta function*, Math. Proc. Cambridge Philos. Soc. **142** (2007), no. 3, 395–405, DOI 10.1017/S0305004107000059. MR2329691

[19] H. Tsumura, *On alternating analogues of Tornheim's double series. II*, Ramanujan J. **18** (2009), no. 1, 81–90, DOI 10.1007/s11139-007-9016-x. MR2471618

[20] P. Turán, *On the zeros of the polynomials of Legendre* (English, with Czech summary), Časopis Pěst. Mat. Fys. **75** (1950), 113–122. MR41284

[21] O. Espinosa and V. H. Moll, *The evaluation of Tornheim double sums. I*, J. Number Theory **116** (2006), no. 1, 200–229, DOI 10.1016/j.jnt.2005.04.008. MR2197867

[22] O. Espinosa and V. H. Moll, *The evaluation of Tornheim double sums. II*, Ramanujan J. **22** (2010), no. 1, 55–99, DOI 10.1007/s11139-009-9181-1. MR2610609

Department of Mathematics, Ramanujan School of Mathematical Sciences, Pondicherry University, Puducherry–605014, India

*Email address*: rakeshparmar27@gmail.com,rakeshparmar@pondiuni.ac.in

Institute of Applied Mathematics, John von Neumann Faculty of Informatics, Óbuda University, Budapest 1034, Hungary; and Faculty of Maritime Studies, University of Rijeka, 51000 Rijeka, Croatia

*Email address*: pogany.tibor@nik.uni-obuda.hu,tibor.poganj@uniri.hr

Department of Mathematics, Ramanujan School of Mathematical Sciences, Pondicherry University, Puducherry–605014, India

*Email address*: roy.agnihotri@gmail.com,roy.agnihotri@pondiuni.ac.in

Contemporary Mathematics
Volume **820**, 2025
https://doi.org/10.1090/conm/820/16414

# On distribution of Rice–Middleton model II. Srivastava–Daoust $\mathscr{S}$ function approach

Dragana Jankov Maširević and Tibor K. Pogány

ABSTRACT. The probability density function of the single sinusoidal signal combined with Gaussian noise is obtained in terms of the Srivastava–Daoust generalized Lauricella $F_D$ hypergeometric function by two different methods applied to an auxiliary integral: firstly, expanding the integrand into a triple Maclaurin series, secondly, by the Cauchy boundary problem of a linear non–homogeneous ordinary differential equation for the considered auxiliary integral.

## 1. Introduction, motivation and preparation

When a broad band random noise is applied to some physical device, such as an electrical network, the statistical properties of the output are often of interest. For example, when the noise is due to shot effect current assuming the physical device is linearly working, Rice described this shot effect noise results from the superposition of a great number of disturbances which occur at random, see e.g. [**7**, p. 282]. The non–negative stochastic process which is modeling this phenomenon consists of a sum of a narrow band signal, which has a standard Arcsine distribution[1] and a broad band signal which is a centered Gaussian rv - both being time dependent, assuming simultaneously the independence of that two signals [**6**, §9.3] and [**5**, p. 74, Eq. (5)]. The model is therefore called as the Rice–Middleton (RC) model, consult [**3**].

Now-a-days, Trifonov [**12**] considered this model, more precisely the non–negative stochastic process

$$v(t) = A\cos(\omega t + \phi) + u(t),$$

where $t$ denotes time, $A, \omega \in (0, \pi)$, $\phi \in (0, 2\pi)$ stand respectively for the amplitude, frequency and phase of the real–valued sinusoid and $u(t)$ is the independent/uncorrelated (with the cosine wave) measurement noise possessing Gaussian

2020 *Mathematics Subject Classification.* Primary 33C65, 33C70, 33C90, 40B05, 40C15, 40H05, 60E05.

*Key words and phrases.* Modified Bessel function of the first kind; Srivastava–Daoust $\mathscr{S}$ function; Neumann series of the second type; sinusoidal signal; Gaussian noise.

The research of TKP was partially supported by the University of Rijeka under the project code `uniri-iskusni-prirod-23-98`.

The second author is the corresponding author.

[1]Assume that the input random variable (rv) $\xi$ is uniformly $U(0, 2\pi A)$ distributed, than the rv $A\cos\xi$ is Arcsine$(-A, A)$ distributed.

$\mathscr{N}(0,\sigma^2)$ distribution. The probability density (pdf) of the model $v(t)$ becomes [**12**, p. 984, Eq. (1)]

$$p(x) = \frac{1}{\sigma\sqrt{2\pi}}\,\mathrm{e}^{-\frac{x^2}{2\sigma^2}} \sum_{n\geq 0}(-1)^n \beta_n I_{2n}\Big(\sqrt{2\psi}\frac{x}{\sigma}\Big), \qquad x \in \mathbb{R}. \tag{1.1}$$

Here the abbreviation $\psi = \frac{1}{2}\,A^2\sigma^{-2}$ is the signal–to–noise ratio and the coefficient

$$\beta_n = k_n \mathrm{e}^{-\frac{\psi}{2}}\, I_n\Big(\frac{\psi}{2}\Big) > 0, \qquad \sum_{n\geq 0}\beta_n = 1.$$

That means that (1.1) presents the pdf of the instantaneous value of a single sinusoidal signal additively combined with Gaussian noise [**12**, p. 984]. Here $I_\mu$ stands for the modified Bessel function of the first kind of the order $\mu$ which power series definition reads [**2**, p. 13]

$$I_\mu(z) = \sum_{k\geq 0}\frac{1}{\Gamma(\mu+k+1)\,k!}\Big(\frac{z}{2}\Big)^{2k+\mu}, \qquad \Re(\mu) > -1,\, z \in \mathbb{C},$$

and

$$k_n = \begin{cases} 1, & n = 0,\\ 2, & n \geq 1.\end{cases}$$

The following transformed description of the pdf (1.1) holds:
(1.2)

$$p(x) = \frac{1}{\sigma\sqrt{2\pi}}\mathrm{e}^{-\frac{x^2}{2\sigma^2}-\frac{\psi}{2}}\left\{I_0\Big(\frac{\psi}{2}\Big)I_0\Big(\sqrt{2\psi}\frac{x}{\sigma}\Big) + 2\sum_{n\geq 1}(-1)^n I_n\left(\frac{\psi}{2}\right) I_{2n}\left(\sqrt{2\psi}\frac{x}{\sigma}\right)\right\}.$$

It is worth to mention that an analytical expression for the appropriate pdf (1.1) pioneered by Rice [**7**–**9**], also Middleton [**6**, p. 422, Eq.(9.67a)] obtained the formula in terms of the Kummer's confluent hypergeometric function ${}_1F_1$ and Ho *et al.* [**5**, p. 74, Eq. (5)] derived the formula for such pdf in terms of the Hermite polynomials of even order $H_{2k}, k \in \mathbb{N}_0$. Trifonov pointed out that his formula (1.1) has the advantage of importantly faster convergence faster than the series by Rice, Middleton and Ho *et al.*

The problem of summing up the second type Neumann series (1.2) (consult [**1**] and/or [**2**, Chapter 2, pp. 27 *et seq.*]) is the central problem of our research in the very recent article by Górska *et al.* [**3**]. In turn, in the article [**3**] the pdf of the RC distribution model is expressed in three different ways: firstly, by an integral expression of the modified Bessel function of the first kind of integer order; secondly, by a hyperbolic cosine differential operator and thirdly, by making use the Grünwald–Letnikov fractional derivative. The cumulative distribution functions (cdf) are also established in all these cases, see [**3**] and the appropriate references therein.

In turn, here we present two new methods for summing up the auxiliary Neumann series (1.2); these are the Srivastava–Daoust $\mathscr{S}$ function and a linear non–homogeneous ODE for the auxiliary integral

$$M(a,b) := \int_0^1 \frac{\mathrm{e}^{-ax^2+bx}}{\sqrt{1-x^2}}\,\mathrm{d}x. \tag{1.3}$$

We begin with transforming the second type Neumann series which arrives in (1.2), *viz.*

(1.4)

$$\begin{aligned}
\mathscr{N}_{\sigma,\psi}(x) &= \sum_{n\geq 1}(-1)^n I_n\Big(\frac{\psi}{2}\Big) I_{2n}\Big(\sqrt{2\psi}\frac{x}{\sigma}\Big) \\
&= \frac{1}{2\pi}\int_{-\pi}^{\pi} \mathrm{e}^{-\sqrt{2\psi}\frac{x}{\sigma}\sin t}\sum_{n\geq 1}\cos(2nt) I_n\Big(\frac{\psi}{2}\Big)\mathrm{d}t \\
&= \frac{\mathrm{e}^{\frac{\psi}{2}}}{2\pi}\int_0^1 \frac{\mathrm{e}^{-\psi u^2}\cosh\Big(\sqrt{2\psi}\frac{x}{\sigma}u\Big)}{\sqrt{1-u^2}}\,\mathrm{d}u - \frac{1}{\pi}I_0\Big(\frac{\psi}{2}\Big)\int_0^{\pi/2}\cosh\Big(\sqrt{2\psi}\frac{x}{\sigma}\sin t\Big)\mathrm{d}t \\
&= \frac{\mathrm{e}^{\frac{\psi}{2}}}{2\pi}\Big[M\Big(\psi,\sqrt{2\psi}\frac{x}{\sigma}\Big) + M\Big(\psi,-\sqrt{2\psi}\frac{x}{\sigma}\Big)\Big] - \frac{1}{2}I_0\Big(\frac{\psi}{2}\Big)\, I_0\Big(\sqrt{2\psi}\frac{x}{\sigma}\Big),
\end{aligned} \tag{1.5}$$

reported as [**3**, p. 5, Eq. (8)] and combined with the appropriate part in the proof of Theorem 3 in [**3**, p. 8], applying the formula [**4**, p. 193, Eq. 512.6a)]

$$I_{2n}(z) = (-1)^n\frac{2}{\pi}\int_0^{\pi/2}\cosh(z\sin x)\cos(2nx)\,\mathrm{d}x, \qquad n\in\mathbb{Z}.$$

So, we see that the central role in expressing the pdf (1.1) belongs to calculation of the auxiliary integral (1.3).

The Neumann series $\mathscr{N}_{\sigma,\psi}(x)$ is presentable in the form of a specific power series of three variables, the so–called Srivastava–Daoust $\mathscr{S}$ function which is a generalized Lauricella hypergeometric function $F_D$, by the $n$-tuple power series [**10**, p. 454]

$$\mathscr{S}^{A:B^{(1)};\cdots;B^{(n)}}_{C:D^{(1)};\cdots;D^{(n)}}\left(\begin{array}{c} [(a):\theta^{(1)},\cdots,\theta^{(n)}]:[(b^{(1)}):\varphi^{(1)}];\cdots;[(b^{(n)}):\varphi^{(n)}] \\ \\ [(c):\psi^{(1)};\cdots;\psi^{(n)}]:[(d^{(1)}):\delta^{(1)}];\cdots;[(d^{(n)}):\delta^{(n)}] \end{array}\middle|\begin{array}{c} x_1 \\ \vdots \\ x_n\end{array}\right)$$

(1.6)

$$= \sum_{\boldsymbol{m}\geq 0}\frac{\prod\limits_{j=1}^{A}(a_j)_{m_1\theta_j^{(1)}+\cdots+m_n\theta_j^{(n)}}\prod\limits_{j=1}^{B^{(1)}}\big(b_j^{(1)}\big)_{m_1\varphi_j^{(1)}}\cdots\prod\limits_{j=1}^{B^{(n)}}\big(b_j^{(n)}\big)_{m_n\varphi_j^{(n)}}}{\prod\limits_{j=1}^{C}(c_j)_{m_1\psi_j^{(1)}+\cdots+m_n\psi_j^{(n)}}\prod\limits_{j=1}^{D^{(1)}}\big(d_j^{(1)}\big)_{m_1\delta_j^{(1)}}\cdots\prod\limits_{j=1}^{D^{(n)}}\big(d_j^{(n)}\big)_{m_n\delta_j^{(n)}}}\frac{x_1^{m_1}}{m_1!}\cdots\frac{x_n^{m_n}}{m_n!},$$

where $\boldsymbol{m} := (m_1,\cdots,m_n)$ and the parameters satisfy

$$\theta_1^{(1)},\cdots,\theta_A^{(1)},\cdots,\delta_1^{(n)},\cdots,\delta_{D^{(n)}}^{(n)} > 0.$$

Here $(a)_n$ stands for the Pochhammer symbol (or the shifted factorial)

$$(a)_n = \frac{\Gamma(a+n)}{\Gamma(a)} = \begin{cases} a(a+1)\cdots(a+n-1), & n \in \mathbb{N},\, a \in \mathbb{C}\setminus\{0\}, \\ 1, \quad n = 0. \end{cases}$$

For convenience, we write (a) to denote the sequence of $A$ parameters $a_1, \cdots, a_A$, with similar interpretations for $(b^{(1)}), \cdots, (c), \cdots, (d^{(n)})$. Empty products should be interpreted as unity. Srivastava and Daoust [**11**, pp. 157–158] reported that the series in (1.6) converges absolutely for all $x_1, \cdots, x_n \in \mathbb{C}$ when

$$\Delta_\ell = 1 + \sum_{j=1}^{C} \psi_j^{(\ell)} + \sum_{j=1}^{D^{(\ell)}} \delta_j^{(\ell)} - \sum_{j=1}^{A} \theta_j^{(\ell)} - \sum_{j=1}^{B^{(\ell)}} \varphi_j^{(\ell)} > 0, \qquad \ell = \overline{1,n}. \tag{1.7}$$

In the case $\Delta_\ell = 0$, $\ell = \overline{1,n}$, the convergence constraints of the series are described by the relation [**11**, p. 157, Eqs. (5.2-5.3); p. 158 (**ii**)]. Finally, when all $\Delta_\ell < 0$, $\mathscr{S}^{A:B^{(1)};\cdots;B^{(n)}}_{C:D^{(1)};\cdots;D^{(n)}}(x_1, \cdots, x_n)$ it diverges except at the origin, that is, this series is formal.

## 2. The pdf $p(x)$ in terms of Srivastava–Daoust $\mathscr{S}$ function

Now, we present our first main result.

THEOREM 2.1. *For all $x \in \mathbb{R}$ and for all $\sigma, \psi > 0$ there holds*

$$p(x) = \frac{e^{-\frac{x^2}{2\sigma^2}}}{\sigma\sqrt{8\pi^3}} \Bigg\{ \mathscr{S}^{1:-;-;1}_{1:-;-;-} \left( \begin{matrix} [1:2,1,2]:-;-;[\frac12:1] \\ \\ [2:2,1,2]:-;-;- \end{matrix} \,\middle|\, \begin{matrix} -\psi \\ \sqrt{2\psi}\frac{x}{\sigma} \\ -1 \end{matrix} \right)$$

$$+ \mathscr{S}^{1:-;-;1}_{1:-;-;-} \left( \begin{matrix} [1:2,1,2]:-;-;[\frac12:1] \\ \\ [2:2,1,2]:-;-;- \end{matrix} \,\middle|\, \begin{matrix} -\psi \\ -\sqrt{2\psi}\frac{x}{\sigma} \\ -1 \end{matrix} \right)$$

$$- \frac{\pi}{e^{\frac{\psi}{2}}} I_0\Big(\frac{\psi}{2}\Big) \cdot I_0\Big(\sqrt{2\psi}\frac{x}{\sigma}\Big) \Bigg\}.$$

PROOF. Expand all three functions in the integrand of (1.3) into their power series. Since all three series converge in the unit interval, we conclude that

$$\begin{aligned} M(a,b) &= \sum_{m,n,k\ge 0} \frac{(-a)^m b^n}{m!\,n!} \binom{-\frac12}{k} \int_0^1 x^{2m+n+2k}\,\mathrm{d}x \\ &= \sum_{m,n,k\ge 0} \frac{(-a)^m b^n \binom{-\frac12}{k}}{m!\,n!\,(2m+n+2k+1)} \\ &= \sum_{m,n,k\ge 0} \frac{(1)_{2m+n+2k}\,(\frac12)_k}{(2)_{2m+n+2k}} \frac{(-a)^m}{m!}\frac{b^n}{n!}\frac{(-1)^k}{k!} \end{aligned} \tag{2.1}$$

which means that

$$M(a,b) = \mathscr{S}^{1:-;-;1}_{1:-;-;-} \left( \begin{matrix} [1:2,1,2]:-;-;[\frac12:1] \\ \\ [2:2,1,2]:-;-;- \end{matrix} \,\middle|\, \begin{matrix} -a \\ b \\ -1 \end{matrix} \right).$$

It remains the convergence check. According to the convergence conditions (1.7), we see that $\Delta_1(m) = \Delta_2(n) = 2 > 0$, whilst $\Delta_3(k) = 0$ so, $I^{(2)}(a,b)$ converges in $(a,b) \in \mathbb{C}^2$, that rather for $a,b > 0$. The related integral (1.4) reads now

$$\frac{1}{2}\big[M(a,b) + M(a,-b)\big] = \frac{1}{2}\left\{ \mathscr{S}^{1:-;-;1}_{1:-;-;-}\left( \begin{matrix} [1:2,1,2]:-;-;[\frac{1}{2}:1] \\ \\ [2:2,1,2]:-;-;- \end{matrix} \,\middle|\, \begin{matrix} -a \\ b \\ -1 \end{matrix} \right) + \mathscr{S}^{1:-;-;1}_{1:-;-;-}\left( \begin{matrix} [1:2,1,2]:-;-;[\frac{1}{2}:1] \\ \\ [2:2,1,2]:-;-;- \end{matrix} \,\middle|\, \begin{matrix} -a \\ -b \\ -1 \end{matrix} \right) \right\},$$

whilst specifying $a = \psi$, $b = \sqrt{2\psi}\,x\,\sigma^{-1}$ we incorporate this expression into the Neumann series $\mathscr{N}_{\sigma,\psi}(x)$ getting

$$\mathscr{N}_{\sigma,\psi}(x) = \frac{\mathrm{e}^{\frac{\psi}{2}}}{4\pi}\Big[M\Big(\psi, \sqrt{2\psi}\frac{x}{\sigma}\Big) + M\Big(\psi, -\sqrt{2\psi}\frac{x}{\sigma}\Big)\Big] - \frac{1}{2}I_0\Big(\frac{\psi}{2}\Big)\cdot I_0\Big(\sqrt{2\psi}\frac{x}{\sigma}\Big),$$

which confirms the statement. □

## 3. The method of linear non–homogeneous ODE

The second method in calculating $M(a,b)$ exposed in (1.3) includes an ordinary linear non–homogeneous differential equation for $M(a)$, associated with the appropriate initial condition which includes $M(0)$.

LEMMA 3.1. *For all $a, b >$ we have*

$$M(a,b) = a\mathrm{e}^{-a}\,\mathscr{S}^{2:-;-;1;-}_{2:-;-;-;-}\left( \begin{matrix} [1:2,1,2,0],[1:1,0,0,1]:-;-;[\frac{1}{2}:1];- \\ \\ [2:2,1,2,0],[2:1;0;0;1]:-;-;-;- \end{matrix} \,\middle|\, \begin{matrix} -a \\ b \\ -1 \\ a \end{matrix} \right)$$

$$+\frac{\pi}{2}\,\mathrm{e}^{-a}\big(\mathbf{L}_0(b) + I_0(b)\big) := M(a), \tag{3.1}$$

*where $\mathbf{L}_0(b)$ denotes the modified Struve function of zeroth order.*

PROOF. Consider the derivative of $M(a)$ with respect to $a$:

$$M'(a) = -\int_0^1 \frac{x^2\,\mathrm{e}^{-ax^2+bx}}{\sqrt{1-x^2}}\,\mathrm{d}x = \int_0^1 \frac{(1-x^2-1)\mathrm{e}^{-ax^2+bx}}{\sqrt{1-x^2}}\,\mathrm{d}x$$

$$= \int_0^1 \sqrt{1-x^2}\,\mathrm{e}^{-ax^2+bx}\,\mathrm{d}x - M(a), \tag{3.2}$$

which is an ordinary linear differential equation

$$M'(a) + M(a) = \mathscr{S}^{1:-;-;1}_{1:-;-;-}\left( \begin{matrix} [1:2,1,2]:-;-;[\frac{1}{2}:1] \\ \\ [2:2,1,2]:-;-;- \end{matrix} \,\middle|\, \begin{matrix} -a \\ b \\ -1 \end{matrix} \right) := \mathscr{S}(-a,b,-1). \tag{3.3}$$

The integral (3.2) we handle by the same procedure as the previous one in (2.1).

Also, consider the associated initial condition

$$M(0) = \int_0^1 \frac{e^{bx}}{\sqrt{1-x^2}}\,dx = \frac{\pi}{2}\big(\mathbf{L}_0(b) + I_0(b)\big). \tag{3.4}$$

Therefore the Cauchy boundary value problem (3.3) & (3.4) possesses the solution

$$M(a) = e^{-a}\Big[\int_0^a e^x\, \mathscr{S}(-x,b,-1)\,dx + \frac{\pi}{2}\big(\mathbf{L}_0(b)+I_0(b)\big)\Big].$$

It remains only to solve the integral, which we denote as $I_{\mathscr{S}}(a)$.

Expanding the exponential term in the integrand into the Maclaurin series we conclude:

$$\begin{aligned}
I_{\mathscr{S}}(a) &= \sum_{j\ge 0} \frac{1}{j!}\int_0^a x^j\, \mathscr{S}(-x,b,-1)\,dx \\
&= \sum_{m,n,k,j\ge 0} \frac{(1)_{2m+n+2k}\,(\frac{1}{2})_k}{(2)_{2m+n+2k}}\,\frac{(-1)^{m+k}\,b^n}{m!\,n!\,k!\,j!}\int_0^a x^{m+j}\,dx \\
&= a\sum_{m,n,k,j\ge 0} \frac{(1)_{2m+n+2k}\,(1)_{m+j}\,(\frac{1}{2})_k}{(2)_{2m+n+2k}\,(2)_{m+j}}\,\frac{(-a)^m}{m!}\,\frac{b^n}{n!}\,\frac{(-1)^k}{k!}\,\frac{a^j}{j!} \\
&= a\,\mathscr{S}^{2:-;-;1;-}_{2:-;-;-;-}\left(\begin{matrix}[1:2,1,2,0],[1:1,0,0,1]:-;-;[\frac{1}{2}:1];-\\[2:2,1,2,0],[2:1;0;0;1]:-;-;-;-\end{matrix}\;\middle|\;\begin{matrix}-a\\ b\\ -1\\ a\end{matrix}\right).
\end{aligned}$$

Collecting all these formulae, we proved the asserted result, since in this case $\Delta_1(m) = \Delta_2(n) = \Delta_4(j) = 1$ and $\Delta_3(k) = 0$; the third case is not connected to a variable, as the summation index $k$ reflects the sum behavior with a constant variable $-1$. □

COROLLARY 3.2. *For all $\sigma, b > 0$ we have*

$$\begin{aligned}
&M(a,b) + M(a,-b) \\
&\quad = a\,e^{-a}\Bigg\{\mathscr{S}^{2:-;-;1;-}_{2:-;-;-;-}\left(\begin{matrix}[1:2,1,2,0],[1:1,0,0,1]:-;-;[\frac{1}{2}:1];-\\[2:2,1,2,0],[2:1;0;0;1]:-;-;-;-\end{matrix}\;\middle|\;\begin{matrix}-a\\ b\\ -1\\ a\end{matrix}\right) \\
&\qquad + \mathscr{S}^{2:-;-;1;-}_{2:-;-;-;-}\left(\begin{matrix}[1:2,1,2,0],[1:1,0,0,1]:-;-;[\frac{1}{2}:1];-\\[2:2,1,2,0],[2:1;0;0;1]:-;-;-;-\end{matrix}\;\middle|\;\begin{matrix}-a\\ -b\\ -1\\ a\end{matrix}\right) \\
&\qquad + \frac{\pi}{a}\,I_0(b)\Bigg\}.
\end{aligned}$$

PROOF. Considering (3.1), the modified Struve function $\mathbf{L}_0$ vanishes from the resulting expression, since it is odd function. □

THEOREM 3.3. *For all $x \in \mathbb{R}$ and for all $\sigma, \psi > 0$ there holds*

$$p(x) = c(x)\Bigg\{ \mathscr{S}^{2:-;-;1;-}_{2:-;-;-;-}\left( \begin{matrix} [1:2,1,2,0],[1:1,0,0,1]:-;-;[\frac12:1];- \\ \\ [2:2,1,2,0],[2:1;0;0;1]:-;-;-;- \end{matrix} \;\middle|\; \begin{matrix} -\psi \\ \sqrt{2\psi}\dfrac{x}{\sigma} \\ -1 \\ \psi \end{matrix} \right)$$

$$+ \mathscr{S}^{2:-;-;1;-}_{2:-;-;-;-}\left( \begin{matrix} [1:2,1,2,0],[1:1,0,0,1]:-;-;[\frac12:1];- \\ \\ [2:2,1,2,0],[2:1;0;0;1]:-;-;-;- \end{matrix} \;\middle|\; \begin{matrix} -\psi \\ -\sqrt{2\psi}\dfrac{x}{\sigma} \\ -1 \\ \psi \end{matrix} \right)$$

$$+ \frac{\pi}{a} I_0\Big(\sqrt{2\psi}\frac{x}{\sigma}\Big)\Bigg\} - \frac12 I_0\Big(\frac{\psi}{2}\Big)\cdot I_0\Big(\sqrt{2\psi}\frac{x}{\sigma}\Big),$$

*where*

$$c(x) = \frac{\psi\, \mathrm{e}^{-\frac{x^2}{2\sigma^2}-\frac{\psi}{2}}}{\sigma\sqrt{8\,\pi^3}}.$$

PROOF. Following the lines of the proof of the previous theorem, it is enough to insert the values $(a,b) = (\psi, \sqrt{2\psi}\,x/\sigma)$ into $M(a,b)+M(a,-b)$ presented in the Corollary 3.2. and then into

$$\mathscr{N}_{\sigma,\psi}(x) = \frac{\mathrm{e}^{\frac{\psi}{2}}}{4\pi}\cdot\Big[M\Big(\psi, \sqrt{2\psi}\frac{x}{\sigma}\Big) + M\Big(\psi, -\sqrt{2\psi}\frac{x}{\sigma}\Big)\Big] - \frac12 I_0\Big(\frac{\psi}{2}\Big)\cdot I_0\Big(\sqrt{2\psi}\frac{x}{\sigma}\Big),$$

finally into the associated expression for

$$p(x) = \frac{2}{\sigma\sqrt{2\pi}}\,\mathrm{e}^{-\frac{x^2}{2\sigma^2}-\frac{\psi}{2}}\cdot\mathscr{N}_{\sigma,\psi}(x),$$

which becomes the stated result. □

Finally, we give a surprising by–product of the proofs of Theorems 2.1 and 23.3, and the Lemma 3.1 in the form of summation for a special case of Srivastava–Daoust three–, and four–variables functions.

COROLLARY 3.4. *For all positive $a, b$ we have the following result:*

$$\mathscr{S}^{1:-;-;1}_{1:-;-;-}\left( \begin{matrix} [1:2,1,2]:-;-;[\frac12:1] \\ \\ [2:2,1,2]:-;-;- \end{matrix} \;\middle|\; \begin{matrix} -a \\ b \\ -1 \end{matrix} \right) - \frac{\pi}{2}\,\mathrm{e}^{-a}\big(\mathbf{L}_0(b) + I_0(b)\big)$$

$$= a\,\mathrm{e}^{-a}\,\mathscr{S}^{2:-;-;1;-}_{2:-;-;-;-}\left( \begin{matrix} [1:2,1,2,0],[1:1,0,0,1]:-;-;[\frac12:1];- \\ \\ [2:2,1,2,0],[2:1;0;0;1]:-;-;-;- \end{matrix} \;\middle|\; \begin{matrix} -a \\ b \\ -1 \\ a \end{matrix} \right).$$

## Acknowledgments

The authors would like to thank the editor and the unknown referee for careful reading and comments which improve the exposition.

## References

[1] Á. Baricz and T. K. Pogány, *Properties of the product of modified Bessel functions*, Analytic number theory, approximation theory, and special functions, Springer, New York, 2014, pp. 809–820. MR3329261

[2] Á. Baricz, D. Jankov Maširević, and T. K. Pogány, *Series of Bessel and Kummer-type functions*, Lecture Notes in Mathematics, vol. 2207, Springer, Cham, 2017, DOI 10.1007/978-3-319-74350-9. MR3753667

[3] K. Górska, A. Horzela, D. Jankov Maširević, and T. K. Pogány, *On distribution of Rice-Middleton model*, Results Math. **79** (2024), no. 3, Paper No. 106, 22, DOI 10.1007/s00025-024-02141-3. MR4709508

[4] W. Gröbner and N. Hofreiter, *Integraltafel. Zweiter Teil. Bestimmte Integrale*, Vierte verbesserte Auflage, Springer-Verlag, GmbH, Wien, 1966.

[5] K.–P. Ho, C.–K. Chant, F. Tong, and L. K. Chen, *Exact analysis of homodyne crosstalk induced penalty in optical networks*, Proceedings of SPIE **3420**, 1998, pp. 72–77.

[6] D. Middleton, *An introduction to statistical communication theory*, International Series in Pure and Applied Physics, McGraw-Hill Book Co., Inc., New York-Toronto-London, 1960. MR118561

[7] S. O. Rice, *Mathematical analysis of random noise*, Bell System Tech. J. **23** (1944), 282–332, DOI 10.1002/j.1538-7305.1944.tb00874.x. MR10932

[8] S. O. Rice, *Mathematical analysis of random noise*, Bell System Tech. J. **24** (1945), 46–156, DOI 10.1002/j.1538-7305.1945.tb00453.x. MR11918

[9] S. O. Rice, *Statistical properties of a sine wave plus random noise*, Bell System Tech. J. **27** (1948), 109–157, DOI 10.1002/j.1538-7305.1948.tb01334.x. MR23483

[10] H. M. Srivastava and M. C. Daoust, *Certain generalized Neumann expansions associated with the Kampé de Fériet function*, Indag. Math. **31** (1969), 449–457. Nederl. Akad. Wetensch. Proc. Ser. A **72**. MR252701

[11] H. M. Srivastava and M. C. Daoust, *A note on the convergence of Kampé de Fériet's double hypergeometric series*, Math. Nachr. **53** (1972), 151–159, DOI 10.1002/mana.19720530114. MR316763

[12] M. Trifonov, *Bessel representation for amplitude distribution of noisy sinusoidal signals*, Statist. Papers **63** (2022), no. 3, 983–988, DOI 10.1007/s00362-021-01262-z. MR4426423

School of Applied Mathematics and Informatics, University of Osijek, 31000 Osijek, Croatia
*Email address*: djankov@mathos.hr

Institute of Applied Mathematics, Óbuda University, 1034 Budapest, Hungary
*Email address*: pogany.tibor@nik.uni-obuda.hu

Faculty of Maritime Studies, University of Rijeka, 51000 Rijeka, Croatia
*Email address*: tibor.poganj@uniri.hr

Contemporary Mathematics
Volume **820**, 2025
https://doi.org/10.1090/conm/820/16415

# Modified Bohr phenomena for some classes of functions

Marrikanti Keerthana Reddy, Saminathan Ponnusamy,
and Karl-Joachim Wirths

ABSTRACT. In this article we consider three problems of recent interest. Firstly, modified Bohr radius for the class of holomorphic functions $f$ such that $\operatorname{Re} f(z) > \alpha$ in the unit disk $|z| < 1$ and $\lim_{z\to 1} f(z) = \infty$ is investigated. In the same spirit, a similar problem for functions satisfying the condition $\operatorname{Re} f(z) < 1$ was also considered. Finally, we determine the classical Bohr radius for the family of holomorphic $\alpha$-Bloch functions $f$ for which the Taylor expansion of $f'$ has the form $f'(z) = \sum_{k=0}^{\infty} b_k z^k$.

## 1. Introduction and statement of results

Throughout this paper $\mathbb{D}$ will denote the open unit disk $\{z\colon |z| < 1\}$ in the complex plane $\mathbb{C}$. Let $\mathcal{P}(\alpha)$ denote the family of holomorphic functions $f$ in $\mathbb{D}$ such that $\operatorname{Re} f(z) > \alpha$ in $\mathbb{D}$, where $\alpha \geq 0$.

Let $F$ be a holomorphic function satisfying $|F(z)| < 1$ in $|z| < 1$ such that the angular limit $\angle \lim_{z\to 1} F(z) = 1$. Then the angular limit $\angle \lim_{z\to 1} \frac{1-F(z)}{1-z}$ exists and is called the angular derivative of $F$ at $z = 1$. We have the following important result [**2**].

LEMMA A (Julia-Wolff-Carathéodory). *Let $F\colon \mathbb{D} \to \mathbb{D}$ be a bounded holomorphic function such that*

$$\alpha := \liminf_{z\to\sigma} \frac{1-|F(z)|}{1-|z|} < +\infty$$

*for some $\sigma \in \partial\mathbb{D}$ and let $\tau$ be the angular limit of $F$ at $\sigma$. Then both the incremental ratio $\frac{\tau-F(z)}{1-z}$ and the derivative $F'(z)$ have angular limit $\alpha\overline{\sigma}\tau$ at $\sigma$.*

---

2020 *Mathematics Subject Classification.* Primary 30B10, 30D45, 30H30; Secondary 30C80, 30A10.

*Key words and phrases.* Bohr phenomena, Bloch functions, angular limit.

The work of the first author was supported by CSIR (No.: 09/0084(11998)/2021-EMR-I)..

The second author is the corresponding author of this article.

In the following we use two theorems due to Komatu [**4**]

THEOREM B ([**4**, Theorem 4]). *Let $F$ be a holomorphic function satisfying $|F(z)| < 1$ in $|z| < 1$. Then there exists a positive real constant $D$, possibly equal to $\infty$, i.e., the so called angular derivative of $F(z)$ at $z = 1$, for which we have*

$$\frac{1-|F(z)|^2}{|1-F(z)|^2} \geq \frac{1}{D}\frac{1-|z|^2}{|1-z|^2} \quad for \quad |z|<1,$$

*$D$ being understood as the least possible constant satisfying this inequality. Here the equality holds at a point if and only if $F$ is a linear function defined by*

$$\frac{1+F(z)}{1-F(z)} = \frac{1}{D}\frac{1+z}{1-z} + i\beta,$$

*$\beta = \mathrm{Im}\,((1+F(0))/(1-F(0)))$ being a real constant; in this case, $D$ is necessarily finite. In general, the function satisfies the limit relation*

$$\frac{1-F(z)}{1-z} \to D,$$

*and, if $D$ is supposed finite, its successive derivatives satisfy the limit relations*

$$F'(z) \to D \quad and \quad (1-z)^{n-1}F^{(n)}(z) \to 0 \quad (n = 2, 3, ..),$$

*every relation holding uniformly as $z$ tends to 1 along any path through a Stolz angle in $|z| < 1$ with the vertex at $z = 1$.*

THEOREM C ([**4**, Theorem 6]). *Let $F$ be a holomorphic function with $|F(z)| < 1$ in $\mathbb{D}$ and $D$ be the angular derivative of $F$ at $z = 1$.Then we have, for any integer $n \geq 1$, an inequality*

$$\left|\frac{1}{n!}\frac{d^n}{dz^n}\frac{1}{1-F(z)} - \frac{1}{D}\frac{1}{(1-z)^{n+1}}\right| \leq \left(\mathrm{Re}\frac{1+F(0)}{1-F(0)} - \frac{1}{D}\right)\frac{1}{(1-|z|)^{n+1}}$$

In Theorem C, we have recalled only the necessary part of the statement of [**4**, Theorem 6] that is relevant for our investigation.

The classical Bohr phenomenon ([**1**]) for unimodular bounded functions deals with functions

$$\omega(z) = \sum_{k=0}^{\infty} a_k z^k, \quad z \in \mathbb{D},$$

that fulfill $|\omega(z)| \leq 1$ for $z \in \mathbb{D}$ and it is the assertion that

$$\sum_{k=0}^{\infty} |a_k| r^k \leq 1 \text{ for } r \leq \frac{1}{3}.$$

The number $1/3$ is optimal. In the recent years there has been a lot of interest on this topic. Refinements, improvements and generalizations of this result are investigated by a number of authors.

Now, we are in a position to state our first main result.

THEOREM 1.1. *Let $f(z) = \sum_{k=0}^{\infty} A_n z^n \in \mathcal{P}(\alpha)$ where $\alpha \geq 0$ and $A_0 > \alpha$ such that $\lim_{z\to 1} f(z) = \infty$, and $d$ be the angular derivative of $f$ at $z = 1$ defined as $d = \lim_{z\to 1}(1-z)f(z)$. Then we have $d \leq 2(A_0 - \alpha)$ and the modified Bohr inequality*

$$\sum_{n=1}^{\infty} |A_n - d| r^n \leq \mathrm{dist}(f(0), I_\alpha) \quad for \quad r \leq \frac{A_0 - \alpha}{3(A_0-\alpha) - d},$$

*where* $I_\alpha$ *denotes the axis* $\{w\colon \operatorname{Re} w = \alpha\}$. *The result is sharp, i.e.,*

(1) *for* $d \in (0, 2]$ *with* $a = d - 1$, *the sharpness is attained for the class of functions*

$$f_a(z) = \frac{2(1 + az)}{1 - z^2} + \alpha - 1,$$

*where* $a \in (-1, 1]$;

(2) *for* $d \geq 2$ *with* $a = (d-1)/(d+1)$, *the sharpness is attained for the class of functions*

$$f_a(z) = \frac{2\left(1 + az^2\right)}{(1-a)\left(1 - z^2\right)} + \alpha - 1,$$

*where* $a \in [1/3, 1)$.

REMARK 1.2. The case $\alpha = 0$ of Theorem 1.1 gives the corresponding results for holomorphic functions with positive real part.

Another result of independent interest is the following and the assumption on $f$ also covers the class of holomorphic self-mappings of the unit disk.

COROLLARY 1.1. *Let* $f(z) = \sum_{n=0}^{\infty} A_n z^n$ *be holomorphic in* $\mathbb{D}$ *such that* $\operatorname{Re} f(z) < 1$ *for* $z \in \mathbb{D}$, $0 < f(0) < 1$, $\lim_{z\to 1} f(z) = \infty$ *and* $d$ *be the angular derivative of* $f$ *at* $z = 1$. *Then the following modified Bohr inequality holds:*

$$\sum_{n=1}^{\infty} |A_n - d| r^n \leq \operatorname{dist}(f(0), I_1) \quad \textit{for} \quad r \leq \frac{1 - A_0}{3(1 - A_0) + d},$$

*where* $I_1$ *denotes the axis* $\{w\colon \operatorname{Re} w = 1\}$. *The result is sharp, i.e.,*

(1) *for* $d \in [-2, 0)$ *with* $a = -(d+1)$, *the sharpness is attained for the class of functions*

$$f_a(z) = \frac{-2z(z + a)}{1 - z^2}, \quad a \in (-1, 1];$$

(2) *for* $d \leq -2$ *with* $a = -(d+1)/(1-d)$, *the sharpness is attained for the class of functions*

$$f_a(z) = \frac{-2\left(a + z^2\right)}{(1-a)\left(1 - z^2\right)}, \quad a \in [1/3, 1).$$

For $\alpha \in [0, \infty)$, we say that a holomorphic function $f$ in the unit disk $\mathbb{D}$ belongs to $\alpha$-Bloch space denoted by $\mathcal{B}_\alpha$ if

$$\sup_{|z|<1} (1 - |z|^2)^\alpha |f'(z)| < \infty.$$

The $\alpha$-Bloch norm is given by

$$\|f\|_\alpha = |f(0)| + \sup_{|z|<1} (1 - |z|^2)^\alpha |f'(z)|.$$

Let the Taylor expansion of $f'$ be

$$f'(z) = \sum_{k=0}^{\infty} b_k z^k. \tag{1.1}$$

It seems natural to ask, whether for functions in the unit ball of $\alpha$-Bloch space with the expansion (1.1) there exists $r_0 \in (0,1)$ such that

$$(1.2)\qquad (1-r^2)^\alpha \sum_{k=0}^{\infty} |b_k| r^k \leq 1 \text{ for } r \leq r_0.$$

More precisely, we let

$$\mathcal{G}_\alpha = \{f \in \mathcal{B}_\alpha \colon \|f\|_\alpha \leq 1\}$$

and ask the above questions for the family $\mathcal{G}_\alpha$ for a given value of $a = |f'(0)|$. We show that when $\alpha \in \mathbb{N}$ and $a = 1$, there doesn't exist an $r_0 > 0$ satisfying the equation (1.2). In the case of $a \in [0,1)$, we present an affirmative answer to this question. We can now state our next result.

THEOREM 1.3. *Let $f \in \mathcal{G}_\alpha$, $\alpha > 0$, where $|f'(0)| = a \in [0,1)$ with $f'(z) = \sum_{k=0}^{\infty} b_k z^k$. Then we have*

$$(1.3)\qquad (1-r^2)^\alpha \sum_{k=0}^{\infty} |b_k| r^k \leq 1 \textit{ for } r \leq r_0(a,\alpha),$$

*where $r_0(a,\alpha)$ is the smallest positive root of the equation*

$$\left(\frac{e}{2\alpha}\right)^\alpha \left(\sum_{k=1}^{\infty} (k+2\alpha)^\alpha r^k\right) = 1-a.$$

*For smaller integer values of $\alpha$, one obtains better estimates for $r_0(a,\alpha)$. More precisely, we obtain*

(1) *$r_0(a,1)$ is the smallest positive root of the equation*

$$(3\sqrt{3}-3e)r^3 + (4e - 6\sqrt{3} - 2 + 2a)r^2 + (3\sqrt{3} + 4 - 4a)r + 2(a-1) = 0;$$

(2) *$r_0(a,2)$ is the smallest positive root of the equation*

$$(36e^2 - 108)r^5 - (83e^2 - 324 + 25\sqrt{5})r^4 + (49e^2 + 16(1-a) - 324 + 75\sqrt{5})r^3 + (108 - 75\sqrt{5} - 48(1-a))r^2 + (25\sqrt{5} + 48(1-a))r - 16(1-a) = 0;$$

(3) *$r_0(a,3)$ is the smallest positive root of the 7-th degree polynomial equation given by*

$$\begin{aligned}&(729e^3 - 2187\sqrt{3})r^7 + (8748\sqrt{3} - 2048 - 2404e^3)r^6\\ &+(8192 - 343\sqrt{7} - 13122\sqrt{3} + 2669e^3)r^5\\ &+(216(1-a) + 1372\sqrt{7} - 12288 + 8748\sqrt{3} - 1000e^3)r^4\\ &+(8192 - 2187\sqrt{3} - 2058\sqrt{7} - 864(1-a))r^3 + (1296(1-a) + 1372\sqrt{7} - 2048)r^2\\ &-(864(1-a) + 343\sqrt{7})r + 216(1-a) = 0.\end{aligned}$$

*Moreover when $\alpha \in \mathbb{N}$ and $a = 1$, there doesn't exist $r_0 > 0$ satisfying the inequality (1.3).*

TABLE 1. The values of $r_0(a, \alpha)$ for $\alpha = 1, 2, 3$ and $a$ are tabulated.

| $a$ | $r_0(a,1)$ | $r_0(a,2)$ | $r_0(a,3)$ |
|---|---|---|---|
| 0 | 0.230155 | 0.177819 | 0.1538 |
| 0.3 | 0.181048 | 0.140376 | 0.1208 |
| 0.5 | 0.141576 | 0.109809 | 0.0938 |
| 0.6 | 0.119116 | 0.0922613 | 0.0784 |
| 0.8 | 0.066796 | 0.0512378 | 0.0430 |
| 0.9 | 0.035699 | 0.0270725 | 0.0226 |

## 2. Proofs of Theorems 1.1 and 1.3

PROOF OF THEOREM 1.1. Let $f \in \mathcal{P}(\alpha)$. Then there exists a unique holomorphic function $F$ which is unimodular bounded in $\mathbb{D}$ such that

$$f(z) = \frac{1+F(z)}{1-F(z)} + \alpha = \frac{2}{1-F(z)} + \alpha - 1 \tag{2.1}$$

where $F(0) = \frac{A_0-1-\alpha}{A_0+1-\alpha} \in (-1,1)$. We assume that $\lim_{z\to 1} F(z) = 1$ in any Stolz angle with vertex at the point 1. From the theorem of Julia - Wolff - Carathéodory, it is well known that there exists the angular derivative

$$D := \lim_{z\to 1} \frac{F(z)-1}{z-1} \tag{2.2}$$

and that the inequality

$$A_0 - \alpha = \frac{1+F(0)}{1-F(0)} \geq \frac{1}{D}$$

is satisfied, by Theorem B. Hence by (2.1) and (2.2) it follows that

$$\lim_{z\to 1}(1-z)f(z) = \frac{2}{D} =: d$$

the angular derivative of $f$ at $z = 1$. From Theorem C, and (2.1), we deduce that

$$|A_n - d| \leq 2(A_0 - \alpha) - d, \quad n \geq 1.$$

Therefore the modified Bohr inequality

$$\sum_{n=1}^{\infty} |A_n - d|\, r^n \leq d(f(0), I_\alpha) = A_0 - \alpha,$$

is satisfied, if

$$[2(A_0 - \alpha) - d]\,\frac{r}{1-r} \leq A_0 - \alpha,$$

Here $I_\alpha$ is as in the statement. This inequality is valid for

$$r \leq \frac{A_0 - \alpha}{3(A_0 - \alpha) - d}.$$

Let us call this quantity the modified Bohr radius. In the sequel, we prove that this modified Bohr radius cannot be improved for any fixed positive value $d$ of the angular derivative of $f$ at $z = 1$. In the cases $F(0) = 0$, where $d \in [0, 2]$, we consider the unimodular bounded functions $F\colon \mathbb{D} \to \mathbb{D}$ by

$$F(z) = z\frac{z+a}{1+az}, \quad a \in (-1, 1].$$

For these functions in view of (2.1) we get

$$f(z)=\frac{2(1+az)}{1-z^2}+\alpha-1=1+\alpha+2\sum_{k=1}^{\infty}z^{2k}+2a\sum_{k=0}^{\infty}z^{2k+1}.$$

Since $\lim_{z\to 1}(1-z)f(z)=1+a=d$, we have

$$|A_n-d|=1-a.$$

and therefore,

$$\sum_{n=1}^{\infty}|A_n-d|\,r^n=(1-a)\frac{r}{1-r}\leq 1=A_0-\alpha$$

if and only if

$$r\leq\frac{1}{2-a}=\frac{A_0-\alpha}{3(A_0-\alpha)-d}$$

These examples show that any $r\in(1/3,1)$ occurs as such sharp modified Bohr radius.

For $d\geq 2$ we consider the unimodular bounded functions $F\colon\mathbb{D}\to\mathbb{D}$ by

$$F(z)=\frac{z^2+a}{1+az^2},\quad a\in[1/3,1).$$

In view of (2.1), the functions with real part greater than $\alpha$ constructed as above are the following

$$f(z)=\frac{2\left(1+az^2\right)}{(1-a)\left(1-z^2\right)}+\alpha-1=\alpha+\frac{1+a}{1-a}\left(\frac{1+z^2}{1-z^2}\right)=\alpha+\frac{1+a}{1-a}\left(1+2\sum_{k=1}^{\infty}z^{2k}\right).$$

It is easy to calculate the equalities $d=(1+a)/(1-a)$ and $|A_n-d|=(1+a)/(1-a)$. Hence, we have

$$\sum_{n=1}^{\infty}|A_n-d|\,r^n=\frac{(1+a)r}{(1-a)(1-r)}\leq\frac{1+a}{1-a}$$

if and only if

$$r\leq\frac{1}{2}=\frac{A_0-\alpha}{3(A_0-\alpha)-d},$$

and we complete the proof. □

PROOF OF COROLLARY 1.1. We are given that $f_0(z):=1-f(z)\in\mathcal{P}(0)$ and $1-A_0>0$. If we set $f_0(z)=\sum_{n=0}^{\infty}B_nz^n$, then we easily have

$$B_0=1-A_0\text{ and }B_n=-A_n\text{ for }n\geq 1,\tag{2.3}$$

and

$$\lim_{z\to 1}(1-z)f_0(z)=\lim_{z\to 1}(1-z)(1-f(z))=-d.$$

By Theorem 1.1, we deduce that

$$\sum_{n=1}^{\infty}|B_n+d|r^n\leq B_0\text{ for }r\leq\frac{B_0}{3B_0+d},$$

which, by (2.3), is equivalent to the required inequality

$$\sum_{n=1}^{\infty}|A_n-d|r^n\leq\operatorname{dist}(f(0),I_1)\text{ for }r\leq\frac{1-A_0}{3(1-A_0)+d}.$$

Following the proof of Theorem 1.1, we may consider

$$f_0(z) = \frac{2(1+az)}{1-z^2} - 1 \text{ and } f(z) = 1 - f_0(z),$$

so that

$$\lim_{z\to 1}(1-z)f_0(z) = 1+a, \quad d = \lim_{z\to 1}(1-z)f(z) = -(1+a)$$

and $|A_n - d| = 1-a$. The rest follows as in the proof of Theorem 1.1.

For $d \leq -2$, we consider

$$f_0(z) = \frac{2\left(1+az^2\right)}{(1-a)\left(1-z^2\right)} - 1 \quad \text{and } f(z) = 1 - f_0(z).$$

so that $\lim_{z\to 1}(1-z)f_0(z) = \frac{1+a}{1-a}$ and thus, $d = \lim_{z\to 1}(1-z)f(z) = -\frac{1+a}{1-a}$ and $|A_n - d| = \frac{1+a}{1-a}$. The remaining part of the proof follows similarly. □

For the proof of Theorem 1.3, we need the following result (see [**3**] and [**5**]).

Lemma D ([**3**]). *Let $f \in \mathcal{B}_\alpha$, $\alpha > 0$ and $\|f\|_\alpha \leq 1$ such that $f'(z) = \sum_{k=0}^{\infty} b_k z^k$. Then we have the estimates*

$$|b_k| \leq \left(\frac{k+2\alpha}{2\alpha}\right)^\alpha \left(\frac{k+2\alpha}{k}\right)^{k/2} \quad \textit{for } k \geq 1.$$

Proof of Theorem 1.3. For $f \in \mathcal{G}_\alpha$, by Lemma D we get the following estimates

$$|b_k| \leq \left(\frac{e(k+2\alpha)}{2\alpha}\right)^\alpha \quad \text{for } k \geq 1 \tag{2.4}$$

Given $a = |b_0| < 1$, it is sufficient to prove the existence of such $r_0(a,\alpha)$ which satisfies the stronger inequality

$$\sum_{k=1}^{\infty} |b_k| r^k \leq 1 - a \text{ for } r \leq r_0(a,\alpha).$$

By (2.4), we have following decent looking inequality

$$\sum_{k=1}^{\infty} |b_k| r^k \leq \left(\frac{e}{2\alpha}\right)^\alpha \sum_{k=1}^{\infty} (k+2\alpha)^\alpha r^k =: M(r).$$

Obviously, $M(0) = 0$, $M(r)$ is monotonically increasing for $r \in [0,1)$, and $\lim_{r\to 1} M(r) = \infty$.Therefore, there exists $r_0(a,\alpha) \in (0,1)$ such that

$$\sum_{k=1}^{\infty} |b_k| r^k \leq M(r_0(a,\alpha)) = 1 - a \quad for \quad r \leq r_0(a,\alpha).$$

Next, we deal with the second part of the conclusion of Theorem 1.3 by considering the case $a = 1$.

When $\alpha \in \mathbb{N}$, we show that for sufficiently small $b$, the function $f$ belongs to $\alpha$-Bloch space , where

$$f'(z) = (1 - z^2 + bz^3)^\alpha, \quad b > 0.$$

Let $L(z)$ be such that

$$L'(z) = 1 - z^2 + bz^3, \quad b > 0.$$

Now we calculate for $z = re^{i\phi}, x = \cos\phi$,

$$|L'(z)|^2 = (1+r^2)^2 + b^2 r^6 + 2r^2 h_r(x), \tag{2.5}$$

where

$$h_r(x) = -2x^2 + 4brx^3 - 3brx - br^3x. \tag{2.6}$$

An elementary discussion of the function $h_r(x), x \in [-1, 1]$ shows that for sufficiently small $b > 0$, it attains a unique maximum at

$$x_m = \frac{1}{6br}\left(1 - \sqrt{1 + 9r^2b^2 + 3r^4b^2}\right).$$

For sufficiently small $b > 0$ we may use the estimates

$$-rb \le x_m \le -\frac{3}{4}rb.$$

If we insert this inequalities into (2.5) and (2.6), we get the rough estimate

$$|L'(z)|^2 \le (1 + r^2)^2 + 7b^2r^4. \tag{2.7}$$

Therefore,

$$(1 - r^2)^2|L'(z)|^2 < 1 - 2r^4 + r^8 + 7b^2r^4 < 1, \tag{2.8}$$

for sufficiently small positive values of $b$. Hence, for such $b$ the function $f$ is a $\alpha$-Bloch function as

$$(1 - r^2)^\alpha|f'(z)| = ((1 - r^2)|L'(z)|)^\alpha < 1$$

For this Bloch function, we get

$$(1 - r^2)^\alpha \sum_{k=0}^{\infty} |b_k|r^k = 1 + \alpha br^3 + r^4P(r) =: H(r)$$

where $P(r)$ is a polynomial in $r$. Since $H(0) = 1$ and

$$H'(r) = r^2(3\alpha b + 4rP'(r)) > 0$$

for some $r$ sufficiently small. This shows that there exist $\alpha$-Bloch functions with $|b_0| = 1$ for which (1.2) is not valid when $\alpha \in \mathbb{N}$.

We give a better estimate for $r_0(a, \alpha)$ when $\alpha = 1, 2, 3$. For $\alpha = 1$, by Lemma D, we get

$$|b_1| \le \frac{3\sqrt{3}}{2} \text{ and } |b_k| \le \left(\frac{k+2}{2}\right)e \quad for \quad k \ge 2.$$

Using these estimates, we obtain that

$$\sum_{k=1}^{\infty} |b_k|r^k \le \frac{3\sqrt{3}}{2}r + \frac{e}{2}\left(\sum_{k=2}^{\infty}(k+2)r^k\right) = \frac{3\sqrt{3}}{2}r + \frac{e}{2}\left(\frac{(4-3r)r^2}{(1-r)^2}\right)$$

which is less than or equal to $1 - a$ whenever $r \le r_0(a, 1)$, where $r_0(a, 1)$ is the smallest positive root of the equation

$$(3\sqrt{3} - 3e)r^3 + (4e - 6\sqrt{3} - 2 + 2a)r^2 + (3\sqrt{3} + 4 - 4a)r + 2(a - 1) = 0.$$

Similarly for $\alpha = 2$, we have

$$|b_1| \le \frac{25\sqrt{5}}{16}, \ |b_2| \le \frac{27}{4} \text{ and } |b_k| \le \left(\frac{k+4}{4}\right)^2 e^2 \quad for \quad k \ge 3.$$

Again, using these coefficient inequalities, we obtain that

$$\begin{aligned}\sum_{k=1}^{\infty}|b_k|r^k &\leq \frac{25\sqrt{5}}{16}r + \frac{27}{4}r^2 + \frac{e^2}{16}\left(\sum_{k=3}^{\infty}(k+4)^2 r^k\right)\\ &= \frac{25\sqrt{5}}{16}r + \frac{27}{4}r^2 + \frac{e^2}{16}\left(\frac{(49-83r+36r^2)r^3}{(1-r)^3}\right)\end{aligned}$$

which is less than or equal to $1-a$ whenever $r \leq r_0(a,2)$, where $r_0(a,2)$ is as in the statement. Finally, for $\alpha = 3$, we have

$$|b_1| \leq \frac{343\sqrt{7}}{216},\ |b_2| \leq \frac{256}{27},\ |b_3| \leq \frac{81\sqrt{3}}{8} \text{ and } |b_k| \leq \left(\frac{k+6}{6}\right)^3 e^3 \text{ for } k \geq 4,$$

and therefore, we obtain that

$$\begin{aligned}\sum_{k=1}^{\infty}|b_k|r^k &\leq \frac{343\sqrt{7}}{216}r + \frac{256}{27}r^2 + \frac{81\sqrt{3}}{8}r^3 + \frac{e^3}{216}\sum_{k=4}^{\infty}(k+6)^3 r^k\\ &= \frac{343\sqrt{7}}{216}r + \frac{256}{27}r^2 + \frac{81\sqrt{3}}{8}r^3\\ &\quad + \frac{e^3}{216}\left(\frac{(1000-2669r+2404r^2-729r^3)r^4}{(1-r)^4}\right)\end{aligned}$$

which is less than or equal to $1-a$ whenever $r \leq r_0(a,3)$, where $r_0(a,3)$ is as in the statement. The proof is complete. □

## Acknowledgments

The authors thank the referee for his/her suggestions. This work was completed in final form during the visit of the second author to "Lomonosov Moscow State University, Moscow Center of Fundamental and Applied Mathematics, Moscow," under the support of "IMU-CDC Individual Research Travel Support"

## References

[1] H. Bohr, *A Theorem Concerning Power Series*, Proc. London Math. Soc. (2) **13** (1914), 1–5, DOI 10.1112/plms/s2-13.1.1. MR1577494

[2] R. B. Burckel, *An introduction to classical complex analysis. Vol. 1*, Pure and Applied Mathematics, vol. 82, Academic Press, Inc. [Harcourt Brace Jovanovich, Publishers], New York-London, 1979. MR555733

[3] I. R. Kayumov and K.-J. Wirths, *Coefficient inequalities for Bloch functions*, Lobachevskii J. Math. **40** (2019), no. 9, 1319–1323, DOI 10.1134/s1995080219090117. MR4021521

[4] Y. Komatu, *On angular derivative*, Kōdai Math. Sem. Rep. **13** (1961), 167–179. MR138767

[5] K.-J. Wirths, *Über holomorphe Funktionen, die einer Wachstumsbeschränkung unterliegen* (German), Arch. Math. (Basel) **30** (1978), no. 6, 606–612, DOI 10.1007/BF01226108. MR492223

Department of Mathematics, Indian Institute of Technology Madras, Chennai-600 036, India.

*Email address*: ma20d020@smail.iitm.ac.in

S. Ponnusamy, Department of Mathematics, Indian Institute of Technology Madras, Chennai-600 036, India.

*Email address*: samy@iitm.ac.in

K.-J. Wirths, Institut für Analysis und Algebra, TU Braunschweig, 38106 Braunschweig, Germany.

*Email address*: kjwirths@tu-bs.de

Contemporary Mathematics
Volume **820**, 2025
https://doi.org/10.1090/conm/820/16416

# Geometry of stereographic (type) projections I

Ritwick Maity and Swadesh Kumar Sahoo

ABSTRACT. The aim of this manuscript is to explore the geometric properties of stereographic projection and stereographic type projections of a subset of $\mathbb{R}^{n+1}$ on a unit sphere in $\mathbb{R}^{n+1}$. In some cases, we examine the existence of the density of the chordal distance. We also illustrate a couple of situations where stereographic type projections preserve the stereographic projection.

## 1. Introduction and preliminaries

Notations of this manuscript are mostly standard as used in [**3**]. Quite a few are listed below. The sphere centered at $a \in \mathbb{R}^{n+1}$ with radius $r > 0$ is denoted by $S(a, r)$. The notation $\overline{\mathbb{R}^{n+1}}$ is used to denote the Möbius space $\mathbb{R}^{n+1} \cup \{\infty\}$. Throughout the manuscript, we use the notation $X_i$, $i = 1, 2, \ldots, n+1$, to represent the coordinate axes of $\mathbb{R}^{n+1}$, $n \geq 1$.

The inversion $\phi\colon \overline{\mathbb{R}^{n+1}} \to \overline{\mathbb{R}^{n+1}}$ in $S(a, r)$ is defined by

$$\phi(x) = \begin{cases} a + \dfrac{r^2}{|x-a|^2}(x-a), & \text{if } x \in \mathbb{R}^{n+1} \setminus \{a\}, \\ \infty, & \text{if } x = a, \\ a, & \text{if } x = \infty. \end{cases} \tag{1.1}$$

A plane in $\overline{\mathbb{R}^{n+1}}$ is defined by $P(a, t) = \{x \in \mathbb{R}^{n+1} : (x \cdot a) = t\} \cup \{\infty\}$, where $a(\neq 0) \in \mathbb{R}^{n+1}$, $(x \cdot a) = \sum_{i=1}^{n+1} a_i x_i$ and $t \in \mathbb{R}$.

The reflection $\psi\colon \overline{\mathbb{R}^{n+1}} \to \overline{\mathbb{R}^{n+1}}$ in the plane $P(a, t)$ is given by

$$\psi(x) = \begin{cases} x - 2[(x \cdot a) - t]\dfrac{a}{|a|^2}, & \text{if } x \in \mathbb{R}^{n+1}, \\ \infty, & \text{if } x = \infty. \end{cases}$$

A Möbius transformation on $\overline{\mathbb{R}^{n+1}}$ is defined as a finite composition of sphere inversions and plane reflections.

In general, the stereographic projection is considered as a one-to-one correspondence between the extended plane and the Riemann sphere. Recent developments

---

2020 *Mathematics Subject Classification.* Primary 51M15; Secondary 51B10, 51F15.

*Key words and phrases.* Stereographic projection, projection sphere, projection line, chordal metric, spherical metric.

The work of the first author is supported by the DST-INSPIRE Fellowship (Code: IF210630). We also acknowledge the support of the DST-FIST Project (File No.: SR/FST/MS I/2018/26) for providing research facilities in the department.

related to stereographic projection may be found in [**4**]. However, this manuscript deals with stereographic (type) projection of a subset $G \subset \mathbb{R}^{n+1}$ on a unit sphere in $\mathbb{R}^{n+1}$. For this purpose, we mainly use the following two important terminologies, namely, *projection sphere* and *projection lines.* They are defined as follows: let $d \in \mathbb{R}$ and $e_{n+1} = (0, 0, \cdots, 0, 1) \in \mathbb{R}^{n+1}$.

- We name the unit sphere $S((d+1)e_{n+1}, 1)$ with the north pole $N$ at $(d+2)e_{n+1}$ as the *projection sphere* (see Figure 1).
- A simple differentiable parametrized curve in $\mathbb{R}^{n+1}$ that joins the north pole $N$, at least one point $x \in G$, and at most one point on the projection sphere other than the north pole $N$ is called a *projection line.* We denote this projection line by $\ell_x$.
- To define stereographic type projection of $G$, we take the projection sphere $S((d+1)e_{n+1}, 1)$ and a family of projection lines, say $\ell = \{\ell_x \colon x \in G\}$, with the property that each point of $G$ is connected to $N$ through a single projection line. Thus, *stereographic type projection* is a function $\pi^{\ell}_{(d+2)e_{n+1}} \colon G \to S((d+1)e_{n+1}, 1)$ described by
$$\pi^{\ell}_{(d+2)e_{n+1}}(x) = \begin{cases} y, & \text{if } \ell_x \cap \{S((d+1)e_{n+1}, 1) \setminus N\} = \{y\}, \\ N, & \text{otherwise;} \end{cases}$$
see Figure 1(a).
- When all $\ell_x$ are straight lines, we call such a stereographic type projection as *a stereographic projection* and denote it by $\pi_{(d+2)e_{n+1}}$ or simply by $\pi$ (see Figure 1(a)).

Stereographic type projection is completely dependent on the chosen family of projection lines. For $x \in G$ and two different families of projection lines $\ell^1 = \{\ell^1_x : x \in G\}$ and $\ell^2 = \{\ell^2_x : x \in G\}$, two stereographic type projections $\pi^{\ell^1}_{(d+2)e_{n+1}}$ and $\pi^{\ell^2}_{(d+2)e_{n+1}}$ of $G$ will be considered different even if $\pi^{\ell^1}_{(d+2)e_{n+1}}(x) = \pi^{\ell^2}_{(d+2)e_{n+1}}(x)$; see for instance Subsection 3.2 where a couple of examples are presented on this.

Also, by the property of a family of projection lines, each point of $G$ has a unique image on $S((d+1)e_{n+1}, 1)$, which establishes the well-definedness of a stereographic type projection. In general, stereographic (type) projection of $G$ is not necessarily one-to-one or onto (see Figure 1).

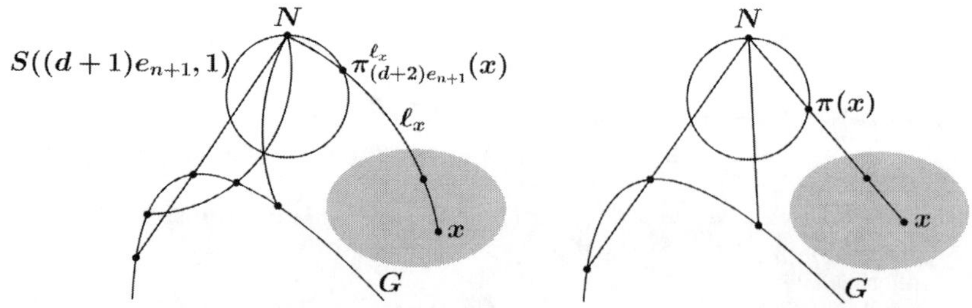


(a) Stereographic type projection (b) Stereographic projection

FIGURE 1. Stereographic type and stereographic projection

If $G$ is unbounded then the image of the point at infinity depends both on $G$ and the chosen family of projection lines. In some cases, the point at infinity may not have a unique image. For this reason, in this article, we consider neither the projection of the point at infinity nor the chordal distance related to it, except Theorem 2.3.

## 2. Stereographic projection

**2.1. Stereographic projection of $\mathbb{R}^n$.** In this subsection, we consider the stereographic projection of $\mathbb{R}^n$ on the projection sphere $S((d+1)e_{n+1}, 1)$ (see Figure 2).

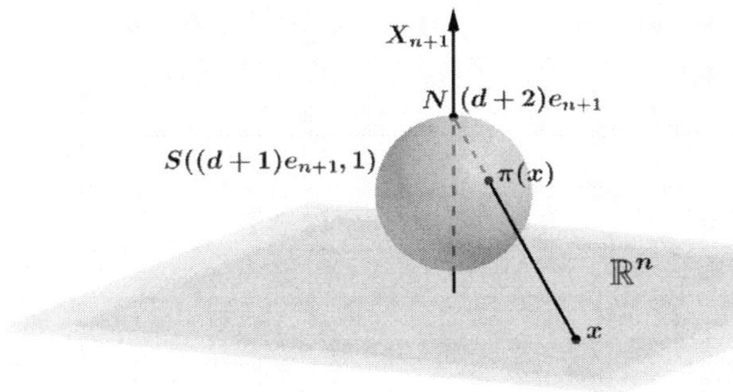


FIGURE 2. Stereographic projection of $\mathbb{R}^n$

The stereographic projection, $\pi(x)$, of $x = (x_1, x_2, \ldots, x_n) \in \mathbb{R}^n$ on the projection sphere is formulated in [**2**] for $n = 2$. A similar computation leads to the formula

$$\pi(x) = \left( \frac{2(d+2)x_1}{|x|^2 + (d+2)^2}, \ldots, \frac{2(d+2)x_n}{|x|^2 + (d+2)^2}, (d+2)\frac{|x|^2 + d(d+2)}{|x|^2 + (d+2)^2} \right) \tag{2.1}$$

for $d \in \mathbb{R}$. Note that if $d+2 = 0$ then $\pi(x) = (0, 0, \cdots, 0) = N$, the north pole. As a result, this situation does not provide any non-trivial projection. For this reason, we now assume $d + 2 \neq 0$.

For the projection sphere $S((d+1)e_{n+1}, 1)$, we denote the chordal distance, $|\pi(x) - \pi(y)|$, between $x$ and $y$ in $\mathbb{R}^n$ by $g_{(d+2)e_{n+1}}(x, y)$. An elementary computation yields

$$g_{(d+2)e_{n+1}}(x, y) = \frac{2|d+2||x-y|}{\sqrt{|x|^2 + (d+2)^2}\sqrt{|y|^2 + (d+2)^2}}, \quad x, y \in \mathbb{R}^n. \tag{2.2}$$

Note that for $d = -1, n = 2$, $g_{(d+2)e_{n+1}}$ becomes $g_{e_3}$, which describes the classical chordal distance; see [**1**] for the plane and [**3**] for the $n$-space (readers are referred to [**6**] for further geometries).

The following lemma describes a relation between chordal distances of a pair of points for two different projection spheres.

LEMMA 2.1. *For $x, y \in \mathbb{R}^n$ and $d, h \in \mathbb{R} \setminus \{-2\}$, if $g_{(h+2)e_{n+1}}(x, y)$ and $g_{(d+2)e_{n+1}}(x, y)$ are respectively the chordal distances for the projection spheres $S((h+1)e_{n+1}, 1)$ and $S((d+1)e_{n+1}, 1)$, then $g_{(h+2)e_{n+1}}(f(x), f(y)) = g_{(d+2)e_{n+1}}(x, y)$ holds, where $f(x) = [(h+2)/(d+2)]x$.*

PROOF. Rewriting the formula (2.2) for $g_{(d+2)e_{n+1}}(x,y)$, $x,y\in\mathbb{R}^n$, we obtain

$$g_{(d+2)e_{n+1}}(x,y)=\frac{2\left|\dfrac{x}{d+2}-\dfrac{y}{d+2}\right|}{\sqrt{\left|\dfrac{x}{d+2}\right|^2+1}\sqrt{\left|\dfrac{y}{d+2}\right|^2+1}},\quad d\in\mathbb{R}\setminus\{-2\}.$$

For $f(x)=[(h+2)/(d+2)]x$, $h\in\mathbb{R}\setminus\{-2\}$, we compute

$$\begin{aligned}g_{(h+2)e_{n+1}}(f(x),f(y))&=\frac{2\left|\dfrac{f(x)}{h+2}-\dfrac{f(y)}{h+2}\right|}{\sqrt{\left|\dfrac{f(x)}{h+2}\right|^2+1}\sqrt{\left|\dfrac{f(y)}{h+2}\right|^2+1}}\\&=\frac{2\left|\dfrac{x}{d+2}-\dfrac{y}{d+2}\right|}{\sqrt{\left|\dfrac{x}{d+2}\right|^2+1}\sqrt{\left|\dfrac{y}{d+2}\right|^2+1}}=g_{(d+2)e_{n+1}}(x,y),\end{aligned}$$

completing the proof. □

REMARKS 2.2. We have the following observations from Lemma 2.1:

(1) for the choices $d=2, h=-1$, Lemma 2.1 is geometrically viewed in Figure 3;
(2) Lemma 2.1 also holds for the map $f(x)=[(h+2)/(d+2)]Ax$, where $A$ is an $n\times n$ orthogonal matrix and $x\in\mathbb{R}^n$.

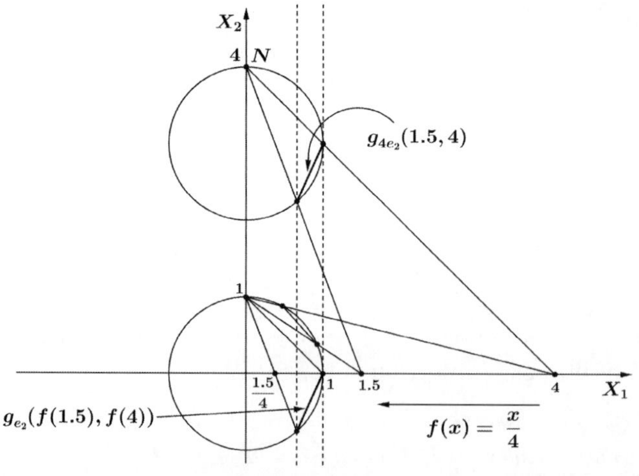


FIGURE 3. Relation of chordal distances for two projection spheres

**2.2. Stereographic projection as a restriction of Möbius transformation.** As described in [**3**, Section 3.4], the stereographic projection $\pi$ of $\overline{\mathbb{R}^n}$ onto $\mathbb{S}^n$ (unit sphere centered at origin in $\mathbb{R}^{n+1}$) can be written as inversion in $S(e_{n+1}, \sqrt{2})$. In a similar fashion, we have the following general statement:

THEOREM 2.3. *The stereographic projection of $\overline{\mathbb{R}^n}$ onto $S((d+1)e_{n+1}, 1)$ can be written as a restriction of a Möbius transformation on $\overline{\mathbb{R}^{n+1}}$.*

Here, the restriction of a Möbius transformation on $\overline{\mathbb{R}^{n+1}}$ is understood as domain restriction on $\overline{\mathbb{R}^n}(\subset \overline{\mathbb{R}^{n+1}})$.

PROOF. We notice that the origin and the point $de_{n+1}$ are the stereographic projections of one another with respect to the projection sphere $S((d+1)e_{n+1}, 1)$. Let they also be inverse points with respect to the sphere $S((d+2)e_{n+1}, r)$ for some $r > 0$. To determine $r$, we use the corresponding inversion formula

$$|(d+2)e_{n+1} - de_{n+1}||(d+2)e_{n+1} - 0| = r^2.$$

This yields $r = \sqrt{2|d+2|}$. Thus, the inversion of $\overline{\mathbb{R}^n}$ in $S((d+2)e_{n+1}, \sqrt{2|d+2|})$ becomes

$$\begin{aligned}\phi(x) &= (d+2)e_{n+1} + 2|d+2|\frac{x-(d+2)e_{n+1}}{|x-(d+2)e_{n+1}|^2}\\ &= \left(\frac{2|d+2|x_1}{|x|^2+(d+2)^2}, \dots, \frac{2|d+2|x_n}{|x|^2+(d+2)^2}, (d+2)\frac{|x|^2+(d+2)^2-2|d+2|}{|x|^2+(d+2)^2}\right),\end{aligned}$$

where $x = (x_1, x_2, \dots, x_n, 0)$, and by (1.1) we have $\phi(\infty) = N = (d+2)e_{n+1}$.

To investigate our required Möbius maps, we consider two cases.

CASE 1 ($d+2 > 0$). In this case, we have (see Figure 4)

$$\begin{aligned}\phi(x) &= \left(\frac{2(d+2)x_1}{|x|^2+(d+2)^2}, \dots, \frac{2(d+2)x_n}{|x|^2+(d+2)^2}, (d+2)\frac{|x|^2+(d+2)^2-2(d+2)}{|x|^2+(d+2)^2}\right)\\ &= \left(\frac{2(d+2)x_1}{|x|^2+(d+2)^2}, \dots, \frac{2(d+2)x_n}{|x|^2+(d+2)^2}, (d+2)\frac{|x|^2+d(d+2)}{|x|^2+(d+2)^2}\right)\\ &= \pi(x),\end{aligned}$$

as obtained in (2.1), and we also have $\phi(\infty) = N = (d+2)e_{n+1}$.

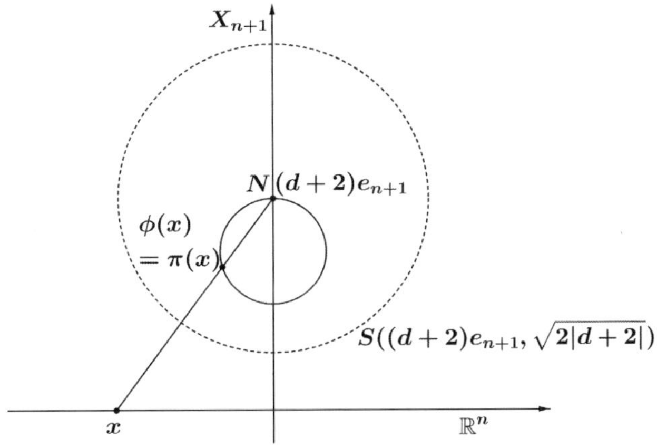


FIGURE 4. Stereographic projection as an inversion

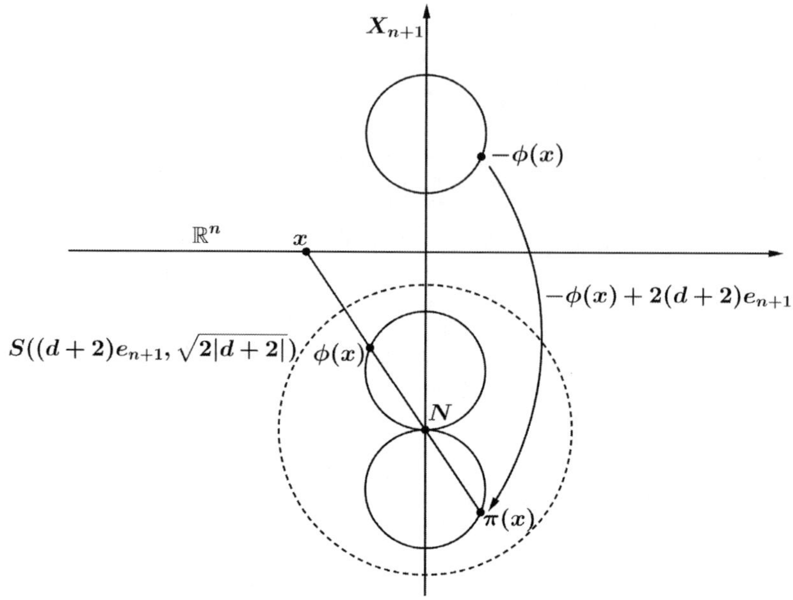


FIGURE 5. Stereographic projection as inversion and plane reflections

CASE 2 ($d+2<0$). For this, we have (see Figure 5)

$$\phi(x)=\left(\frac{-2(d+2)x_1}{|x|^2+(d+2)^2},\ldots,\frac{-2(d+2)x_n}{|x|^2+(d+2)^2},(d+2)\frac{|x|^2+(d+2)^2+2(d+2)}{|x|^2+(d+2)^2}\right),$$

and $\phi(\infty)=(d+2)e_{n+1}$. It follows that

$$\begin{aligned}-\phi(x)&+2(d+2)e_{n+1}\\&=\left(\frac{2(d+2)x_1}{|x|^2+(d+2)^2},\ldots,\frac{2(d+2)x_n}{|x|^2+(d+2)^2},(d+2)\frac{|x|^2+d(d+2)}{|x|^2+(d+2)^2}\right)\\&=\pi(x),\end{aligned}$$

and $-\phi(\infty)+2(d+2)e_{n+1}=(d+2)e_{n+1}$.

We see that $x\mapsto\phi(x)$ and $x\mapsto-\phi(x)+2(d+2)e_{n+1}$ both are Möbius maps on $\overline{\mathbb{R}^{n+1}}$. Stereographic projection of $\overline{\mathbb{R}^n}$ is obtained by restricting these maps to $\overline{\mathbb{R}^n}$.

□

REMARK 2.4. According to [**5**, p. 35], stereographic projection of $S(ae_{n+1},|a|)$ is restriction of inversion in $S(2ae_{n+1},2|a|)$, $a>0$. Therefore, there is a scope to generalize Theorem 2.3 for a projection sphere of arbitrary radius.

**2.3. Stereographic projection of a subset of $\mathbb{R}^{n+1}$.** Let $G\subset\mathbb{R}^{n+1}$ and $x=(x_1,x_2,\ldots,x_n,x_{n+1})\in G$. Let $\pi(x)=(\pi_1(x),\ldots,\pi_n(x),\pi_{n+1}(x))$ be the stereographic projection of $x$ on $S((d+1)e_{n+1},1)$; see Figure 1(b). Thus,

$$\pi(x)=x+t((d+2)e_{n+1}-x);\ t\in[0,1].$$

Note that if $x=(d+2)e_{n+1}\in G$, then trivially $\pi(x)=(d+2)e_{n+1}$, i.e. $\pi(N)=N$.

We now compute $\pi(x)$, when $x \neq (d+2)e_{n+1}$. As $\pi(x)$ lies on the projection sphere, it satisfies $|\pi(x) - (d+1)e_{n+1}| = 1$. This leads to

$$|x + t((d+2)e_{n+1} - x) - (d+1)e_{n+1}|^2 = 1,$$

which is equivalent to

$$t^2|(d+2)e_{n+1} - x|^2 + 2t((d+2)e_{n+1} - x)\cdot(x - (d+1)e_{n+1}) + |x - (d+1)e_{n+1}|^2 = 1.$$

This produces $t = 1$ and $t = [|x - (d+1)e_{n+1}|^2 - 1]/|(d+2)e_{n+1} - x|^2$.

Trivially, $t = 1$ corresponds to the north pole. The second choice of $t$ gives

$$\begin{aligned}
(2.3)\quad \pi(x) &= x + \frac{|x-(d+1)e_{n+1}|^2 - 1}{|(d+2)e_{n+1} - x|^2}((d+2)e_{n+1} - x)\\
&= \frac{2(d+2-x_{n+1})}{|(d+2)e_{n+1} - x|^2}x + \frac{|x-(d+1)e_{n+1}|^2 - 1}{|(d+2)e_{n+1} - x|^2}(d+2)e_{n+1}\\
&= \Bigg(\frac{2(d+2-x_{n+1})}{|(d+2)e_{n+1} - x|^2}x_1, \ldots, \frac{2(d+2-x_{n+1})}{|(d+2)e_{n+1} - x|^2}x_n,\\
&\qquad \frac{2(d+2-x_{n+1})}{|(d+2)e_{n+1} - x|^2}x_{n+1} + \frac{|x-(d+1)e_{n+1}|^2 - 1}{|(d+2)e_{n+1} - x|^2}(d+2)\Bigg).\\
&= \Bigg(\frac{2(d+2-x_{n+1})}{|(d+2)e_{n+1} - x|^2}x_1, \ldots, \frac{2(d+2-x_{n+1})}{|(d+2)e_{n+1} - x|^2}x_n,\\
&\qquad (d+2) - \frac{2(d+2-x_{n+1})^2}{|(d+2)e_{n+1} - x|^2}\Bigg).
\end{aligned}$$

It's evident that every point in $G \setminus \{N\}$, that is collinear with the north pole $N$ and $x$, corresponds to $\pi(x)$.

Clearly, for $x_{n+1} = 0$, the above expression of $\pi(x)$ gives the formula (2.1).

REMARK 2.5. For the projection sphere $S((d+2)e_2, 1)$ and a curve $G(\subset \mathbb{R}^2)$, if a projection line (straight line) is tangent to $G$ at $x \in G$, then in a neighbourhood of $x$, injectivity of the stereographic projection is destroyed (see Figure 6). Similar observation will occur when a surface in $\mathbb{R}^{n+1}$ is stereographically projected.

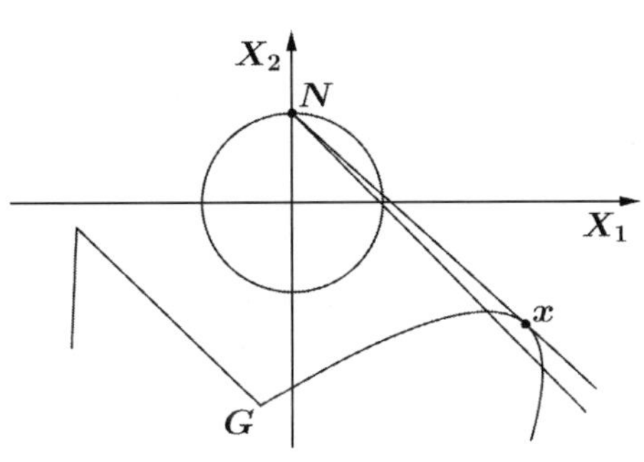


FIGURE 6. Destruction of one-to-one property of stereographic projection

It is now natural to compute the chordal distance between two points of $G$.

**The chordal distance.** Let $x = (x_1, x_2, \cdots, x_{n+1})$ and $y = (y_1, y_2, \cdots, y_{n+1})$ be two points in $G$. Since $\pi(x)$ and $\pi(y)$ lie on $S((d+1)e_{n+1}, 1)$, the relations

$$|\pi(x) - (d+1)e_{n+1}| = 1 \text{ and } |\pi(y) - (d+1)e_{n+1}| = 1$$

provide

$$\begin{cases} \sum_{i=1}^{n+1} \pi_i(x)^2 & = 2(d+1)\pi_{n+1}(x) - d(d+2), \\ \sum_{i=1}^{n+1} \pi_i(y)^2 & = 2(d+1)\pi_{n+1}(y) - d(d+2). \end{cases} \tag{2.4}$$

To compute the chordal distance between $x$ and $y$, we consider two cases: (i) $x \neq (d+2)e_{n+1},\ y \neq (d+2)e_{n+1}$ and (ii) either $x$ or $y$ is $N = (d+2)e_{n+1}$, if $N \in G$.

Case (i) ($x \neq (d+2)e_{n+1},\ y \neq (d+2)e_{n+1}$). Referring to the formula (2.3) for $\pi(x)$, the chordal distance becomes

$$\begin{aligned} g_{(d+2)e_{n+1}}(x,y) &= |\pi(x) - \pi(y)| \\ &= \left[\sum_{i=1}^{n+1} (\pi_i(x) - \pi_i(y))^2\right]^{\frac{1}{2}} \\ &= \left[\sum_{i=1}^{n+1} \pi_i(x)^2 + \sum_{i=1}^{n+1} \pi_i(y)^2 - 2\sum_{i=1}^{n+1} \pi_i(x)\pi_i(y)\right]^{\frac{1}{2}} \\ &= \left[2\left\{(d+1)\pi_{n+1}(x) + (d+1)\pi_{n+1}(y) - d(d+2) - \sum_{i=1}^{n+1} \pi_i(x)\pi_i(y)\right\}\right]^{\frac{1}{2}} \\ &= \frac{2\left[\sum_{i=1}^{n} \left[x_i(y_{n+1} - (d+2)) - y_i(x_{n+1} - (d+2))\right]^2\right]^{\frac{1}{2}}}{|x - (d+2)e_{n+1}||y - (d+2)e_{n+1}|}. \end{aligned}$$

Case (ii) ($x \neq (d+2)e_{n+1},\ y = (d+2)e_{n+1} = N$.). If $N \in G$, $\pi(N) = N$. By the help of (2.3) and (2.4), it follows that

$$\begin{aligned} g_{(d+2)e_{n+1}}(x,N) &= |\pi(x) - \pi(N)| \\ &= |\pi(x) - (d+2)e_{n+1}| \\ &= \left[\sum_{i=1}^{n} \pi_i(x)^2 + (\pi_{n+1}(x) - (d+2))^2\right]^{\frac{1}{2}} \\ &= \sqrt{2}[(d+2) - \pi_{n+1}(x)]^{\frac{1}{2}} \\ &= 2\frac{|x_{n+1} - (d+2)|}{|x - (d+2)e_{n+1}|}. \end{aligned}$$

In particular, the choices $n = 2, (x_1, x_2) = z, (y_1, y_2) = w, d = -1$ lead to the standard chordal distance between $z$ and $w$ on the complex plane. Indeed, it has the formula

$$g(z,w) := \frac{2|z-w|}{\sqrt{(1+|z|^2)(1+|w|^2)}}.$$

Its infinitesimal form $2|dz|/(1+|z|^2)$ is obtained through the limit

$$\lim_{w\to z}\frac{g(z,w)}{|z-w|}=\frac{2}{1+|z|^2},$$

which represents the density of the spherical metric on the plane. In a similar fashion, we define the *density of the chordal distance* for $G$ as

$$\lim_{x\to y}\frac{g_{(d+2)e_{n+1}}(x,y)}{|x-y|},\quad x,y\in G.$$

For arbitrary $G$, this limit may not exist. We intend to investigate the existence of this limit when $G$ is a plane. For an easy computation, we consider this situation for $n=2$ and arrive at the following result:

THEOREM 2.6. *For a plane $P$: $aX_1+bX_2+cX_3=k$, the density of the chordal distance exists if and only if $P$ is of the form $X_3=k$, where $a,b,c,k\in\mathbb{R}$.*

PROOF. Consider the general equation of a plane $aX_1+bX_2+cX_3=k$ and two distinct points $x=(x_1,x_2,x_3)$, $y=(y_1,y_2,y_3)$ on it. It produces

$$\begin{cases} ax_1+bx_2+cx_3 &= k,\\ ay_1+by_2+cy_3 &= k.\end{cases}\tag{2.5}$$

We assume that $x\neq(d+2)e_3\neq y$. To compute $\lim_{x\to y}g_{(d+2)e_3}(x,y)/|x-y|$, it is sufficient to consider the limit

$$\lim_{x\to y}\frac{\sum_{i=1}^{2}\left[x_i(y_3-(d+2))-y_i(x_3-(d+2))\right]^2}{|x-y|^2}.\tag{2.6}$$

If $c\neq 0$, (2.5) can be written as

$$x_3=\frac{1}{c}(k-ax_1-bx_2)=:f(x_1,x_2)$$

and

$$y_3=\frac{1}{c}(k-ay_1-by_2)=:f(y_1,y_2).$$

Now, we compute

$$\begin{aligned}x_1(y_3-(d+2))&-y_1(x_3-(d+2))\\&=\left(\frac{k}{c}-(d+2)-\frac{b}{c}y_2\right)(x_1-y_1)+\frac{b}{c}y_1(x_2-y_2).\end{aligned}$$

Similarly,

$$\begin{aligned}x_2(y_3-&(d+2))-y_2(x_3-(d+2))\\&=\left(\frac{k}{c}-(d+2)-\frac{a}{c}y_1\right)(x_2-y_2)+\frac{a}{c}y_2(x_1-y_1),\end{aligned}$$

and

$$x_3-y_3=-\frac{a}{c}(x_1-y_1)-\frac{b}{c}(x_2-y_2).$$

Taking the limit $x = (x_1, x_2, f(x_1, x_2)) \to y = (y_1, y_2, f(y_1, y_2))$, (2.6) yields

$$\lim_{x\to y} \frac{(x_1(y_3-(d+2)) - y_1(x_3-(d+2)))^2 + (x_2(y_3-(d+2)) - y_2(x_3-(d+2)))^2}{(x_1-y_1)^2+(x_2-y_2)^2+(x_3-y_3)^2}$$

$$= \lim_{(x_1,x_2)\to(y_1,y_2)} \frac{\left[\left(\frac{k}{c}-(d+2)-\frac{b}{c}y_2\right)(x_1-y_1)+\frac{b}{c}y_1(x_2-y_2)\right]^2}{(x_1-y_1)^2+(x_2-y_2)^2+\left(-\frac{a}{c}(x_1-y_1)-\frac{b}{c}(x_2-y_2)\right)^2}$$

$$+ \lim_{(x_1,x_2)\to(y_1,y_2)} \frac{\left[\left(\frac{k}{c}-(d+2)-\frac{a}{c}y_1\right)(x_2-y_2)+\frac{a}{c}y_2(x_1-y_1)\right]^2}{(x_1-y_1)^2+(x_2-y_2)^2+\left(-\frac{a}{c}(x_1-y_1)-\frac{b}{c}(x_2-y_2)\right)^2}$$

and along the straight line paths $(X_2-y_2) = m(X_1-y_1)$ on $X_1X_2$-plane, it results

$$\frac{\left[\frac{k}{c}-(d+2)-\frac{b}{c}(y_2-my_1)\right]^2+\left[\left(\frac{k}{c}-(d+2)\right)m+\frac{a}{c}(y_2-my_1)\right]^2}{1+m^2+\left(\frac{a+bm}{c}\right)^2},$$

for $m \in \mathbb{R}$. This quantity depends upon $m$ for either of the cases $a = 0, b \neq 0, c \neq 0$; $a \neq 0, b = 0, c \neq 0$; and $a \neq 0, b \neq 0, c \neq 0$. Hence, the limit does not exist for these cases.

Assuming $c = 0, b \neq 0$, (2.5) simplifies to

$$ax_1 + bx_2 = k \iff x_2 = \frac{1}{b}(k - ax_1) =: f(x_1)$$

and

$$ay_1 + by_2 = k \iff y_2 = \frac{1}{b}(k - ay_1) =: f(y_1).$$

Proceeding as above, taking the limit $x = (x_1, f(x_1), x_3) \to y = (y_1, f(y_1), y_3)$ along the straight line paths $X_3 - y_3 = m(X_1 - y_1)$ on $X_1X_3$-plane, the limit in (2.6) simplifies to

$$\frac{[-(d+2)+(y_3-my_1)]^2+\left[-\frac{km}{b}+\frac{a}{b}((d+2)-(y_3-my_1))\right]^2}{1+\frac{a^2}{b^2}+m^2}.$$

It depends on $m$ in both the cases $a \neq 0$ and $a = 0$.

If $c = 0, b = 0$ and $a \neq 0$, then (2.5) gives $ax_1 = k$ and $ay_1 = k$. Then the limit in (2.6) along the straight line paths $X_3 - y_3 = m(X_2 - y_2)$ on $X_2X_3$-plane gives

$$\frac{(km/a)^2 + (-(d+2)+(y_3-my_2))^2}{1+m^2},$$

which clearly depends on $m$ and hence the limit does not exist in this case as well.

Finally, we can see that for $a = 0, b = 0, c \neq 0$ i.e. for $X_3 = k/c$, the limit in (2.6) equals $((k/c) - (d+2))^2$ and hence we have

$$\lim_{x \to y} \frac{g_{(d+2)e_3}(x,y)}{|x-y|} = \frac{2\left|\dfrac{k}{c} - (d+2)\right|}{y_1^2 + y_2^2 + \left(\dfrac{k}{c} - (d+2)\right)^2}. \tag{2.7}$$

Moreover, for the plane $X_3 = k/c = d+2$, the limit $\lim_{x\to y} g_{(d+2)e_3}(x,y)/|x-y|$ exists and equals zero as $g_{(d+2)e_3}(x,y) = 0$ for all $x, y \in P$, which completes the proof. □

REMARK 2.7. It is easy to compute that the density found in (2.7) induces a metric on the plane $X_3 = k/c \neq d+2$ with curvature $+1$.

Next we consider the stereographic projection of a sphere. Here the sphere $S((d+1)e_3, 1)$ is the projection sphere and a sphere passing through $N(0,0,d+2)$ serves as $G$; see Figure 7. Then we have the following result:

THEOREM 2.8. *The density of the chordal distance exists for a sphere passing through the north pole if and only if its center lies on $X_3$-axis. Moreover, the density is equal to the reciprocal of the radius of the sphere.*

PROOF. Equation of a sphere passing through the north pole $N(0,0,d+2)$ with center $(a,b,c)$ is given by

$$(X_1 - a)^2 + (X_2 - b)^2 + (X_3 - c)^2 = a^2 + b^2 + (d+2-c)^2$$

i.e.

$$X_1^2 + X_2^2 + X_3^2 = 2aX_1 + 2bX_2 + 2cX_3 - 2c(d+2) + (d+2)^2. \tag{2.8}$$

Let $x = (x_1, x_2, x_3)$ be a point on the sphere (2.8). We are interested to check the existence of the limit

$$\lim_{x \to (d+2)e_3} \frac{g_{(d+2)e_3}(x, (d+2)e_3)}{|x - (d+2)e_3|} = \lim_{x \to (d+2)e_3} \frac{2\,|x_3 - (d+2)|}{|x - (d+2)e_3|^2}.$$

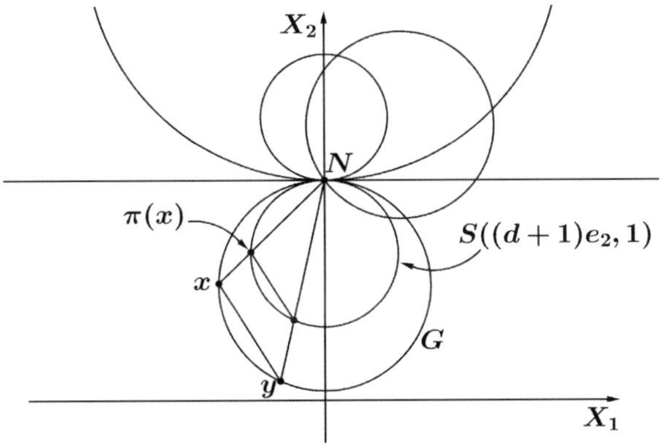


FIGURE 7. Spheres passing through $N$

With the help of (2.8) the above limit simplifies to

$$\lim_{(x_1,x_2,x_3)\to(0,0,d+2)} \frac{|x_3-(d+2)|}{ax_1+bx_2+(c-(d+2))(x_3-(d+2))}. \tag{2.9}$$

If we consider the planes $X_3-(d+2)=m[aX_1+bX_2+(c-(d+2))(X_3-(d+2))]$ i.e. $maX_1+mbX_2+(m(c-(d+2))-1)(X_3-(d+2))=0$, where $m\in\mathbb{R}$, and take the limit along the paths given by intersection of the sphere (2.8) and these planes, then the value of the limit in (2.9) is $|m|$. This implies that the limit does not exist when either $a\neq 0$ or $b\neq 0$. When $a=0$ and $b=0$, the value of the limit in (2.9) is $1/|d+2-c|$, $d+2-c\neq 0$.

For $a=0, b=0, d+2-c=0$, the sphere (2.8) becomes the point $N$ and in limiting sense we can say that the density for the north pole is infinity.

Conversely, suppose the center of the sphere (2.8) lies on $X_3$-axis. Then it has the equation

$$X_1^2+X_2^2+(X_3-c)^2=(d+2-c)^2,\ d+2-c\neq 0. \tag{2.10}$$

We transform the Cartesian coordinates into spherical coordinates by the formula:

$$\begin{aligned} X_1 &= r\sin\theta\cos\phi \\ X_2 &= r\sin\theta\sin\phi \\ X_3 &= r\cos\theta+d+2, \end{aligned}$$

where $r\geq 0, 0\leq\theta\leq\pi, 0\leq\phi<2\pi$, which converts (2.10) to $r+2(d+2-c)\cos\theta=0$.

If $x\neq(d+2)e_3\neq y$, consider any two points

$$x=(x_1,x_2,x_3)=(r_1\sin\theta_1\cos\phi_1, r_1\sin\theta_1\sin\phi_1, r_1\cos\theta_1+d+2)$$

and

$$y=(y_1,y_2,y_3)=(r_2\sin\theta_2\cos\phi_2, r_2\sin\theta_2\sin\phi_2, r_2\cos\theta_2+d+2)$$

on the sphere in (2.10), with $r_1+2(d+2-c)\cos\theta_1=0$ and $r_2+2(d+2-c)\cos\theta_2=0$. Then we compute

$$\frac{\left[\sum_{i=1}^2\left[x_i(y_3-(d+2))-y_i(x_3-(d+2))\right]^2\right]^{\frac{1}{2}}}{|x-y|}=\frac{r_1r_2}{2|d+2-c|}.$$

Hence, the density is obtained by

$$\lim_{x\to y}\frac{g_{(d+2)e_3}(x,y)}{|x-y|}=\lim_{x\to y}\frac{2}{r_1r_2}\times\frac{r_1r_2}{2|d+2-c|}=\frac{1}{|d+2-c|}.$$

If $x\neq(d+2)e_3=y$, then referring to the first part of the proof for the case $a=0=b$, we obtain the same limit. □

REMARK 2.9. In limiting sense, the plane $X_3=d+2$ is also a sphere passing through $N$ with center at infinity in the direction of $X_3$-axis. Theorem 2.8 says that the density of the chordal distance for $X_3=d+2$ is zero and this agrees with the density of $X_3=d+2$ as a plane (see the last line of the proof of Theorem 2.6).

## 3. Certain stereographic type projections

This section deals with stereographic type projections with some particular family of projection lines.

**3.1. Stereographic type projection through circular arcs.** In this subsection we take the plane $X_3 = 0$ as $G$ and study the stereographic type projection, $\pi^c$, of it where the projection lines $\ell_x$ are circular arcs touching $X_3 = 0$. For this purpose, we consider the projection sphere

$$X_1^2 + X_2^2 + (X_3 - (d+1))^2 = 1. \tag{3.1}$$

The projection line $\ell_x$ corresponding to $x = (a, b, 0) \neq (0, 0, 0)$ is the circular arc $\widehat{N\pi^c(x)x}$, (see Figure 8) given by

$$(X_1 - a)^2 + (X_2 - b)^2 + (X_3 - c)^2 = c^2 \tag{3.2}$$

and

$$X_1 = at, X_2 = bt \text{ i.e. } bX_1 - aX_2 = 0, \tag{3.3}$$

with

$$c = \frac{a^2 + b^2 + (d+2)^2}{2(d+2)}, \quad d + 2 \neq 0; \tag{3.4}$$

where $a, b, c, t$ are real. Here, we omit $d+2 = 0$, because it doesn't fulfil our purpose, that is, the circular arcs are always cutting $X_3 = 0$ at the north pole.

Solving (3.1) and (3.2) and using (3.4), we obtain

$$aX_1 + bX_2 + (c - (d+1))(X_3 - (d+2)) = 0. \tag{3.5}$$

For $c - (d+1) \neq 0$, the equations (3.3) and (3.5) together give

$$X_1 = at, X_2 = bt, X_3 = d + 2 - \frac{t(a^2 + b^2)}{c - (d+1)}. \tag{3.6}$$

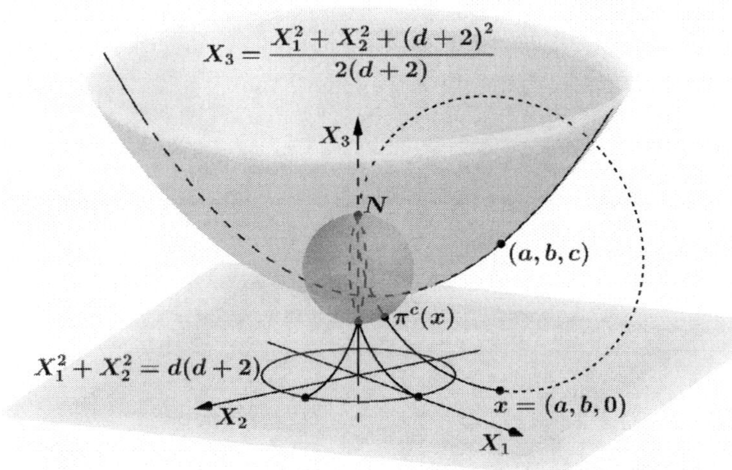


FIGURE 8. Projection through circular arcs

Putting these values of $X_1, X_2, X_3$ in (3.1), we get a quadratic equation in $t$ which produces

$$t = 0, t = \frac{4(d+2)(a^2+b^2-d(d+2))}{8(d+2)(a^2+b^2)+(a^2+b^2+d(d+2))^2}.$$

The choice $t = 0$ gives $(0, 0, d+2)$. For the second choice of $t$, (3.6) yields

$$\pi^c_{(d+2)e_3}(a, b, 0) := (p, q, r)$$

as the stereographic type projection of $(a, b, 0)$ on $S((d+1)e_3, 1)$, where

$$\begin{aligned} p &= \frac{4a(d+2)(a^2+b^2-d(d+2))}{8(d+2)(a^2+b^2)+(a^2+b^2+d(d+2))^2}, \\ q &= \frac{4b(d+2)(a^2+b^2-d(d+2))}{8(d+2)(a^2+b^2)+(a^2+b^2+d(d+2))^2}, \\ r &= \frac{(d+2)(a^2+b^2+d(d+2))^2}{8(d+2)(a^2+b^2)+(a^2+b^2+d(d+2))^2}. \end{aligned}$$

We observe that when $c-(d+1) = 0$ i.e. $a^2+b^2 = d(d+2)$, the equations (3.3) and (3.5) show that $\{(a, b, 0) : a^2+b^2 = d(d+2)\}$ maps to $de_3$; see Figure 8. Clearly, one can easily verify that $(0, 0, 0)$ maps to $de_3$ if $d = 0$, whereas $(0, 0, 0)$ maps to $N$ if $d \neq 0$.

It is evident that this projection is one-to-one if and only if $d = 0$, and in fact, only the point at infinity of $X_3 = 0$ corresponds to the north pole $N$. Moreover, for this case, the projection $\pi^c_{2e_3}$ of $x$ is the Möbius map $\pi_{2e_3}(x/2)$, i.e. $\pi^c_{2e_3}(x) = \pi_{2e_3}(x/2)$.

Let the chordal distance corresponding to $\pi^c_{(d+2)e_3}$ be denoted and defined by

$$g^c_{(d+2)e_3}(x, y) := |\pi^c_{(d+2)e_3}(x) - \pi^c_{(d+2)e_3}(y)|.$$

Then we observe the following result:

LEMMA 3.1. *For $x, y \in \mathbb{R}^2$ and $d+2 \neq 0$,*

$$g^c_{2e_3}(x, y) = g_{4e_3}(x, y) = g_{(d+2)e_3}((d+2)x/4, (d+2)y/4)$$

*holds.*

PROOF. Since we consider $G = \mathbb{R}^2 \subset \mathbb{R}^3$, denoting $x = (a_1, b_1, 0)$ and $y = (a_2, b_2, 0)$, a straightforward calculation yields

$$\pi^c_{2e_3}(a_1, b_1, 0) = \left(\frac{8a_1}{a_1^2+b_1^2+16}, \frac{8b_1}{a_1^2+b_1^2+16}, \frac{2(a_1^2+b_1^2)}{a_1^2+b_1^2+16}\right) = \pi_{2e_3}(a_1/2, b_1/2, 0).$$

Hence, it follows that

$$\begin{aligned} g^c_{2e_3}(x, y) &= |\pi^c_{2e_3}(a_1, b_1, 0) - \pi^c_{2e_3}(a_2, b_2, 0)| \\ &= |\pi_{2e_3}(a_1/2, b_1/2, 0) - \pi_{2e_3}(a_2/2, b_2/2, 0)| \\ &= g_{2e_3}(x/2, y/2) \\ &= g_{4e_3}(x, y) \\ &= g_{(d+2)e_3}((d+2)x/4, (d+2)y/4), \end{aligned}$$

where the last two identities hold by Lemma 2.1; see Figure 9 for an illustration in $\mathbb{R}^2$. □

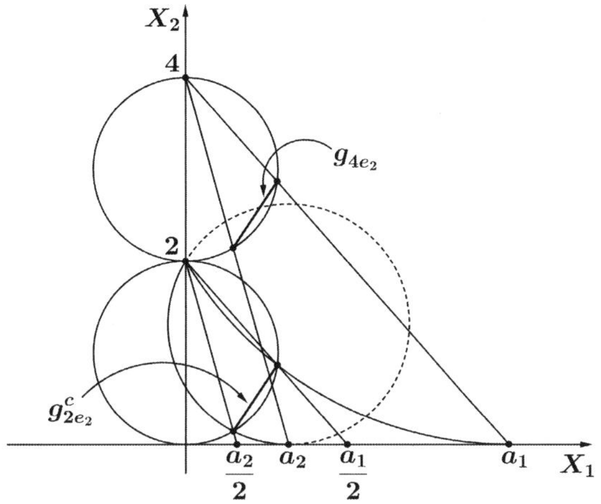


FIGURE 9. Equality of the chordal distances $g^c_{2e_2}$ and $g_{4e_2}$ in $\mathbb{R}^2$

**3.2. Projections that preserve the stereographic projection.** A stereographic type projection that preserves the stereographic projection is that whose projection line $\ell_x$ joins the north pole $N$, $\pi(x)$ and the point $x$ of $G$. In this subsection, we construct two such examples.

EXAMPLE 3.2. Let us consider the stereographic projection of $X_2 = 0$ on $S(e_2, 1)$. We denote $D$ as the interior of $S(e_2, 1)$ and $H$ as $\{(x_1, x_2) : x_2 > 0\} \cap D^c$. Now, for the point $x = (a, 0)$, we draw two circles: one centered at $(2/a, 2)$ passing through $N$ and $\pi(x)$; and the other centered at $(a/2, 0)$ passing through $\pi(x)$ and $x$; see Figure 10. We take $\ell_x$ as the concatenation of the circular arcs $\widehat{N\pi(x)}$ and $\widehat{\pi(x)x}$. Clearly, this maps $x$ to $\pi(x)$ and the projection line $\ell_x$ is differentiable at $\pi(x)$. Thus, our constructed stereographic type projection preserves the stereographic projection.

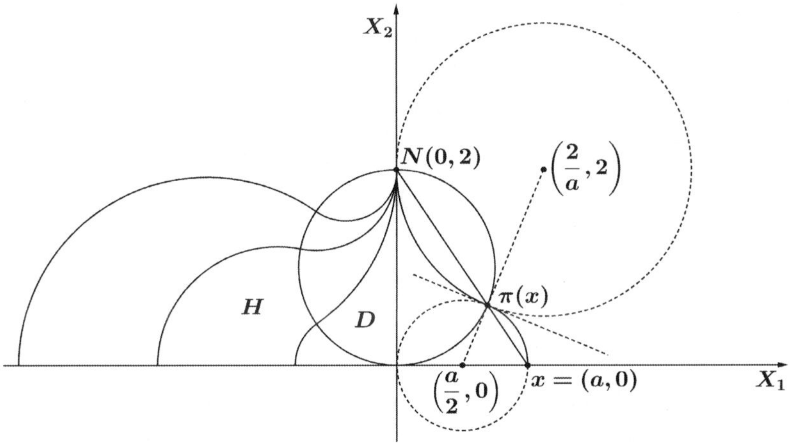


FIGURE 10. Projection through concatenated circular arcs

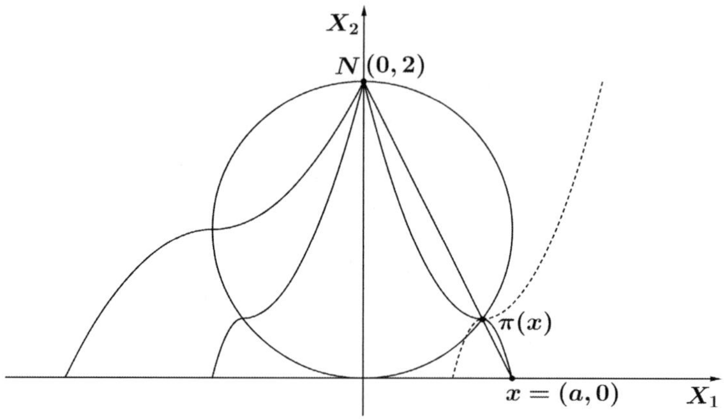


FIGURE 11. Projection through concatenated polynomial segments for $k = 2$

One can similarly construct many other stereographic type projections which preserve the stereographic projection. The next example provides another such construction.

EXAMPLE 3.3. Concatenation of two polynomial curve segments as projection line can be constructed as follows: we assume that $X_2 = 0$ is projected stereographically on $S(e_2, 1)$, and $\pi(x)$ is the stereographic projection of $x = (a, 0)$; see Figure 11. Now, for an integer $k \geq 1$, we consider $\ell_x$ as the concatenation of the segment of

$$\left(x - \frac{4a}{a^2+4}\right)^k = \frac{(-4a)^k}{8(a^2+4)^{k-1}}\left(y - \frac{2a^2}{a^2+4}\right)$$

joining $N$ and $\pi(x)$; and the segment of

$$\left(x - \frac{4a}{a^2+4}\right)^k = -\frac{(a^3)^k}{2a^2(a^2+4)^{k-1}}\left(y - \frac{2a^2}{a^2+4}\right)$$

joining $\pi(x)$ and $x$. Obviously, this constructed projection line $\ell_x$ preserves the stereographic projection. For $k = 1$, we get the straight line segment joining $N, \pi(x)$ and $x$, which represents the stereographic projection.

## 4. Concluding remarks

In the construction with reference to Example 3.2, we observe that the projection line $\ell_x$ is a combination of hyperbolic geodesics in $D$ and $H$ in the following way: the quantity $2/(1 - x_1^2 - (x_2 - 1)^2)$ defines the hyperbolic metric on $D$ and the arc $\widehat{N\pi(x)}$ is a geodesic. The arc $\widehat{\pi(x)x}$ is a geodesic of the metric space $H$ as a subspace of the upper half plane $x_2 > 0$ with the classical hyperbolic metric. Moreover, the projection lines cover the whole upper half plane if we consider the projection of $\{\pm\infty\}$ of $X_2 = 0$.

The following question is of our interest: *does there exist a metric on a space containing the upper half-plane $x_2 > 0$ where $\ell_x$ (as constructed in Example 3.2) is a geodesic and the metric is closely related to the hyperbolic metric both on $H$ as a subspace of the upper half-plane and on $D$?*

One of the reasons to ask this question is that, we want to identify a stereographic projection preserving space $(X, \rho)$ in the following way: first we take $G \subset \mathbb{R}^2$ and a projection sphere $S((d+1)e_2, 1)$. Now, we consider a family of projection lines $\{\ell_x : x \in G\}$ which preserves the stereographic projection of $G$. Finally we want to find a metric $\rho$ on a space $X \subseteq \mathbb{R}^2$ containing $\{\ell_x : x \in G\}$ such that $\{\ell_x : x \in G\}$ is a geodesic family of $(X, \rho)$. The above question demands not only one such $\rho$, but also a connection of $\rho$ to the hyperbolic metrics.

We are in the process of developing further geometries in this direction. It is expected that these will have connections with hyperbolic-type geometries and possibly have applications in Astrophysics.

## Acknowledgment

The authors are thankful to the referee for providing constructive suggestions for improving the manuscript.

## References

[1] L. V. Ahlfors, *Complex analysis*, 3rd ed., International Series in Pure and Applied Mathematics, McGraw-Hill Book Co., New York, 1978. An introduction to the theory of analytic functions of one complex variable. MR510197

[2] Md. N. Akhtar and A. Hossain, *Stereographic metric and dimensions of fractals on the sphere*, Results Math. **77** (2022), no. 6, Paper No. 213, 31, DOI 10.1007/s00025-022-01745-x. MR4483575

[3] A. F. Beardon, *The geometry of discrete groups*, Graduate Texts in Mathematics, vol. 91, Springer-Verlag, New York, 1983, DOI 10.1007/978-1-4612-1146-4. MR698777

[4] M. Fujimura, O. Rainio, and M. Vuorinen, *Collinearity of points on Poincaré unit disk and Riemann sphere*, Publ. Math. Debrecen **105** (2024), no. 1-2, 141–169. MR4787116

[5] T. Iwaniec and G. Martin, *Geometric function theory and non-linear analysis*, Oxford Mathematical Monographs, The Clarendon Press, Oxford University Press, New York, 2001. MR1859913

[6] M. Vuorinen, *Conformal geometry and quasiregular mappings*, Lecture Notes in Mathematics, vol. 1319, Springer-Verlag, Berlin, 1988, DOI 10.1007/BFb0077904. MR950174

Department of Mathematics, Indian Institute of Technology Indore, Simrol, Indore 452020, India

*Email address*: `ritwickmaitym@gmail.com`

Department of Mathematics, Indian Institute of Technology Indore, Simrol, Indore 452020, India

*Email address*: `swadesh.sahoo@iiti.ac.in`

Contemporary Mathematics
Volume **820**, 2025
https://doi.org/10.1090/conm/820/16417

# Some results on a subclass of analytic univalent functions

Yao Liang Chung, Maisarah Haji Mohd, and See Keong Lee

ABSTRACT. Let $\mathcal{A}$ be the class of normalized analytic functions in the open unit disc $\mathbb{D} := \{z \in \mathbb{C} : |z| < 1\}$ and $\mathcal{S}$ be a subset of $\mathcal{A}$ consisting of functions that are univalent in $\mathbb{D}$. Let $\lambda > 0$ and $\mu \in \mathbb{C}$ such that $|1-\mu| < \lambda \leq 1$ and $\lambda + |1-\mu| \leq 1$. A subclass of functions $f$ in $\mathcal{S}$ satisfying $|f'(z)(z/f(z))^2 - \mu| < \lambda$, $z \in \mathbb{D}$, is considered in this article. By using the bound of the second coefficient of $f$ in this subclass, several coefficients bounds are being explored for this subclass and sharp results are obtained.

## 1. Introduction

Let $\mathcal{A}$ be the class of normalized analytic functions $f$ in $\mathbb{D} := \{z \in \mathbb{C} : |z| < 1\}$ of the form

$$f(z) = z + \sum_{n=2}^{\infty} a_n z^n \tag{1.1}$$

and $\mathcal{S}$ be a subset of $\mathcal{A}$ consisting of functions that are univalent (namely one-to-one) in $\mathbb{D}$. The univalence of analytic functions is an important property in geometric function theory. For this reason, many authors have provided several types of sufficient conditions for $f$ in $\mathcal{A}$ to be univalent in $\mathbb{D}$. Among these, Ozaki and Nunokawa [**22**, Theorem 2] and Aksent'ev [**3**] independently showed that if $f$ is a function in $\mathcal{A}$ satisfying

$$\left| \left(\frac{z}{f(z)}\right)^2 f'(z) - 1 \right| < 1, \tag{1.2}$$

then $f$ is univalent in $\mathbb{D}$. The class of all normalized analytic functions satisfying inequality (1.2) in $\mathbb{D}$ is denoted by $\mathcal{U}$. Many interesting properties of this class can be seen in [**18**]. In [**16**], Obradović found a function $f$ in $\mathcal{A}$ that satisfies

$$\left| \left(\frac{z}{f(z)}\right)^2 f'(z) - 1 \right| < \lambda, \tag{1.3}$$

2020 *Mathematics Subject Classification.* Primary 30C50.

*Key words and phrases.* Univalent, Hankel determinant, Zalcman functional, Krushkal inequality.

The authors are grateful for the support of the Universiti Sains Malaysia Research University Grant (1001/PMATHS/8011015).

where $\lambda$ is a value between 0 and 1, but $\operatorname{Re}\left(zf'(z)/f(z)\right) < 0$, that is, a function in the class $\mathcal{U}$ but not a starlike function. This motivated Obradović [**16**] to define and investigate the class $\mathcal{U}(\lambda)$, $0 < \lambda \leq 1$:

$$\mathcal{U}(\lambda) = \left\{ f \in \mathcal{A} : \left| \left(\frac{z}{f(z)}\right)^2 f'(z) - 1 \right| < \lambda \right\}.$$

Since then, many works have been done on this subclass, such as [**1, 17, 19, 29**].

Recently, the subclass

$$\mathcal{U}(\lambda, \mu) = \left\{ f \in \mathcal{A} : \left| \left(\frac{z}{f(z)}\right)^2 f'(z) - \mu \right| < \lambda \right\}, \tag{1.4}$$

where $\lambda > 0$ and $\mu \in \mathbb{C}$ are such that $|1-\mu| < \lambda \leq 1$ and $\lambda + |1-\mu| \leq 1$, is introduced [**2**]. It is obvious that $\mathcal{U}(1,1) := \mathcal{U}$ and $\mathcal{U}(\lambda, 1) := \mathcal{U}(\lambda)$. The univalence of functions in $\mathcal{U}(\lambda, \mu)$ and some geometrical properties have been investigated. Meromorphic functions that satisfy condition in (1.4) have also been investigated in [**14**].

In literature, the integral representation of an analytic function plays a vital role in obtaining some interesting results. The integral representation of $f$ in $\mathcal{U}(\lambda, \mu)$ is given by

$$\frac{z}{f(z)} = 1 - a_2 z - \lambda(1-|a|^2)z \int_0^z \frac{\phi(t)}{t^2(1+\overline{a}\phi(t))}\, dt, \quad a = \frac{1-\mu}{\lambda}, \tag{1.5}$$

where $|a| < 1$ and $\phi$ is analytic in $\mathbb{D}$ with $\phi(0) = 0 = \phi'(0)$ and $|\phi(z)| < 1$ for all $z \in \mathbb{D}$. The equation (1.5) is also a sufficient condition for $f \in \mathcal{A}$ to be in $\mathcal{U}(\lambda, \mu)$, see [**2**, Theorem 2]. By using the integral representation (1.5), the sharp bound of second coefficient for $f \in \mathcal{U}(\lambda, \mu)$ is obtained, that is,

$$|a_2| \leq 1 + \lambda + |1-\mu|, \tag{1.6}$$

see [**2**, Theorem 5]. The equality is attained by the function

$$f(z) = z + (1 + \lambda + |1-\mu|)z^2 + \left(\lambda(1-|a|) + (1 + \lambda + |1-\mu|)^2\right)z^3 + \cdots,$$

where $|a| = |1-\mu|/\lambda < 1$. Just as for functions in $\mathcal{S}$, the bound of this $a_2$ is instrumental in determining some properties of functions in $\mathcal{U}(\lambda, \mu)$. In particular, the bounds for several coefficients and functionals related to the coefficients will be discussed in the next sections.

## 2. Results

It is well-known that the class $\mathcal{S}$ is normal compact [**6**]. For a continuous functional $\phi$ on a subclass $\mathcal{F}$ of $\mathcal{A}$, if $\mathcal{F}$ is normal compact, then the supremum $\sup_{f \in \mathcal{F}} |\phi(f)|$ exists, that is, there exists $g \in \mathcal{F}$ such that $|\phi(f)| \leq |\phi(g)|$ (see, [**6, 8**]). So, first we show that the class $\mathcal{U}(\lambda, \mu)$ is compact.

THEOREM 2.1. *Let $\lambda > 0$ and $\mu$ be a complex number such that $|1-\mu| < \lambda \leq 1$ and $\lambda + |1-\mu| \leq 1$. Then, the class $\mathcal{U}(\lambda, \mu)$, as defined in (1.4), is compact.*

PROOF. It is enough to show that $\mathcal{U}(\lambda, \mu)$ is closed and locally bounded. Let $\{f_n\}$ be a sequence in $\mathcal{U}(\lambda, \mu)$ which converges to $f$ uniformly on every compact subset of $\mathbb{D}$. We have $f(0) = 0 = f'(0) - 1$ and $f_n'$ converges to $f'$ uniformly on every compact subset of $\mathbb{D}$. Let

$$g_n(z) = \left(\frac{z}{f_n(z)}\right)^2 f_n'(z) - \mu \quad \text{and} \quad g(z) = \left(\frac{z}{f(z)}\right)^2 f'(z) - \mu.$$

We need to prove $g_n \to g$ uniformly on every compact subset of $\mathbb{D}$. Let

$$h_n(z) = \frac{f_n(z)}{z} \quad \text{and} \quad h(z) = \frac{f(z)}{z}.$$

Then $h_n(z) \neq 0$ and $h(z) \neq 0$, for $z \in \mathbb{D}$ and $n \in \mathbb{N}$. Let

$$M_n = \sup |h_n(z) - h(z)| = \sup \frac{|f_n(z) - f(z)|}{|z|}, z \in \mathbb{D}.$$

Since $h_n(z) - h(z)$ is analytic in $\mathbb{D}$ and continuous on $\overline{\mathbb{D}}$, so by the maximum modulus theorem

$$\max\{|h_n(z) - h(z)| : z \in \overline{\mathbb{D}}\} = \max\{|h_n(z) - h(z)| : z \in \partial D\}.$$

Since $f_n$ converges to $f$ uniformly on $\mathbb{D}$, it follows that $M_n$ converges to 0 as $n \to \infty$. Hence $h_n$ converges to $h$ uniformly on $\mathbb{D}$. Therefore, $h_n$ converges to $h$ uniformly on every compact subset of $\mathbb{D}$. Since $h_n \to h$ uniformly on every compact subset of $\mathbb{D}$ and $h_n(z) \neq 0, h(z) \neq 0$, for $z \in \mathbb{D}$ and $n \in \mathbb{N}$, it follows that $1/h_n$ converges to $1/h$ uniformly on every compact subset of $\mathbb{D}$. Also, $(1/h_n)^2 \to (1/h)^2$ uniformly on every compact subset of $\mathbb{D}$. Also, $f_n'$ converges to $f'$ uniformly on every compact subset of $\mathbb{D}$. Hence $(1/h_n)^2 f_n'$ converges to $(1/h)^2 f'$ uniformly on every compact subset of $\mathbb{D}$. Therefore $g_n$ converges to $g$ uniformly on every compact subset of $\mathbb{D}$.

Since $|g_n(z)| < \lambda$ for each $n \in \mathbb{N}$, we prove that $|g(z)| < \lambda$. Suppose not, then there exists $z_0 \in \mathbb{D}$ such that $|g(z_0)| > \lambda$. Let $\epsilon = |g(z_0)| - \lambda$. Then there exists $N \in \mathbb{N}$ such that

$$|g_n(z) - g(z)| < \epsilon = |g(z_0)| - \lambda$$

for $n \geq N$ and $z \in \mathbb{D}$. In particular,

$$|g_N(z_0) - g(z_0)| < |g(z_0)| - \lambda$$

Therefore,

$$|g(z_0)| - |g_n(z_0)| < |g_N(z_0) - g(z_0)| < |g(z_0)| - \lambda$$

shows that $|g_n(z_0)| > \lambda$, which is a contradiction and hence $|g(z)| \leq \lambda$ in $\mathbb{D}$. If there exists some point $z_0 \in D$ such that $|g(z_0)| = \lambda$, then by the maximum modulus theorem, $g$ must be a constant function, which is a contradiction. Therefore, $|g(z)| < \lambda$ for $z \in \mathbb{D}$ and hence $f \in \mathcal{U}(\lambda, \mu)$. This shows that $\mathcal{U}(\lambda, \mu)$ is closed. In view of the growth theorem for class $\mathcal{S}$, we have that the class $\mathcal{S}$ is locally bounded. Since the class $\mathcal{U}(\lambda, \mu) \subset \mathcal{S}$, it follows that class $\mathcal{U}(\lambda, \mu)$ is also locally bounded. Therefore, $\mathcal{U}(\lambda, \mu)$ is compact. $\square$

The rest of the results in this article are on coefficient problems for functions in $\mathcal{U}(\lambda, \mu)$. The compactness of $\mathcal{U}(\lambda, \mu)$ guarantees the sharpness of the bound, being investigated.

THEOREM 2.2. *Let* $f(z) = z + \sum_{n=2}^{\infty} a_n z^n$ *be in* $\mathcal{U}(\lambda, \mu)$ *and* $|a| = |1-\mu|/\lambda < 1$. *Then*

(i) $|a_3| \leq (1+\lambda+|1-\mu|)^2 + \dfrac{\lambda^2 - |1-\mu|^2}{\lambda}$,

(ii) $|a_4| \leq (1+\lambda+|1-\mu|)^3 + \dfrac{2(\lambda^2 - |1-\mu|^2)(1+\lambda+|1-\mu|)}{\lambda}$,

(iii) $|a_5| \leq 2(1+\lambda+|1-\mu|)((1+\lambda+|1-\mu|)^3 + 2\lambda(1-|a|^2)(1+\lambda+|1-\mu|))$
$+ ((1+\lambda+|1-\mu|)^2 + \frac{\lambda^2-|1-\mu|^2}{\lambda})^2 + (1+\lambda+|1-\mu|)^4$
$+ 3(1+\lambda+|1-\mu|)^2((1+\lambda+|1-\mu|)^2 + \frac{\lambda^2-|1-\mu|^2}{\lambda}) + \frac{\lambda(1-|1-\mu|^2/\lambda^2)}{3}$.

*Equalities in* (i) *and* (ii) *are respectively attained by the function*

$$
\begin{aligned}
f(z) &= \frac{z}{1-(1+\lambda+|1-\mu|)z - \lambda(1-|a|^2)z\int_0^z \frac{1}{1+\overline{a}t^2}\,dt} \\
&= z + (1+\lambda+|1-\mu|)z^2 + \left[(1+\lambda+|1-\mu|)^2 + \frac{\lambda^2-|1-\mu|^2}{\lambda}\right]z^3 \\
&\quad + \left[(1+\lambda+|1-\mu|)^3 + \frac{2(\lambda^2-|1-\mu|^2)(1+\lambda+|1-\mu|)}{\lambda}\right]z^4 + \cdots .
\end{aligned} \tag{2.1}
$$

PROOF. Since $f(z) = z + a_2z^2 + a_3z^3 + \cdots$, it follows that

$$
\begin{aligned}
\frac{z}{f(z)} &= 1 - a_2z + (a_2^2 - a_3)z^2 + (2a_2a_3 - a_2^3 - a_4)z^3 \\
&\quad + (a_2^4 - 3a_2^2a_3 + 2a_2a_4 + a_3^2 - a_5)z^4 + \cdots .
\end{aligned} \tag{2.2}
$$

Recall from (1.5), we have

$$\frac{z}{f(z)} = 1 - a_2z - \lambda(1-|a|^2)z\int_0^z \frac{\phi(t)}{t^2(1+\overline{a}\phi(t))}\,dt, \quad a = \frac{1-\mu}{\lambda},$$

where $|a| < 1$ and $\phi$ is analytic in $\mathbb{D}$ with $\phi(0) = 0 = \phi'(0)$ and $|\phi(z)| < 1$ for all $z \in \mathbb{D}$. Since $\phi(0) = \phi'(0) = 0$ and $|\phi(z)| < 1$, by Schwarz's lemma, we have $|\phi(z)| \leq |z|^2$. Taking $w(z) = \phi(z)/z^2$ in (1.5) yields

$$\frac{z}{f(z)} = 1 - a_2z - \lambda(1-|a|^2)z\int_0^z \frac{w(t)}{1+\overline{a}t^2w(t)}\,dt, \quad a = \frac{1-\mu}{\lambda},$$

where $w$ is analytic in $\mathbb{D}$ and $|w(z)| \leq 1$. The function $w$ has the form $w(z) = c_1 + c_2z + c_3z^2 + \cdots$. It follows that

$$
\begin{aligned}
\int_0^z \frac{w(t)}{1+\overline{a}t^2w(t)}\,dt &= \int_0^z \left(\sum_{k=0}^{\infty}(-1)^k(\overline{a})^k t^{2k} w^{k+1}(t)\right)dt \\
&= c_1z + \frac{1}{2}c_2z^2 + \frac{1}{3}\left(c_3 - \overline{a}c_1^2\right)z^3 + \cdots .
\end{aligned}
$$

Hence,

$$\frac{z}{f(z)} = 1 - a_2z - \lambda(1-|a|^2)c_1z^2 - \frac{\lambda(1-|a|^2)}{2}c_2z^3 - \frac{1}{3}(\lambda(1-|a|^2))(c_3 - \overline{a}c_1^2)z^4 + \cdots . \tag{2.3}$$

By comparing the coefficient of $z^2$ in (2.2) and (2.3), we obtain $a_2^2 - a_3 = -\lambda(1-|a|^2)c_1$ or

$$a_3 = a_2^2 + \lambda(1-|a|^2)c_1. \tag{2.4}$$

By taking the absolute value on both sides of (2.4) and using $|c_1| \leq 1$ would yield $|a_3| \leq |a_2|^2 + \lambda(1-|a|^2)$. Then, by using (1.6), we obtain

$$\begin{aligned} |a_3| &\leq (1+\lambda+|1-\mu|)^2 + \lambda(1-|a|^2) \\ &= (1+\lambda+|1-\mu|)^2 + \frac{\lambda^2-|1-\mu|^2}{\lambda}. \end{aligned} \tag{2.5}$$

Next, by comparing the coefficient of $z^3$ in (2.2) and (2.3), we obtain

$$a_4 = 2a_2a_3 - a_2^3 + \frac{\lambda(1-|a|^2)}{2}c_2. \tag{2.6}$$

With (2.4), it follows that

$$\begin{aligned} a_4 &= 2a_2(a_2^2+\lambda(1-|a|^2)c_1) - a_2^3 + \frac{\lambda(1-|a|^2)}{2}c_2 \\ &= a_2^3 + 2\lambda(1-|a|^2)c_1a_2 + \frac{\lambda(1-|a|^2)}{2}c_2. \end{aligned}$$

By using (1.6) and $|c_2| \leq 1-|c_1|^2$, we get

$$\begin{aligned} |a_4| &\leq |a_2|^3 + 2\lambda(1-|a|^2)|c_1||a_2| + \frac{\lambda(1-|a|^2)}{2}|c_2| \\ &\leq (1+\lambda+|1-\mu|)^3 + 2\lambda(1-|a|^2)(1+\lambda+|1-\mu|)|c_1| \\ &\quad + \frac{\lambda(1-|a|^2)}{2}(1-|c_1|^2). \end{aligned} \tag{2.7}$$

In view of (2.7), let $x = |c_1|$ and consider

$$h(x) = (1+\lambda+|1-\mu|)^3 + 2\lambda(1-|a|^2)(1+\lambda+|1-\mu|)x + \frac{\lambda(1-|a|^2)}{2}(1-x^2).$$

Since $x = |c_1| \leq 1$, it follows that $h$ is nonnegative in $[0,1]$. Also

$$h'(x) = 2\lambda(1-|a|^2)(1+\lambda+|1-\mu|) - \lambda(1-|a|^2)x$$

is nonnegative in $[0,1]$. Therefore, $h$ is an increasing function and so

$$h(1) = (1+\lambda+|1-\mu|)^3 + 2\lambda(1-|a|^2)(1+\lambda+|1-\mu|)$$

is the maximum value. Hence,

$$|a_4| \leq (1+\lambda+|1-\mu|)^3 + 2\lambda(1-|a|^2)(1+\lambda+|1-\mu|).$$

Lastly, by comparing the coefficient of $z^4$ in (2.2) and (2.3), we obtain

$$a_5 = 2a_2a_4 + a_3^2 + a_2^4 - 3a_2^2a_3 + \frac{1}{3}\lambda(1-|a|^2)(c_3 - \overline{a}c_1^2). \tag{2.8}$$

It follows that

$$\begin{aligned} |a_5| &\leq 2|a_2||a_4| + |a_3|^2 + |a_2|^4 + 3|a_2|^2|a_3| + \frac{\lambda(1-|a|^2)|\overline{a}||c_1|^2}{3} \\ &\leq 2(1+\lambda+|1-\mu|)((1+\lambda+|1-\mu|)^3 + 2\lambda(1-|a|^2)(1+\lambda+|1-\mu|)) \\ &\quad + ((1+\lambda+|1-\mu|)^2 + \lambda - |1-\mu|^2/\lambda)^2 + (1+\lambda+|1-\mu|)^4 \\ &\quad + 3(1+\lambda+|1-\mu|)^2\left((1+\lambda+|1-\mu|)^2 + \lambda - \frac{|1-\mu|^2}{\lambda}\right) \\ &\quad + \frac{\lambda(1-\frac{|1-\mu|^2}{\lambda^2})}{3}. \end{aligned}$$

To show the equality, consider the function $f$ in (2.1). It follows that

$$\frac{z}{f(z)} = 1 - (1+\lambda+|1-\mu|)z - \lambda(1-|a|^2)z\int_0^z \frac{1}{1+\overline{a}t^2}\,dt.$$

In view of (1.5), we see that $f$ is in $\mathcal{U}(\lambda,\mu)$. Note that

$$\int_0^z \frac{1}{1+\overline{a}t^2}\,t = \int_0^z \sum_{k=0}^{\infty}(-1)^k(\overline{a})^k t^{2k}\,dt = \sum_{k=0}^{\infty}(-1)^k(\overline{a})^k\frac{z^{2k+1}}{2k+1}.$$

The series expansion of (2.1) is given by

$$\begin{aligned} f(z) &= z\Bigg[1+\left\{(1+\lambda+|1-\mu|)z+\lambda(1-|a|^2)z^2-\frac{1}{3}\overline{a}\lambda(1-|a|^2)z^4+\cdots\right\} \\ &\qquad + \left\{(1+\lambda+|1-\mu|)z+\lambda(1-|a|^2)z^2+\cdots\right\}^2+\cdots\Bigg] \\ &= z+(1+\lambda+|1-\mu|)z^2+\Big(\lambda(1-|1-\mu|^2/\lambda^2)+(1+\lambda+|1-\mu|)^2\Big)z^3 \\ &\qquad + \Big[(1+\lambda+|1-\mu|)^3+2\lambda(1-|1-\mu|^2/\lambda^2)(1+\lambda+|1-\mu|)\Big]z^4+\cdots. \end{aligned}$$

Here, we see that

$$\begin{aligned} a_3 &= (1+\lambda+|1-\mu|)^2+\frac{\lambda^2-|1-\mu|^2}{\lambda}, \\ a_4 &= (1+\lambda+|1-\mu|)^3+2\lambda\Big(1-|1-\mu|^2/\lambda^2\Big)(1+\lambda+|1-\mu|). \end{aligned}$$

Therefore, equality occurs in (i) and (ii). □

**2.1. Fekete-Szegö problem.** For any $f(z) = z+\sum_{n=2}^{\infty}a_nz^n$ in $\mathcal{S}$, the inequality

$$|a_3-a_2^2| \le 1 \tag{2.9}$$

holds and equality occurs when $f$ is the Koebe function, $k(z) = z/(1-z)^2$ or its rotations. A classical theorem of Fekete and Szegö [**7**] states that, for $f$ in $\mathcal{S}$ and real $\delta$,

$$|a_3-\delta a_2^2| \le \begin{cases} 3-4\delta, & \text{if } \delta\le 0, \\ 1+2e^{-2\delta/(1-\delta)}, & \text{if } 0\le\delta<1, \\ 4\delta-3, & \text{if } \delta\ge 1. \end{cases} \tag{2.10}$$

Note that (2.9) is a special case of (2.10) when $\delta = 1$. The determination of upper bound for the functional $|a_3-\delta a_2^2|$ for any subclass $\mathcal{F}$ of normalized analytic functions is popularly known as the Fekete-Szegö functional problem for $\mathcal{F}$. Our next result is regarding the Fekete-Szegö functional problem for $\mathcal{U}(\lambda,\mu)$.

THEOREM 2.3. *Let $f(z) = z+\sum_{n=2}^{\infty}a_nz^n$ be in $\mathcal{U}(\lambda,\mu)$ and $|a| = |1-\mu|/\lambda < 1$. For $\delta\le 1$,*

$$|a_3-\delta a_2^2| \le \frac{\lambda^2-|1-\mu|^2}{\lambda}+(1-\delta)\big(1+\lambda+|1-\mu|\big)^2. \tag{2.11}$$

*Equality occurs for the function defined in* (2.1).

PROOF. Note that for $\delta \in \mathbb{R}$,

$$|a_3 - \delta a_2^2| = |a_3 - \delta a_2^2 + a_2^2 - a_2^2| \le |a_3 - a_2^2| + |1-\delta||a_2|^2. \tag{2.12}$$

From (2.4), we have

$$a_3 = a_2^2 + \lambda(1-|a|^2)c_1.$$

By using $|c_1| \le 1$, it follows that

$$|a_3 - a_2^2| = \lambda(1-|a|^2)|c_1| \le \lambda(1-|a|^2) = \frac{\lambda^2 - |1-\mu|^2}{\lambda}.$$

Therefore,

$$|a_3 - \delta a_2^2| \le \frac{\lambda^2 - |1-\mu|^2}{\lambda} + |1-\delta||a_2|^2.$$

By (1.6), we have

$$|a_3 - \delta a_2^2| \le \frac{\lambda^2 - |1-\mu|^2}{\lambda} + |1-\delta|\big(1+\lambda+|1-\mu|\big)^2.$$

To show the equality, we use the function $f$ in (2.1). For this function $f$, note that

$$a_2 = 1+\lambda+|1-\mu| \quad \text{and} \quad a_3 = \big(1+\lambda+|1-\mu|\big)^2 + \frac{\lambda^2 - |1-\mu|^2}{\lambda},$$

and so, for $\delta \le 1$,

$$\begin{aligned} |a_3 - \delta a_2^2| &= \left|\frac{\lambda^2 - |1-\mu|^2}{\lambda} + (1-\delta)\big(1+\lambda+|1-\mu|\big)^2\right| \\ &= \frac{\lambda^2 - |1-\mu|^2}{\lambda} + (1-\delta)\big(1+\lambda+|1-\mu|\big)^2. \end{aligned}$$ □

**2.2. Hankel determinants for the class $\mathcal{U}(\lambda,\mu)$.** Another popular bound of functional being investigated is related to the Hankel determinant. Let $f$ be in $\mathcal{A}$ and of the form (1.1). For $q \ge 1$ and $n \ge 1$, the Hankel determinants of $f$ are defined by

$$H_q(n) = \begin{vmatrix} a_n & a_{n+1} & \cdots & a_{n+q-1} \\ a_{n+1} & a_{n+2} & \cdots & a_{n+q} \\ \vdots & \vdots & \vdots & \vdots \\ a_{n+q-1} & a_{n+q} & \cdots & a_{n+2q-2} \end{vmatrix}, \quad a_1 = 1.$$

Pommerenke [**23, 24**] initiated the investigation of Hankel determinants on analytic univalent functions. Before that, there were also some works on zeros of the Hankel determinants done by Ullman [**26, 27**]. For Hankel determinants of analytic functions, it is interesting to see that

$$H_2(1) = \begin{vmatrix} a_1 & a_2 \\ a_2 & a_3 \end{vmatrix} = a_3 - a_2^2,$$

which is the Fekete-Szegö functional discussed above.

In the case $q = 2 = n$, we have

$$H_2(2) = \begin{vmatrix} a_2 & a_3 \\ a_3 & a_4 \end{vmatrix} = a_2a_4 - a_3^2,$$

which is called the second Hankel determinant. The work on second Hankel determinant for subclasses of $\mathcal{S}$ was first initiated by Janteng *et al.* [**9, 10**]. They considered the functional $|H_2(2)|$ for the subclasses

$$\mathcal{S}^* = \left\{ f \in \mathcal{S} : \operatorname{Re}\left(\frac{zf'(z)}{f(z)}\right) > 0, z \in \mathbb{D}\right\},$$

$$\mathcal{C} = \left\{ f \in \mathcal{S} : \operatorname{Re}\left(1 + \frac{zf''(z)}{f'(z)}\right) > 0, z \in \mathbb{D}\right\},$$

and

$$\mathcal{R} = \{ f \in \mathcal{S} : \operatorname{Re}(f'(z)) > 0, z \in \mathbb{D}\}.$$

These are the classes of starlike functions, convex functions and functions whose derivative has positive real part, respectively. They proved that $|H_2(2)|$ is bounded by 1, 1/8 and 4/9, respectively, (see [**10**, Theorem 3.1 and Theorem 3.2] and [**9**, Theorem 3.1]), and the results are sharp.

When $q = 3$ and $n = 1$, we have

$$H_3(1) = \begin{vmatrix} a_1 & a_2 & a_3 \\ a_2 & a_3 & a_4 \\ a_3 & a_4 & a_5 \end{vmatrix} = \begin{vmatrix} 1 & a_2 & a_3 \\ a_2 & a_3 & a_4 \\ a_3 & a_4 & a_5 \end{vmatrix},$$

which is called the third Hankel determinant. By expanding via the third column, we have

$$H_3(1) = a_3(a_2a_4 - a_3^2) - a_4(a_4 - a_2a_3) + a_5(a_3 - a_2^2),$$

Bound for the third Hankel determinant for subclasses $\mathcal{S}^*, \mathcal{C}$ and $\mathcal{R}$ was first studied by Babalola [**4**]. He proved that $|H_3(1)|$ is bounded by $16, (32+33\sqrt{3})/72\sqrt{3}$ and $(2736\sqrt{3} + 675\sqrt{5})/4860\sqrt{3}$, respectively. All these results are not sharp and so this third Hankel determinant problem attracts many researchers to find the sharp bounds. Recently, sharp bounds have been obtained for $\mathcal{S}^*$, $\mathcal{C}$ and $\mathcal{R}$, see [**13**], [**12**, Theorem 2.2] and [**11**, Theorem 1].

For the class $\mathcal{U}$, Obradovic and Tuneski [**20**] proved that $|H_2(2)| \leq 1$ and $|H_3(1)| \leq 1/4$. One method to obtain the bound of second and third Hankel determinant for class $\mathcal{U}(\lambda, \mu)$ is directly using the equations (2.4), (2.6) and (2.8). Here, we use another method to obtain the results. Before that, we look at a lemma.

LEMMA 2.4. *Let $f(z) = z + \sum_{n=2}^{\infty} a_n z^n$ in the class $\mathcal{U}(\lambda, \mu)$. Then there exists an analytic function $w_1$ in $\mathbb{D}$ satisfying $|w_1(z)| \leq |z| < 1$ and $|w_1'(z)| \leq 1$ with*

$$\frac{z}{f(z)} = 1 - a_2 z - \lambda(1 - |a|^2) z w_1(z), \quad z \in \mathbb{D}. \tag{2.13}$$

PROOF. Recall from (1.5), the integral representation for $f$ in $\mathcal{U}(\lambda, \mu)$ is given by

$$\frac{z}{f(z)} = 1 - a_2 z - \lambda(1 - |a|^2) z \int_0^z \frac{\phi(t)}{t^2(1 + \overline{a}\phi(t))}\, dt, \quad a = \frac{1-\mu}{\lambda}, \tag{2.14}$$

where $|a| < 1$ and $\phi$ is analytic in $\mathbb{D}$ with $\phi(0) = 0 = \phi'(0)$ and $|\phi(z)| < 1$ for all $z \in \mathbb{D}$. In view of (2.14), set

$$w_1(z) = \int_0^z \frac{\phi(t)}{t^2(1+\overline{a}\phi(t))}\,dt = \int_0^1 \frac{\phi(tz)}{t^2z^2(1+\overline{a}\phi(tz))} z\,dt.$$

So, (2.14) can be written as in (2.13). Further, we have

$$|w_1(z)| = \left|\int_0^1 \frac{\phi(tz)}{t^2z^2(1+\overline{a}\phi(tz))} zdt\right| \leq \frac{|z|}{|1+\overline{a}\phi(tz)|} \leq |z| < 1$$

and

$$|w_1'(z)| = \left|\frac{\phi(z)}{z^2(1+\overline{a}\phi(z))}\right| \leq \frac{1}{|1+\overline{a}\phi(z)|} < 1. \qquad \square$$

The series representation of the function $w_1$ in Lemma 2.4 is given by $w_1(z) = c_1z + c_2z^2 + \cdots$. It follows that $w_1'(z) = c_1 + 2c_2z + 3c_3z^2 + \cdots$. From a result by Prokhorov and Szynal [**25**], we have

$$|c_1| \leq 1, \tag{2.15}$$

$$|c_2| \leq \frac{1}{2}(1-|c_1|^2), \tag{2.16}$$

$$|c_3| \leq \frac{1}{3}\left(1-|c_1|^2 - \frac{4|c_2|^2}{1+|c_1|}\right). \tag{2.17}$$

Let $\psi = \lambda(1-|a|^2)$ and by using $w_1(z) = c_1z + c_2z^2 + \cdots$ in (2.13), we obtain

$$\begin{aligned} z &= (1 - a_2z - \psi c_1z^2 - \psi c_2z^3 - \cdots)(z + a_2z^2 + a_3z^3 + a_4z^4 + \cdots) \\ &= z + (a_3 - a_2^2 - \psi c_1)z^3 + (a_4 - a_2a_3z^4 - \psi a_2c_1z^4 - \psi c_2)z^4 + \\ &\quad (a_5 - a_2a_4 - \psi a_3c_1 - \psi a_2c_2 - \psi c_3)z^5 + \cdots. \end{aligned}$$

By comparing the coefficients of $z^3$ and $z^4$, we have

$$a_3 = \psi c_1 + a_2^2$$

and

$$a_4 = \psi c_2 + \psi a_2c_1 + a_2a_3 = \psi c_2 + 2\psi a_2c_1 + a_2^3,$$

respectively. By comparing the coefficients of $z^5$,

$$\begin{aligned} a_5 &= \psi c_3 + \psi a_2c_2 + \psi a_3c_1 + a_2a_4 \\ &= \psi c_3 + \psi a_2c_2 + \psi^2c_1^2 + \psi a_2^2c_1 + 2\psi a_2^2c_1 + a_2^4 \\ &= \psi c_3 + 2\psi a_2c_2 + \psi^2c_1^2 + 3\psi a_2^2c_1 + a_2^4. \end{aligned}$$

In short, we have

$$a_3 = \psi c_1 + a_2^2 = a_2^2 + \lambda(1-|a|^2)c_1, \tag{2.18}$$

$$\begin{aligned} a_4 &= \psi c_2 + 2\psi a_2c_1 + a_2^3 \\ &= 2\lambda(1-|a|^2)a_2c_1 + a_2^3 + \lambda(1-|a|^2)c_2, \end{aligned} \tag{2.19}$$

$$a_5 = \psi c_3 + 2\psi a_2c_2 + \psi^2c_1^2 + 3\psi a_2^2c_1 + a_2^4, \tag{2.20}$$

where $\psi = \lambda(1-|a|^2)$. Note that (2.18) is the same as (2.4). Also, it can be shown that both the methods mentioned above give the same bound for second Hankel determinant for class $\mathcal{U}(\lambda,\mu)$.

Now, we obtain the bound of second and third Hankel determinant for class $\mathcal{U}(\lambda,\mu)$ by using the second method, that is, by using (2.15)–(2.20).

THEOREM 2.5. *Let* $f(z)=z+\sum_{n=2}^{\infty}a_nz^n\in\mathcal{U}(\lambda,\mu)$ *and* $|a|=|1-\mu|/\lambda<1$. *Then*

(i) $|H_2(2)|=|a_2a_4-a_3^2|\le\dfrac{(\lambda^2-|1-\mu|^2)(1+\lambda+|1-\mu|)}{2\lambda}$,

(ii) $|H_3(1)|=|a_3(a_2a_4-a_3^2)-a_4(a_4-a_2a_3)+a_5(a_3-a_2^2)|\le\dfrac{(\lambda^2-|1-\mu|^2)^2}{4\lambda^2}$.

*Equality for (ii) is attained for the function*

$$f(z)=\frac{z}{1-\left(\frac{\lambda(1-|a|^2)}{2}\right)z^3+\frac{\lambda(1-|a|^2)\bar{a}}{5}z^6+\cdots}\tag{2.21}$$

$$=z+\frac{\lambda(1-|a|^2)}{2}z^4+\left(\frac{\lambda^2(1-|a|^2)^2}{4}-\frac{\lambda(1-|a|^2)\bar{a}}{5}\right)z^7+\cdots.$$

REMARK 2.6. It is an open problem to find the best bound for $H_2(2)$ for the class $\mathcal{U}(\lambda,\mu)$.

When $q=2$ and $n=3$, we have

$$H_2(3)=\begin{vmatrix}a_3 & a_4\\ a_4 & a_5\end{vmatrix}=a_3a_5-a_4^2.$$

The problem of determining the bound of $H_2(3)$ for class $\mathcal{S}$ and some of it subclasses: $\mathcal{S}^*$, $\mathcal{C}$ and $R$, has been considered by Zaprawa [**30**]. Motivated by Zaprawa's results, the nonsharp bound of $|H_2(3)|$ for functions in $\mathcal{U}(\lambda,\mu)$ is obtained.

THEOREM 2.7. *If* $f(z)=z+\sum_{n=2}^{\infty}a_nz^n$ *is in* $\mathcal{U}(\lambda,\mu)$, *then*

$$|H_2(3)|\le\psi\Big(\psi^2+\frac{7}{12}+1+\lambda+|1-\mu|+\frac{1}{3}(1+\lambda+|1-\mu|^2)\Big),$$

*where* $\psi=(\lambda^2-|1-\mu|^2)/\lambda$.

PROOF. The result follows by using (2.15)–(2.20). □

**2.3. Zalcman and generalized Zalcman conjecture for class $\mathcal{U}(\lambda,\mu)$.** In an attempt to solve the Bieberbach conjecture, in 1970[1], Lawrence Zalcman conjectured that that every $f(z)=z+\sum_{n=2}^{\infty}a_nz^n$ in $\mathcal{S}$ satisfies the more general inequality, that is,

$$|a_n^2-a_{2n-1}|\le(n-1)^2,\quad n\ge2.\tag{2.22}$$

Equality holds when $f$ is the Koebe function, $k(z)=z/(1-z)^2$ or its rotations. If (2.22) is true, then it can imply the Bieberbach conjecture, see [**5**]. Also for $n=2$ in (2.22), we have the Fekete-Szegö inequality $|a_2^2-a_3|\le1$.

Brown and Tsao [**5**] solved the Zalcman conjecture for class $\mathcal{S}^*$ of starlike functions. Obradovic and Tuneski [**21**] proved the results for the class $\mathcal{U}$ in the cases $n=2$ and $n=3$, and the complete result for the class $\mathcal{U}$ was proved by Vasudevarao and Pandey [**28**] by using the extreme point theory. Motivated by

[1]In literature, many wrote that Prof. L. Zalcman conjectured this in 1960. A private conversion between Prof. Zalcman and a well-known professor revealed that the conjecture was made around end of 1960's and early 1970's.

their work, we prove now the Zalcman conjecture for class $\mathcal{U}(\lambda,\mu)$ but for $n=2$ and $n=3$ only. For the case $n=2$, by taking $\delta=1$ in (2.11) yields

THEOREM 2.8. *If $f(z)=z+a_2z^2+a_3z^3+\cdots$ is in $\mathcal{U}(\lambda,\mu)$, then*

$$|a_2^2-a_3|\leq\begin{cases}\dfrac{\lambda^2-|1-\mu|^2}{\lambda}, & \text{if } \mu\neq 1,\\ \lambda, & \text{if } \mu=1.\end{cases}$$

*Equality occurs for the function defined in* (2.1).

Now, we discuss the case $n=3$.

THEOREM 2.9. *If $f(z)=z+\sum_{n=2}^{\infty}a_nz^n$ in $\mathcal{U}(\lambda,\mu)$ , then*

$$|a_3^2-a_5|\leq\frac{(\lambda^2-|1-\mu|^2)(1+\lambda+|1-\mu|)^2}{\lambda}+\frac{2(\lambda^2-|1-\mu|^2)^2}{\lambda^2}.$$

PROOF. By using (2.18) and (2.20), we have

$$\begin{aligned}|a_3^2-a_5| &= \left|(\psi c_1+a_2^2)^2-\psi c_3-2\psi a_2c_2-\psi^2c_1^2-3\psi a_2^2c_1-a_2^4\right|\\ &=\psi\left|c_3+2a_2c_2+a_2^2c_1\right|\\ &=\psi\left|c_3+2a_2c_2-\psi c_1^2+c_1(\psi c_1+a_2^2)\right|\\ &=\psi\left|c_3+2a_2c_2-\psi c_1^2+c_1a_3\right|\\ &\leq\psi\Big[|c_3|+2|a_2||c_2|+\psi|c_1|^2+|c_1||a_3|\Big],\end{aligned}$$

where $\psi=\lambda(1-|a|^2)=(\lambda^2-|1-\mu|^2)/\lambda$. The bounds of $|a_2|$ and $|a_3|$ in (1.6) and (2.5), and also (2.15)–(2.17) would yield

$$\begin{aligned}|a_3^2-a_5|\leq\psi\Bigg[&\frac{1}{3}\left(1-|c_1|^2-\frac{4|c_2|^2}{1+|c_1|}\right)+2|c_2|\left(1+\lambda+|1-\mu|\right)+\psi|c_1|^2\\ &+\left((1+\lambda+|1-\mu|)^2+\frac{\lambda^2-|1-\mu|^2}{\lambda}\right)|c_1|\Bigg].\end{aligned}$$

Let

$$\begin{aligned}g_1(x,y)=\;&\frac{1}{3}\left(1-x^2-\frac{4y^2}{1+x}\right)+2\left(1+\lambda+|1-\mu|\right)y+\psi x^2\\ &+\left((1+\lambda+|1-\mu|)^2+\psi\right)x,\end{aligned}$$

where $0\leq x=|c_1|\leq 1$ and $0\leq y=|c_2|\leq\frac{1}{2}(1-x^2)$. For $x=0$, we have $0\leq y\leq 1/2$ and

$$\begin{aligned}g_1(0,y)&=\frac{1}{3}(1-4y^2)+2(1+\lambda+|1-\mu|)y\\ &=\frac{1}{3}\Big[-4y^2+6(1+\lambda+|1-\mu|)y+1\Big]\\ &\leq 1+\lambda+|1-\mu|.\end{aligned}$$

For $y = 0$, we have $0 \le x \le 1$ and

$$\begin{aligned} g_1(x,0) &= \frac{1}{3}(1-x^2) + \psi x^2 + \left((1+\lambda+|1-\mu|)^2 + \psi\right)x, \\ &= \left(\psi - \frac{1}{3}\right)x^2 + (1+\lambda+|1-\mu|)^2 x + \psi x + \frac{1}{3} \\ &\le (1+\lambda+|1-\mu|)^2 + 2\psi. \end{aligned}$$

Therefore,

$$\begin{aligned} |a_3^2 - a_5| &\le \psi((1+\lambda+|1-\mu|)^2 + 2\psi) \\ &= \frac{(\lambda^2 - |1-\mu|^2)(1+\lambda+|1-\mu|)^2}{\lambda} + \frac{2(\lambda^2-|1-\mu|^2)^2}{\lambda^2}. \end{aligned}$$ □

In 1999, Ma [**15**] proposed a generalized Zalcman conjecture for $f(z) = z + \sum_{n=2}^{\infty} a_n z^n$ in $\mathcal{S}$ that

$$(2.23) \qquad |a_n a_m - a_{n+m-1}| \le (n-1)(m-1), \quad (n \ge 2, m \ge 2).$$

When $n = m$, the inequality (2.23) becomes (2.22). Ma proved the generalized Zalcman conjecture holds for the class $\mathcal{S}^*$ of starlike function. Obradovic and Tuneski [**21**] proved the generalized Zalcman conjecture for the class $\mathcal{U}$ for the case $n = 2$ and $m = 3$. Motivated by their work, we find the generalized Zalcman conjecture for the class $\mathcal{U}(\lambda, \mu)$ for the case $n = 2$ and $m = 3$.

THEOREM 2.10. *Let $f(z) = z + a_2 z^2 + a_3 z^3 + a_4 z^4 + \cdots$ be in $\mathcal{U}(\lambda,\mu)$. Then*

$$|a_2 a_3 - a_4| \le \frac{(\lambda^2 - |1-\mu|^2)(1+\lambda+|1-\mu|)}{\lambda}.$$

*Equality is attained by the function* (2.1).

PROOF. By using (2.18) and (2.19), we get

$$\begin{aligned} |a_2 a_3 - a_4| &= |\psi a_2 c_1 + a_2^3 - \psi c_2 - 2\psi a_2 c_1 - a_2^3| \\ &= |-\psi c_2 - \psi a_2 c_1| \\ &\le \psi\left(|c_2| + |a_2||c_1|\right), \end{aligned}$$

where $\psi = \lambda(1-|a|^2) = \lambda(1-|1-\mu|^2/\lambda^2)$. By using (2.15) and (2.16), we have

$$\begin{aligned} |a_2 a_3 - a_4| &\le \psi\left(\frac{1}{2}(1-|c_1|^2) + |c_1|(1+\lambda+|1-\mu|)\right) \\ &= \frac{1}{2}\psi\left(1 - |c_1|^2 + 2|c_1|(1+\lambda+|1-\mu|)\right). \end{aligned}$$

Let $x = |c_1|$ and $|c_1| \le 1$. Consider

$$h(x) = 1 - x^2 + 2x(1+\lambda+|1-\mu|).$$

Clearly, $h$ is nonnegative in $[0,1]$ and

$$h'(x) = 2(1+\lambda+|1-\mu|) - 2x$$

is also nonnegative in $[0,1]$. Therefore, $h$ is an increasing function and hence

$$h(1) = 2\left(1+\lambda+|1-\mu|\right)$$

is the maximum value. Hence,

$$1 - |c_1|^2 + 2|c_1|\left(1+\lambda+|1-\mu|\right) \le 2\left(1+\lambda+|1-\mu|\right).$$

So,

$$\begin{aligned}|a_2a_3 - a_4| &\le \lambda(1-|a|^2)\,(1+\lambda+|1-\mu|)\\ &= \frac{(\lambda^2-|1-\mu|^2)(1+\lambda+|1-\mu|)}{\lambda}.\end{aligned}$$

To show equality, consider the function $f$ in (2.1). Since

$$\begin{aligned}a_2 &= 1+\lambda+|1-\mu|,\\ a_3 &= (1+\lambda+|1-\mu|)^2 + \frac{\lambda^2-|1-\mu|^2}{\lambda},\\ a_4 &= (1+\lambda+|1-\mu|)^3 + \frac{2(\lambda^2-|1-\mu|^2)\,(1+\lambda+|1-\mu|)}{\lambda},\end{aligned}$$

it follows that

$$|a_2a_3 - a_4| = \frac{(\lambda^2-|1-\mu|^2)\,(1+\lambda+|1-\mu|)}{\lambda}. \qquad \square$$

**2.4. Krushkal inequality for the class $\mathcal{U}(\lambda,\mu)$.** In this section, the generalization of the inequality (2.9) to higher coefficients is studied. For $n > 3$ and $p \ge 1$, Krushkal showed that for $f(z) = z + \sum_{n=2}^{\infty} a_n z^n$ in $\mathcal{S}$, the inequality

$$|a_n^p - a_2^{p(n-1)}| \le 2^{p(n-1)} - n^p \tag{2.24}$$

holds. Note that when $n = 3$ and $p = 1$, inequality (2.24) becomes (2.9). Here, we give a direct proof of the inequality (2.24) over the class $\mathcal{U}(\lambda,\mu)$ for the case $n = 4$ and $p = 1$.

THEOREM 2.11. *Let $f(z) = z + \sum_{n=2}^{\infty} a_n z^n$ in the class $\mathcal{U}(\lambda,\mu)$. Then*

$$\left|a_4 - a_2^3\right| \le \frac{2(1+\lambda+|1-\mu|)(\lambda^2-|1-\mu|^2)}{\lambda}.$$

*Equality is attained by the function in* (2.1).

PROOF. By using (2.19), we have

$$\begin{aligned}\left|a_4 - a_2^3\right| &= |\psi c_2 + 2\psi a_2 c_1 + a_2^3 - a_2^3|\\ &\le \psi(|c_2| + 2|a_2||c_1|),\end{aligned}$$

where $\psi = (\lambda^2 - |1-\mu|^2)/\lambda$. The bound of $|a_2|$ and $|c_2|$ in (1.6) and (2.16) would yield

$$\begin{aligned}|a_4 - a_2^3| &\le \psi\left(\frac{1}{2}(1-|c_1|^2 + 2(1+\lambda+|1-\mu|)|c_1|\right)\\ &= \frac{1}{2}\psi(1+4(1+\lambda+|1-\mu|)|c_1| - |c_1|^2.\end{aligned}$$

Let $x = |c_1| \le 1$ and consider the function

$$h(x) = 1 + 4(1+\lambda+|1-\mu|)x - x^2.$$

Here, $h$ is nonnegative in $[0,1]$. Note that

$$h'(x) = 4(1+\lambda+|1-\mu|) - 2x$$

is also nonnegative in $[0,1]$. Therefore, $h$ is an increasing function. So,

$$h(1) = 4(1+\lambda+|1-\mu|)$$

is the maximum value and

$$1 + 4(1+\lambda+|1-\mu|)|c_1| - |c_1|^2 \leq 4(1+\lambda+|1-\mu|).$$

Hence,

$$\begin{aligned}|a_4 - a_2^3| &\leq 2\psi(1+\lambda+|1-\mu|)\\ &= 2\lambda(1-|1-\mu|^2/\lambda^2)(1+\lambda+|1-\mu|)\\ &= \frac{2(1+\lambda+|1-\mu|)(\lambda^2-|1-\mu|^2)}{\lambda}.\end{aligned}$$

To show equality, consider the function $f$ in (2.1). Since

$$\begin{aligned}a_2 &= 1+\lambda+|1-\mu|,\\ a_4 &= (1+\lambda+|1-\mu|)^3 + \frac{2(\lambda^2-|1-\mu|^2)(1+\lambda+|1-\mu|)}{\lambda},\end{aligned}$$

it follows that

$$|a_4 - a_2^3| = \frac{2(1+\lambda+|1-\mu|)(\lambda^2-|1-\mu|^2)}{\lambda}.$$

Hence, equality occurs. □

## 3. Conclusion

In this work, further properties of the class $\mathcal{U}(\lambda,\mu)$ are investigated. Particularly, several types of coefficients bound for the class have been determined and most of them are sharp bounds.


## Acknowledgment

This work benefited greatly from the discussions held with Prof. Rosihan M. Ali and Prof. S. Ponnusamy.


## References


[1] M. F. Ali, V. Allu, and H. Yanagihara, *On a class of univalent functions defined by a differential inequality*, J. Ramanujan Math. Soc. **35** (2020), no. 3, 217–226. MR4166709

[2] R. M. Ali, Y. L. Chung, and S. K. Lee, *On a subclass of analytic functions satisfying a differential inequality*, Mediterr. J. Math. **20** (2023), no. 3, Paper No. 121, 17, DOI 10.1007/s00009-023-02326-0. MR4549898

[3] L. A. Aksent′ev, *Sufficient conditions for univalence of regular functions* (Russian), Izv. Vysš. Učebn. Zaved. Matematika **1958**, no. 3(4), 3–7. MR143895

[4] K. O. Babalola, *On $H_3(1)$ Hankel determinant for some classes of univalent function*, Inequality Theory and Application (edited by C. Kim and S. Dragomir), Nova Science Publishers Inc., New York, 2010, 1–7.

[5] J. E. Brown and A. Tsao, *On the Zalcman conjecture for starlike and typically real functions*, Math. Z. **191** (1986), no. 3, 467–474, DOI 10.1007/BF01162720. MR824446

[6] P. L. Duren, *Univalent functions*, Grundlehren der mathematischen Wissenschaften [Fundamental Principles of Mathematical Sciences], vol. 259, Springer-Verlag, New York, 1983. MR708494

[7] M. Fekete and G. Szegö, *Eine Bemerkung Uber Ungerade Schlichte Funktionen*, J. London Math. Soc. **8** (1933), no. 2, 85–89, DOI 10.1112/jlms/s1-8.2.85. MR1574865

[8] D. J. Hallenbeck and T. H. MacGregor, *Linear problems and convexity techniques in geometric function theory*, Monographs and Studies in Mathematics, vol. 22, Pitman (Advanced Publishing Program), Boston, MA, 1984. MR768747

[9] A. Janteng, S. A. Halim, and M. Darus, *Coefficient inequality for a function whose derivative has a positive real part*, J. Inequal. Pure Appl. Math. **7** (2006), no. 2, Article 50, 5. MR2221331

[10] A. Janteng, S. A. Halim, and M. Darus, *Hankel determinant for starlike and convex functions*, Int. J. Math. Anal. (Ruse) **1** (2007), no. 13-16, 619–625. MR2370200

[11] B. Kowalczyk and A. Lecko, *The sharp bound of the third Hankel determinant for functions of bounded turning*, Bol. Soc. Mat. Mex. (3) **27** (2021), no. 3, Paper No. 69, 13, DOI 10.1007/s40590-021-00383-7. MR4309819

[12] B. Kowalczyk, A. Lecko, and Y. J. Sim, *The sharp bound for the Hankel determinant of the third kind for convex functions*, Bull. Aust. Math. Soc. **97** (2018), no. 3, 435–445, DOI 10.1017/S0004972717001125. MR3802458

[13] B. Kowalczyk, A. Lecko, and D. K. Thomas, *The sharp bound of the third Hankel determinant for starlike functions*, Forum Math. **34** (2022), no. 5, 1249–1254, DOI 10.1515/forum-2021-0308. MR4472881

[14] S. K. Lee, S. Ponnusamy, and K.-J. Wirths, *On classes of meromorphic locally univalent functions defined by differential inequalities*, Bull. Iranian Math. Soc. **46** (2020), no. 1, 149–158, DOI 10.1007/s41980-019-00247-x. MR4053496

[15] W. Ma, *Generalized Zalcman conjecture for starlike and typically real functions*, J. Math. Anal. Appl. **234** (1999), no. 1, 328–339, DOI 10.1006/jmaa.1999.6378. MR1694809

[16] M. Obradović, *Starlikeness and certain class of rational functions*, Math. Nachr. **175** (1995), 263–268, DOI 10.1002/mana.19951750114. MR1355021

[17] M. Obradović and S. Ponnusamy, *New criteria and distortion theorems for univalent functions*, Complex Variables Theory Appl. **44** (2001), no. 3, 173–191, DOI 10.1080/17476930108815354. MR1909496

[18] M. Obradović and S. Ponnusamy, *On the class $U$*, Proceedings of the 21st Annual Conference of the Jammu Mathematical Society and a National Seminar on Analysis and its Application, (2011), 11–26.

[19] M. Obradović, S. Ponnusamy, and K.-J. Wirths, *Geometric studies on the class $\mathcal{U}(\lambda)$*, Bull. Malays. Math. Sci. Soc. **39** (2016), no. 3, 1259–1284, DOI 10.1007/s40840-015-0263-5. MR3515079

[20] M. Obradović and N. Tuneski, *Some properties of the class $\mathcal{U}$*, Ann. Univ. Mariae Curie-Skłodowska Sect. A **73** (2019), no. 1, 49–56. MR4048261

[21] M. Obradović and N. Tuneski, *Zalcman and generalized Zalcman conjecture for a subclass of univalent functions*, Novi Sad J. Math. **52**(1) (2022), 185–190.

[22] S. Ozaki and M. Nunokawa, *The Schwarzian derivative and univalent functions*, Proc. Amer. Math. Soc. **33** (1972), 392–394, DOI 10.2307/2038067. MR299773

[23] Ch. Pommerenke, *On the coefficients and Hankel determinants of univalent functions*, J. London Math. Soc. **41** (1966), 111–122, DOI 10.1112/jlms/s1-41.1.111. MR185105

[24] Ch. Pommerenke, *On the Hankel determinants of univalent functions*, Mathematika **14** (1967), 108–112, DOI 10.1112/S002557930000807X. MR215976

[25] D. V. Prokhorov and J. Szynal, *Inverse coefficients for $(\alpha, \beta)$-convex functions* (English, with Russian and Polish summaries), Ann. Univ. Mariae Curie-Skłodowska Sect. A **35** (1981), 125–143 (1984). MR763927

[26] J. L. Ullman, *Hankel determinants whose elements are sections of a Taylor series. I*, Duke Math. J. **18** (1951), 751–756. MR43200

[27] J. L. Ullman, *Hankel determinants whose elements are sections of a Taylor series. II*, Duke Math. J. **19** (1952), 155–164. MR47764

[28] A. Vasudevarao and A. Pandey, *The Zalcman conjecture for certain analytic and univalent functions*, J. Math. Anal. Appl. **492** (2020), no. 2, 124466, 12, DOI 10.1016/j.jmaa.2020.124466. MR4144258

[29] A. Vasudevarao and H. Yanagihara, *On the growth of analytic functions in the class $\mathcal{U}(\lambda)$*, Comput. Methods Funct. Theory **13** (2013), no. 4, 613–634, DOI 10.1007/s40315-013-0045-8. MR3138356

[30] P. Zaprawa, *On Hankel determinant $H_2(3)$ for univalent functions*, Results Math. **73** (2018), no. 3, Paper No. 89, 12, DOI 10.1007/s00025-018-0854-1. MR3813616

School of Mathematical Sciences, Universiti Sains Malaysia, 11800 USM, Penang, Malaysia

*Current address*: Centre of Excellence for Pre-University Studies, INTI International College Penang, 11900 Bayan Lepas, Penang, Malaysia

*Email address*: yaoliang.chung@newinti.edu.my, chungyaoliang@gmail.com

School of Mathematical Sciences, Universiti Sains Malaysia, 11800 USM, Penang, Malaysia

*Email address*: maisarah_hjmohd@usm.my

School of Mathematical Sciences, Universiti Sains Malaysia, 11800 USM, Penang, Malaysia

*Email address*: sklee@usm.my

Contemporary Mathematics
Volume **820**, 2025
https://doi.org/10.1090/conm/820/16418

# Mathematical approach to the theory of topological insulators

Armen Sergeev

ABSTRACT. The paper is devoted to the theory of topological insulators. Topological insulators are solid states having wide energy gap, stable under small deformations, which motivates the usage of topological methods for their study. A key role in this study is played by the symmetry groups of these objects. The description of possible symmetry types goes back to Kitaev who proposed a classification of topological insulators based on the investigation of the symmetry groups and their representations. In this paper we pay main attention to the topological insulators invariant under time reversion. We introduce the involutive spaces being mathematical equivalent of the momentum spaces of insulators invariant under time reversion. An example of such systems is given by the quantum spin Hall insulator which has a non-trivial topological $\mathbb{Z}_2$-invariant introduced by Kane and Mele.

## Introduction

The topological insulators are the solid bodies with a broad energy gap, stable under small deformations. This motivates the usage of topological methods for their investigation. Apart from the importance of topological insulators for condensed matter physics their theory enjoys numerous and deep connections with different mathematical disciplines such as $C^*$-algebras, K-theory and noncommutative geometry.

A key role in the classification of topological objects in the theory of solid states is played by the study of their symmetry groups. The description of possible symmetry types goes back to Kitaev [**6**] who proposed a classification of topological insulators based on the representation theory of their symmetry groups.

In this paper[1] we pay main attention to the topological insulators invariant under time reversion. An example of such systems is provided by the quantum spin Hall insulator which has a non-trivial topological $\mathbb{Z}_2$-invariant introduced by Kane and Mele [**5**].

---

2020 *Mathematics Subject Classification.* Primary 70G40,70H12.

While preparing this paper the author was supported by SIMC ctnter of Steklov Mathematical Institute.

[1]This paper is based on my talk at the International Conference on Finite and Infinite-Dimensional Complex Analysis and its Applications held in Pondicherry University, India, in August 2023.

## 1. Bloch theory

Classical Bloch theory (cf. [**1**], [**7**]) describes the properties of solid bodies having the crystal lattice called the *Bravais lattice*. From mathematical point of view it is a discrete Abelian group $\Gamma$ in the space $\mathbb{R}^d$ with $d = 3$ isomorphic to $\mathbb{Z}^d$ and acting on $\mathbb{R}^d$ by translations $T_\gamma$ generated by vectors $\gamma \in \Gamma$.

The behavior of free electrons in a solid body is governed by the so called one-particle *Schrödinger equation* which eigenfunctions satisfy the equation

$$H\psi := (-\Delta + V)\psi = E\psi \tag{1.1}$$

with periodic potential $V$ invariant under the action of $\Gamma$. Such operator commutes with all operators $T_\gamma$.

Denote by $\Gamma'$ the *dual lattice* in the momentum space $(\mathbf{R}^d)'$ defined in the following way:

$$\Gamma' = \{k \in (\mathbb{R}^d)' : k \cdot \gamma \in 2\pi\mathbb{Z} \quad \text{for all } \gamma \in \Gamma\}.$$

The fundamental domain (unit cell) of the lattice $\Gamma'$ is called the *Brillouin zone* $Br_d$.

The functions, invariant under $\Gamma$, may be considered as functions on the torus $\mathbb{T}^d = \mathbb{R}^d/\Gamma$. Denote by $\mathcal{H}_0$ the Hilbert space

$$\mathcal{H}_0 = L^2(\mathbb{T}^d) = L^2(\mathbb{R}^d/\Gamma)$$

with respect to the measure on $\mathbb{R}^d/\Gamma$ induced by the Lebesgue measure $dx$ on $\mathbb{R}^d$. The exponential $e_k = e^{ik\cdot x}$ belongs to $\mathcal{H}_0$ if $k \in \Gamma'$. Moreover, such functions form an orthonormal basis in $\mathcal{H}_0$.

The smooth functions of the form

$$\psi(x) = e^{ik\cdot x}\varphi(x)$$

are called the *Bloch functions*. Here, the vector $k$, called the *quasimomentum*, belongs to the Brillouin zone $\mathrm{Br}_d$ and the $C^\infty$-smooth function $\varphi$ is invariant under $\Gamma$, so it may be considered as a function from $C^\infty(\mathbb{R}^d/\Gamma)$. The space of Bloch functions with quasimomentum $k$ is denoted by $L_k$.

The Schrödinger operator (1.1) acts on Bloch functions by the rule

$$H\left(e^{ik\cdot x}\varphi(x)\right) = e^{ik\cdot x}H_k\varphi(x). \tag{1.2}$$

The operator $H_k$, called the *effective* or *Bloch Hamiltonian*, has the form

$$H_k\varphi = (\frac{1}{i}\nabla + k)^2\varphi + V\varphi$$

and maps the space $C^\infty(\mathbb{R}^d/\Gamma)$ to itself. In particular, it follows from the formula (1.2) that the original Schrödinger operator $H = H_0$ maps the space of Bloch functions with quasimomentum $k$ to itself.

If we denote by $I_k$ the operator of multiplication by $e^{ik\cdot x}$ then we can rewrite the formula (1.2) in the form

$$I_k^{-1} \circ H \circ I_k = H_k.$$

In other words,

$$H|_{L_k} = I_k \circ H_k|_{L_0} \circ I_k^{-1}.$$

so the study of the operator $H|_{L_k}$ is reduced to the study of operator $H_k|_{L_0}$.

Denote by $H(k)$ the closure of the operator $H_k|_{L_0}$ in the space $\mathcal{H}_0$. Its definition domain coincides with the subspace

$$D(H(k)) = \{\varphi : \varphi(x) = \sum_{\gamma' \in \Gamma'} c_{\gamma'} e^{i(\gamma' \cdot x)} \text{ such that}$$

$$\sum_{\gamma' \in \Gamma'} (1 + |\gamma'|^2)|c_{\gamma'}|^2 < \infty\}.$$

The operator $H(k)$ has the discrete spectrum and its eigenfunctions $\varphi_m(k)$ are solutions of the equation

$$H(k)\varphi_m(k) = E_m(k)\varphi_m(k). \tag{1.3}$$

Note that the functions $\varphi_m(k)$ are the functions of $x$, depending on two parameters — discrete parameter $m \in \mathbb{N}$ and continuous parameter $k \in \mathrm{Br}_d$. For any fixed $k$ they form a complete orthogonal system un the Hilbert space $\mathcal{H}_0$.

Consider now the physical interpretation of this mathematical picture. In the tight-binding approximation, justified by physical arguments, we can suppose that there is only a finite number $n$ of active one-electron levels corresponding to the eigenfunctions $E_m(k)$. The collection of all energy levels, corresponding to the functions $E_m(k)$ with fixed $m$, is called the *energy zone*.

The ground state (i.e. the state with minimal energy) of a system of $n$ Bloch electrons has the following structure. There is a number of completely occupied one-electron levels with energies not exceeding the quantity $E_F$, called the *Fermi energy*, with only empty (non-occupied) levels lying above it. The interval of energies between the highest occupied level and the lowest empty level is called the *energy gap* or *forbidden zone*. The solid bodies with wide energy gap are called the *insulators*.

The Bloch functions

$$\psi_m(k) = e^{ik \cdot x}\varphi_m(k) \tag{1.4}$$

are the eigenfunctions of the original Schrödinger operator $H$.

Denote by $\mathcal{H}_k$ the completion of the space $L_k$ with respect to the norm determined by the isomorphism $I_k : L_0 \to L_k$ extending to the isometry $I_k : \mathcal{H}_0 \to \mathcal{H}_k$. Consider the Hilbert vector bundle $\pi : \mathfrak{H} \to \mathrm{Br}_d$ with fibre $\mathcal{H}_k$ at the point $k \in \mathrm{Br}_d$ and denote by $\mathcal{H} = L^2(\mathfrak{H})$ the Hilbert space of square integrable sections of $\mathfrak{H}$ with the inner product

$$(s_1, s_2) = \int_{\mathrm{Br}_d} (s_1(k), s_2(k))dk$$

where $(s_1(k), s_2(k))$ is the inner product in $\mathcal{H}_k$.

## 2. Topological invariants and T-symmetry

In the above Bloch picture it is not at all clear where the topology comes from. To explain it consider the Bloch Hamiltonian $H_0(k)$ describing the insulator with forbidden zone on the level of Fermi energy $E_F$. We can diagonalize it by the unitary conjugation:

$$U_0^*(k)H_0(k)U_0(k) = \mathrm{diag}(E_1(k), \dots, E_n(k)).$$

Suppose that the first $p$ levels $E_i(k) > E_F$, $i = 1, \dots, p$, are empty and the occupied levels correspond to the energies $E_i(k) < E_F$ with $i = p + 1, \dots, n$.

Consider the *adiabatic deformation* of this Hamiltonian, i.e. its continuous deformation in the class of considered Hamiltonians not affecting the forbidden zone, which transforms $H_0$ into the Hamiltonian $H_1 \equiv \operatorname{sgn} H_0$ which has only two eigenvalues equal to $\pm 1$. In other words, it exists a continuous map $U_1 : \mathrm{Br}_d \to \mathrm{U}(n)$ such that

$$U_1^*(k)H_1(k)U_1(k) = \operatorname{diag}\left(\mathbf{1}_{p\times p}, -\mathbf{1}_{(n-p)\times(n-p)}\right). \tag{2.1}$$

(The existence of such map can be easily proved using the spectral decomposition for $H_0$.) Note that the matrix $U_1(k)$ in this equation is defined only up to transformations of the form

$$U_1(k) \longmapsto U_1(k)\operatorname{diag}\left(U_{p\times p}, U_{(n-p)\times(n-p)}\right).$$

Hence, the matrix $U_1(k)$, yielding a solution of the equation (2.1), determines a map from the Brillouin zone $\mathrm{Br}_d = \mathbb{T}^d$ to the Grassmannian

$$\mathrm{Gr}_{p,n} = \mathrm{U}(n)/\,\mathrm{U}(p) \times \mathrm{U}(n-p).$$

So we have arrived at the problem of description of homotopy classes $[\mathbb{T}^d, \mathrm{Gr}_{p,n}]$ of continuous maps from the torus $\mathbb{T}^d$ to Grassmann manifolds.

The general problem of description of the homotopy classes $[\mathbb{T}^d, X]$ and the corresponding toric homotopy groups of topological spaces $X$ was studied by Fox [**3**]. It follows from his results that the homotopy classes $[\mathbb{T}^d, X]$ are classified by the homomorphisms

$$\Omega_I^r : \pi_r(X) \to [\mathbb{T}^d, X], r \le d,$$

corresponding to the subsets $I = \{i_1 < \ldots < i_{r-1}\}$ in the set of indices $\{1, \ldots, d\}$. In other words, the homotopy classes $[\mathbb{T}^d, X]$ are parameterized by the collections of $d$ elements from the group $\pi_1(X)$, $\frac{d(d-1)}{2}$ elements from $\pi_2(X), \ldots, C_d^j$ elements from $\pi_j(X)$ and so on.

Applying this result to the space $X = \mathrm{Gr}_{p,n}$, we conclude that for $d = 1$ we have $[\mathbb{T}^1, \mathrm{Gr}_{p,n}] = \pi_1(\mathrm{Gr}_{p,n}) = 0$, for $d = 2$ the homotopy classes of the maps $\mathbb{T}^2 \to \mathrm{Gr}_{p,n}$ are characterized by a unique invariant since $\pi_2(\mathrm{Gr}_{p.n}) = \mathbb{Z}$, and for $d = 3$ the homotopy classes $[\mathbb{T}^3, \mathrm{Gr}_{p,n}]$ are classified by three elements from the group $\pi_2(\mathrm{Gr}_{p,n}) = \mathbb{Z}$ since $\pi_1(\mathrm{Gr}_{p.n}) = \pi_3(\mathrm{Gr}_{p,n}) = 0$.

The non-trivial classes in dimensions $d = 2, 3$ are described in the following way. The space $[\mathbb{T}^d, \mathrm{Gr}_{p,n}]$ may be considered as the classifying space for the bundles over $\mathbb{T}^d$. In dimension $d = 2$ there is a non-trivial bundle over $\mathbb{T}^2$ which is an analog of the *Hopf bundle* $\mathbb{S}^3 \to \mathbb{S}^2$ over $\mathbb{S}^2$. From the physical point of view this case corresponds to the *quantum spin Hall state.* In dimension $d = 3$ there are three bundles over $\mathbb{T}^3$ which are the pullbacks of the Hopf bundle with respect to three different projections $\mathbb{T}^3 = \mathbb{S}^1 \times \mathbb{S}^1 \times \mathbb{S}^1 \mapsto \mathbb{T}^2 = \mathbb{S}^1 \times \mathbb{S}^1$.

We turn now to the time reversal symmetry. This symmetry is given by the *T-transform*, i.e. by an operator $T$ which satisfies the condition $T^2 = -\mathrm{id}$ and is anti-unitary in the sense that

$$(T\varphi, T\psi) = (\psi, \varphi)$$

for any states $\varphi, \psi$. It implies, in particular, that a state $\varphi$ and its T-partner $T\varphi$ are orthogonal to each other: $(\varphi, T\varphi) = 0$. Indeed,

$$(T\varphi, \varphi) = -(T\varphi, T^2\varphi) = -(T\varphi, \varphi),$$

where the last equality follows from the anti-unitarity of $T$. Hence, $(T\varphi, \varphi) = 0$. Since these states have the same energy it means that any energy eigenstate is doubly degenerated. This effect is called the *Kramers degeneration.*

The symmetry with respect to T-transform for the Bloch Hamiltonian $H(k)$ means that

$$TH(k)T^{-1} = H(-k).$$

## 3. Involutive spaces

The topological insulators with time reversal symmetry are described mathematically in terms of involutive spaces.

DEFINITION 3.1. The *involutive space* $(X, \tau)$ is a topological space $X$ provided with the *involution*, i.e. a homeomorphism $\tau : X \to X$ with the square $\tau^2 = \mathrm{id}_X$.

Such space will be called the *R-space.* Here is an example of R-space

$$\mathbb{R}^{p,q} := \mathbb{R}^q + i\mathbb{R}^p$$

with involution $\tau : (x, y) \mapsto (x, -y)$.

The T-symmetry on the involutive space $X$ is usually given by the reversion of signs of local coordinates. If the sphere $\mathbb{S}^d$ is realized as the sphere in the space $\mathbb{R}^{d+1}$ with coordinates $(x_0, x_1, \ldots, x_d)$ such that $x_0^2 + x_1^2 + \ldots + x_d^2 = 1$ then the T-symmetry on it is given by the map $(x_0, x_1, \ldots, x_d) \mapsto (x_0, -x_1, \ldots, -x_d)$. For the torus $\mathbb{T}^d = \mathbb{S}^1 \times \ldots \times \mathbb{S}^1$ with coordinates $(e^{i\theta_1}, \ldots, e^{i\theta_d})$ the T-symmetry is given by the map: $(e^{i\theta_1}, \ldots, e^{i\theta_d}) \mapsto (e^{-i\theta_1}, \ldots, e^{-i\theta_d})$.

The *set of fixed points* of involution $\tau$ on $X$ is the set

$$X^\tau = \{x \in X : \tau(x) = x\}.$$

We suppose that it consists of a finite number of points. In the above example of the sphere $\mathbb{S}^d$ the involution $\tau$ has two fixed points $(\pm 1, 0, \ldots, 0)$ while the involution on the torus $\mathbb{T}^d$ has $2^d$ fixed points given by the collections $(\varepsilon_1, \ldots, \varepsilon_d)$ of the numbers $\varepsilon_i = \pm 1$.

DEFINITION 3.2. The *quaternion vector bundle* or *Q-bundle* $(E, \rho)$ over the involutive space $(X, \tau)$ is a complex vector bundle $E \to X$ provided with the *anti-involution* $\rho : E \to E$ compatible with $\tau$. Here, the anti-involution $\rho$ is given by the anti-linear bundle isomorphism $\rho : E \to E$ such that $\rho^2 = -\mathrm{id}_E$. It is compatible with $\tau$ if $\pi \circ \rho = \tau \circ \pi$.

At fixed points $x \in X^\tau$ the map $\rho_x : E_x \to E_x$ is an anti-linear map with $\rho_x^2 = -\mathrm{id}_{E_x}$. So $\rho_x$ defines a quaternionic structure (action of quaternions) on $E_x$.

If we replace in this definition the anti-involution $\rho$ by the involution $\sigma$ such that $\sigma^2 = \mathrm{id}_E$ then we shall obtain the definition of the *R-bundle* $(E, \sigma)$ over $(X, \tau)$.

## 4. Hilbert bundle of the topological insulator

We return back to the Hilbert bundle $\pi : \mathfrak{H} \to X$ over the momentum space $X = \mathrm{Br}_d$. Its fiber over a point $x \in X$ is the Hilbert space $\mathcal{H}_x$ and we denote by $\mathcal{H} = L^2(\mathfrak{H})$ the Hilbert space of its square integrable sections. Such sections are identified with the physical states with eigenvalues $E(x)$. The *time reversal transform* on the space $\mathcal{H}$ is given by the operator $\Theta$ which is an anti-unitary operator, satisfying the conditions $\Theta^2 = -1$ and $\Theta^* = -\Theta$.

The Hilbert bundle $\pi : \mathfrak{H} \to X$, provided with the operator $\Theta$, becomes the quaternionic vector bundle $\pi : (\mathfrak{H}, \Theta) \to (X, \tau)$. So the action of the involution $\tau$ is pulled up to the action of the anti-involution $\Theta$ on the Hilbert bundle $\pi : \mathfrak{H} \to X$.

The Hamiltonian $H(x)$, invariant under time reversal, should satisfy the condition

$$\Theta H(x)\Theta^* = H(\tau(x)), \; x \in X.$$

If $\varphi$ is an eigenstate of $H$, i.e. $H(x)\varphi(x) = E(x)\varphi(x)$, then $\Theta\varphi$ is an eigenstate of $\Theta H \Theta^*$ with the same energy $E$:

$$[\Theta H \Theta^*]\Theta\varphi(x) = E(\tau(x))\Theta\varphi(x).$$

In other words, the states $\varphi$ and $\Theta\varphi$ belong to the same zone with energy $E(x)$ and each zone under T-symmetry is doubly degenerated so that $(\varphi, \Theta\varphi)$ form the *Kramers pair.*

We shall suppose that $\mathfrak{H}$ is the direct sum

$$\mathfrak{H} = \bigoplus_{i=1}^{N} \mathfrak{H}_i$$

of subbundles $\mathfrak{H}_i$ of rank 2 so that the spaces $\mathcal{H}_i = L^2(\mathfrak{H}_i)$ of square integrable sections of $\mathfrak{H}_i$ are generated by the Kramers sections $\varphi_i$ and $\Theta\varphi_i$.

Denote by $w_i : X \to \mathrm{U}(2)$ the transition function of the bundle $\mathfrak{H}_i$. The $T$-transform on $\mathfrak{H}_i$ changes $w_i$ to $w_i \circ C$ where $C$ is the complex conjugation. In terms of local coordinates $(x, v)$ on $\mathfrak{H}_i$ the operator $\Theta$ acts by the formula: $(x, v) \mapsto (\tau(x), w_i(x)\bar{v})$. Using the relation $\Theta^2 = -\mathrm{id}_{\mathfrak{H}_i}$, we arrive at the equality

$$w_i(\tau(x))\bar{w}_i(x) = -\mathrm{id}_{\mathcal{H}_i}, \text{ i.e. } w_i^t(\tau(x)) = -w_i(x).$$

In particular, at fixed points $x \in X^\tau$ the matrix $w_i(x)$ is skew-symmetric.

Consider the case when the Hilbert bundle $\pi : \mathfrak{H} \to X$ has rank 2 and its space of sections is generated by the Kramers sections $\varphi$ and $\Theta\varphi$. This pair of sections, having the same energy, may be considered as a quasiparticle composed of $\varphi$ and $\Theta\varphi$. Its effective Hamlltonian has the form

$$\tilde{H}(x) = \begin{pmatrix} 0 & \Theta H(x)\Theta^* \\ H(x) & 0 \end{pmatrix}.$$

It is T-invariant and the equation for the eigenvctor $\varphi$ of the Hamiltonian $H$ with eigenvalue $E$ takes in terms of $\tilde{H}$ the following matrix form

$$\begin{pmatrix} 0 & \Theta H(x)\Theta^* \\ H(x) & 0 \end{pmatrix} \begin{pmatrix} \varphi(x) \\ \Theta\varphi(x) \end{pmatrix} = \begin{pmatrix} E(\tau(x))\Theta\varphi(x) \\ E(x)\varphi(x) \end{pmatrix}.$$

## 5. Analytical $\mathbb{Z}_2$-index

Suppose that the operator $H$, hence also $\tilde{H}$, is Fredholm. In a neighborhood of the fixed point $x_0$ the operator $\tilde{H}$ may be approximated by the Dirac operator $D$ with quadratic error term which does not affect the index of $\tilde{H}$ (cf. [**8**]). The Dirac operator in the matrix form corresponds to the operator

$$\mathcal{D} = \begin{pmatrix} 0 & -D \\ D & 0 \end{pmatrix}.$$

So the operator $\tilde{H}$ in a neighborhood of a fixed point is equal to the sum of the skew-symmetric operator $\mathcal{D}$ plus a quadratic error term not affecting the index of $\tilde{H}$.

For real skew-symmetric elliptic operators $P$ Atiyah and Singer (cf. [**2**]) defined their analytical $\mathbb{Z}_2$-index as

$$\mathrm{ind}_a P = \dim_{\mathbb{R}} \mathrm{Ker}\, P \bmod 2.$$

So it is natural to define the $\mathbb{Z}_2$*-invariant of a topological insulator* $\nu$ as the sum

$$\nu = \sum_{x \in X^\tau} \mathrm{ind}_a \tilde{H}(x) = \sum_{x \in X^\tau} \dim_{\mathbb{R}} \mathrm{Ker}\, \tilde{H} \bmod 2$$

of local indices over the set of fixed points.

## 6. Topological $\mathbb{Z}_2$-index

We describe now the construction of the topological $\mathbb{Z}_2$-index for the 3-dimensional involutive spaces $X$. Suppose that the Hilbert bundle $\pi : \mathfrak{H} \to X$ has rank 2 so that the gauge group coincides with $\mathrm{U}(2)$. The anti-involution $\Theta$ on the Hilbert bundle $(\mathfrak{H}, \Theta) \to (X, \tau)$ is compatible with the natural involution $\vartheta$ on $\mathrm{U}(2)$ given by the formula: $\vartheta(g) = -g^t$, in the sense that $w \circ \tau = \vartheta \circ w$ where $w$ is the gauge function. In other words, $w$ defines the equivariant map $w : (X, \tau) \to (\mathrm{U}(2), \vartheta)$.

The *odd Chern character* of a smooth map $g : X \to \mathrm{U}(n)$ from an odd-dimensional space $X^d$ into the group $\mathrm{U}(n)$ is defined by the formula (cf. [**4**])

$$\mathrm{Ch}(g) = \sum_{k=0}^{(d-1)/2} \mathrm{Ch}_{2k+1}(g) = \sum_{k=0}^{(d-1)/2} (-1)^k \frac{k!}{(2k+1)!} \mathrm{tr}\left[(g^{-1}dg)^{2k+1}\right].$$

Using this formula in the 3-dimensional case we can introduce the *topological index* by the formula

$$\mathrm{ind}_t g = \frac{1}{4\pi^2} \int_X \mathrm{Ch}_3(g) = \frac{1}{24\pi^2} \int_X \mathrm{tr}[(g^{-1}dg)^3].$$

Due to $T$-symmetry the *topological index* $\mathrm{ind}_t w$ of the gauge function $w$ is defined modulo $\mathbb{Z}_2$.

The assertion that the topological $\mathbb{Z}_2$-index for 3-dimensional topological insulators coincides with the above analytical index, i.e.

$$\nu = \mathrm{ind}_t w,$$

may be considered as an analog of Atiyah-Singer index theorem in our situation.

## References


[1] N. W. Ashcroft and N. D. Mermin, *Solid state physics*, Saunders, New York, 1976.

[2] M. F. Atiyah and I. M. Singer, *Index theory for skew-adjoint Fredholm operators*, Inst. Hautes Études Sci. Publ. Math. **37** (1969), 5–26. MR285033

[3] R. H. Fox, *Homotopy groups and torus homotopy groups*, Ann. of Math. (2) **49** (1948), 471–510, DOI 10.2307/1969292. MR27143

[4] E. Getzler, *The odd Chern character in cyclic homology and spectral flow*, Topology **32** (1993), no. 3, 489–507, DOI 10.1016/0040-9383(93)90002-D. MR1231957

[5] C. L. Kane and E. J. Mele, $\mathbb{Z}_2$ *topological order and the quantum spin Hall effect*, Phys. Rev. Lett. **95** (2005), 95:146802.

[6] A. Kitaev, *Periodic table for topological insulators and superconductors*, Adv. Theor. Phys. AIP Conf. Proc. **1134** (2009), 22–30.

[7] E. M. Lifschitz and L. P. Pitaevskii, *Statistical physics, Part 2. Theory of condensed matter*, Pergamon Press, 1980.
[8] S. Q. Shen, *Topological insulators: Dirac operators in condensed matters*, Berlin, Springer, 2013.

Steklov Mathematical Institute, Gubkina 8, 119997, Moscow, Russia
*Email address*: `sergeev@mi-ras.ru`

Contemporary Mathematics
Volume **820**, 2025
https://doi.org/10.1090/conm/820/16419

# On the Dirichlet-to-Neumann map for the $p$-Laplacian on a metric measure space

Ryan Gibara and Nageswari Shanmugalingam

ABSTRACT. In this note, we construct a Dirichlet-to-Neumann map, from a Besov space of functions, to the dual of this class. The Besov spaces are of functions on the boundary of a bounded, locally compact uniform domain equipped with a doubling measure supporting a $p$-Poincaré inequality so that this boundary is also equipped with a Radon measure that has a codimensional relationship with the measure on the domain. We construct this map via the following recipe. We show first that solutions to Dirichlet problem for the $p$-Laplacian on the domain with prescribed boundary data in the Besov space induce an operator that lives in the dual of the Besov space. Conversely, we show that there is a solution, in the homogeneous Newton-Sobolev space, to the Neumann problem for the $p$-Laplacian with the Neumann boundary data given by a continuous linear functional belonging to the dual of the Besov space. We also obtain bounds on its operator norm in terms of the norms of trace and extension operators that relate Newton-Sobolev functions on the domain to Besov functions on the boundary.

## 1. Introduction

Classically, for a sufficiently smooth bounded domain $\Omega$ in $\mathbb{R}^n$, the Dirichlet-to-Neumann map for the Laplacian is the operator $\mathrm{DtN} : H^{1/2}(\partial\Omega) \to H^{-1/2}(\partial\Omega)$ given by $\mathrm{DtN}(f) := \partial_\eta u := \eta \cdot \nabla u$, where $\eta$ is the unit outer normal vector to the domain $\Omega$ and $u$ is the unique harmonic extension of $f$ from the boundary $\partial\Omega$ to $\Omega$. Here, $H^{1/2}(\partial\Omega)$ is the classical fractional Sobolev space, with $\partial\Omega$ equipped with its $(n-1)$-dimensional Hausdorff measure, and $H^{-1/2}(\partial\Omega)$ is its dual. For more general domains, where $\eta$ may not be well defined and $\nabla u$ may not exist pointwise, one may define the outer normal derivative of $u \in H^1(\Omega)$ in the weak sense that

$$\int_\Omega (\Delta u)(v) + \int_\Omega \nabla u \cdot \nabla v = \int_{\partial\Omega} (\partial_\eta u)(\mathrm{Tr}\, v)$$

for all $v \in H^1(\Omega)$. Here $\partial_\eta u$ should be thought of as a function on $\partial\Omega$ that serves the role of a weak outer normal derivative of $u$ at $\partial\Omega$; such function may not always exist. It should be clear that the role of $\partial_\eta u$ in the above is reminiscent of

2020 *Mathematics Subject Classification.* Primary 31E05; Secondary 46E36, 31B20, 45Q05.
*Key words and phrases.* boundary-value problems, Besov space, Newton-Sobolev space, doubling measure, homogeneous spaces, Poincaré inequality, upper gradients, Cheeger differential structure, Dirichlet-to-Neumann map.
The second author's research was partially supported by NSF (US) grant DMS#2054960.

that of the usual outer normal derivative in Green's formula. The corresponding Dirichlet-to-Neumann map has been studied for other elliptic operators such as the divergence-form operator $L = \operatorname{div}(\gamma \nabla u)$ for a smooth function $\gamma$, the Helmholtz operator, or Schrödinger operator, see [**1, 7, 13, 38**]. The study of the Dirichlet-to-Neumann map has also been extended to the setting of Riemannian manifolds for the Laplace-Beltrami operator and generalized to differential forms [**9, 18**]. The recent paper [**22**] gave a construction of a Dirichlet-to-Neumann map for the nonlinear operator $\Delta_p$, i.e. the $p$-Laplacian, in the setting of bounded Lipschitz domains in Euclidean spaces.

The goal of the present paper is to generalize this problem even further to the setting of a metric measure space $(\Omega, d, \mu)$ that is bounded, locally compact, supporting a doubling measure $\mu$ and satisfies a $p$-Poincaré inequality for a fixed $1 < p < \infty$, so that $\Omega$ is also a uniform domain in its completion $\overline{\Omega}$. We consider Cheeger-type operators that are analogs of the $p$-Laplacian, and endow the boundary $\partial\Omega$ with a measure $\nu$ that is in a codimensional relationship with $\mu$.

The first main result of this note, contained within Theorem 3.13, is the following, in which we demonstrate the existence of a solution in the homogeneous Newton-Sobolev space $HN^{1,p}(\Omega)$ to the Neumann problem with boundary data taken from the dual of the homogeneous Besov space $HB^\theta_{p,p}(\partial\Omega)$. Here, $\theta$ depends on $p$ and the codimensionality of the boundary measure $\nu$.

THEOREM. *Given an operator $L \in (HB^\theta_{p,p}(\partial\Omega))^*$, there exists $u_L \in HN^{1,p}(\Omega)$ satisfying*

$$\int_\Omega |\nabla u_L|^{p-2} \langle \nabla u_L, \nabla v \rangle \, d\mu = L(\operatorname{Tr}(v))$$

*for all $v \in HN^{1,p}(\Omega)$.*

In the setting of this paper, results of Malý [**34**] imply that there exist a bounded linear trace operator $\operatorname{Tr} : HN^{1,p}(\Omega) \to HB^\theta_{p,p}(\partial\Omega)$ and a bounded linear extension operator $E : HB^\theta_{p,p}(\partial\Omega) \to HN^{1,p}(\Omega)$ with $\operatorname{Tr} \circ E$ the identity operator. We denote by $\|\operatorname{Tr}\|$ and $\|E\|$ their respective operator norms.

The second main result of the this note, see Corollaries 4.3 and 4.5 and Theorem 4.7, is the existence of the Dirichlet-to-Neumann map and its inverse, the Neumann-to-Dirichlet map, in this setting. For $f \in HB^\theta_{p,p}(\partial\Omega)$, we denote by $u_f$ the solution to the Dirichlet problem with boundary data $f$. This allows us to create a functional $L_f \in (HB^\theta_{p,p}(\partial\Omega))^*$ given by

$$L_f(g) := \int_\Omega |\nabla u_f|^{p-2} \langle \nabla u_f, \nabla E g \rangle \, d\mu$$

for any $g \in HB^\theta_{p,p}(\partial\Omega)$

THEOREM. *The Dirichlet-to-Neumann map* $\operatorname{DtN}(f) := L_f$ *and the Neumann-to-Dirichlet map* $\operatorname{NtD}(L) := \operatorname{Tr}(u_L)$ *are well-defined and bounded operators. Moreover, they are inverses of one another and the operator norm of* DtN *satisfies*

$$c_p \, \|\operatorname{Tr}\|^{\frac{p}{1-p}} \le \|DtN\| \le \|E\|^p,$$

*where*

$$c_p := \left[ (2p)^{\frac{1}{p-1}} + \frac{2(p-1)}{p} \right]^{-1}.$$

The Dirichlet-to-Neumann operator plays an important role in the theory of inverse problems and has real-world applications. In particular, in medical imaging, one is interested in obtaining tomographic representations of the inside of organs by measuring the electrical impedance of a current traveling through the tissue of the organ. With knowledge of the Dirichlet-to-Neumann map, also called the voltage-to-current map, this can be accomplished non-invasively. For a survey on inverse problems and applications, we refer the reader to [**27, 39**]. The generalization of the operator to the setting of metric measure spaces permit us to then consider applications in situations where we do not have access to smooth structures, for example in the case where the ambient material corresponds to highly nonsmooth weights modifying measures on Euclidean domain, and also allow us to consider structures in the non-manifold setting of Carnot-Carathéodory spaces.

The construction of the Dirichlet-to-Neumann map, as described in this note, while developed independently of [**22**], follows a similar method as [**22**], but avoids the need for the domain to be Lipschitz regular.

## 2. Background

**2.1. Setting.** In this section, we will review the necessary definitions related to the analysis on metric spaces and set the notation for the rest of this note. Let $(\Omega, d)$ be a locally compact, non-complete metric space with completion $\overline{\Omega}$ and boundary $\partial\Omega := \overline{\Omega} \setminus \Omega$.

In the study of boundary-value problems on such a space, one needs to be able to make sense of notions such as trace to the boundary. At this level of generality, where there is no notion of smoothness in the classical sense, one must rely on geometric conditions instead. In this paper we require the geometric condition called uniform domain condition. First introduced by Martio and Sarvas in [**35**], uniform domains are studied extensively in relation to quasiconformal maps [**4, 15, 33, 35, 40, 41**], potential theory [**2, 3, 16, 17, 30**], and extension domains [**2, 5, 26, 28, 29**], and the references therein. The class of uniform domains includes all bounded Euclidean Lipschitz domains and cardioid domains, but also domains with non-rectifiable boundaries such as the von Koch snowflake domain.

DEFINITION 2.1. Given a number $C \geq 1$ and points $x, y \in \Omega$, we say that a curve $\gamma$ with end points $x$ and $y$ is a *C-uniform curve* connecting $x$ to $y$ if the length $\ell(\gamma)$ of $\gamma$ satisfies the quasiconvexity condition $\ell(\gamma) \leq C\, d(x, y)$ and in addition, for each $z$ in the trajectory of $\gamma$ and for each pair of subcurves $\gamma_{x,z}$ and $\gamma_{z,y}$ of $\gamma$ with end points $x, z$ and $z, y$ respectively, we have that

$$\min\{\ell(\gamma_{x,z}), \ell(\gamma_{z,y})\} \leq C \operatorname{dist}(z, \partial\Omega).$$

The latter condition above is also known as a twisted double cone condition in some literature. We say that $\Omega$ is a *uniform domain* if there is a constant $C \geq 1$ for which each pair $x, y \in \Omega$ with $x \neq y$ can be connected by a $C$-uniform curve $\gamma$.

A triple $(\Omega, d, \mu)$ is called a *metric measure space* if $\mu$ is a Borel regular measure satisfying $0 < \mu(B) < \infty$ for all balls $B \subset \Omega$. We will say that $\mu$ is *doubling* if there exists a constant $C \geq 1$ for which $\mu(2B) \leq C\,\mu(B)$ holds for all balls $B \subset \Omega$, where $2B$ denotes the ball concentric with $B$ having radius twice that of $B$. For references on analysis on metric spaces, we refer the readers to [**23, 25**].

The setting of a metric measure space supporting a doubling measure allows for a significant amount of classical zeroth-order analysis to carry through, see [**11**].

When progressing to first-order analysis, however, one needs a replacement for the classical notion of differentiability. We define here the notion of an upper gradient for a function, which serves as an analogue of the magnitude of the gradient within the context of the fundamental theorem of calculus, see [**24, 32, 36**].

DEFINITION 2.2. We say that a Borel function $g : \Omega \to [0, \infty]$ is an *upper gradient* of a function $u : \Omega \to \mathbb{R}$ if

$$|u(y) - u(x)| \leq \int_\gamma g\, ds \tag{2.3}$$

whenever $\gamma$ is a non-constant compact rectifiable curve in $\Omega$ joining the points $x, y \in \Omega$. We interpret this inequality in the sense that $\int_\gamma g\, ds = \infty$ whenever at least one of $u(x)$ or $u(y)$ is not finite.

When a Borel function $g$ satisfies (2.3) for all rectifiable curves except for a collection $\Gamma$ of curves for which there exists a Borel function $0 \leq \rho \in L^p(\Omega)$ satisfying $\int_\gamma \rho\, ds = \infty$ for all $\gamma \in \Gamma$, we say that $g$ is a *p-weak upper gradient*. For a function $u$, we denote by $D_p(u)$ the collection of all $p$-weak upper gradients of $u$ that are in $L^p(\Omega)$. If it is non-empty, then there is an element that is minimal in the sense of $L^p$-norm and pointwise almost everywhere, see for instance [**20**] or [**25**, Theorem 6.3.20]. This minimal element is called a *minimal p-weak upper gradient*, and plays the role of the norm of gradients of differentiable functions.

With the notion of upper gradient, we are able to define what is meant by a Poincaré inequality in the setting of a metric measure space.

DEFINITION 2.4. For $1 \leq q, p < \infty$, we say that a metric measure space $(\Omega, d, \mu)$ supports a $(q,p)$-*Poincaré inequality* if there are constants $C > 0$ and $\lambda \geq 1$ such that

$$\left( \fint_B |u - u_B|^q \, d\mu \right)^{1/q} \leq C \operatorname{rad}(B) \left( \fint_{\lambda B} g^p \, d\mu \right)^{1/p}$$

for all function-upper gradient pairs $(u, g)$, where $\operatorname{rad}(B)$ is the radius of $B$ and $u_B := \fint_B u\, d\mu$.

It follows from Hölder's inequality that if a space supports a $(q,p)$-Poincaré inequality for some $1 \leq q, p < \infty$, then it satisfies a $(q', p')$-Poincaré inequality for all $q' \leq q$ and $p' \geq p$. Conversely, if $(\Omega, d, \mu)$ supports a $(1,p)$-Poincaré inequality and $\mu$ is doubling, then it supports a $(p,p)$-Poincaré inequality, see for instance [**25**, Theorem 9.1.2] or [**21**, Theorem 5.1]. We will often shorten the notation and say that $\Omega$ supports a $p$-Poincaré inequality when $\Omega$ supports a $(1,p)$-Poincaré inequality.

A metric measure space supporting a Poincaré inequality enjoys strong geometric connectivity properties; in particular, the space is quasiconvex, see for instance [**10**] or [**25**] for the case of $X$ being complete in addition to $\mu$ being doubling, and [**12**] for the case that $\mu$ is doubling but $X$ is only locally compact.

In the study of boundary-value problems, we are interested in the behavior of functions on $\Omega$ near or on its boundary $\partial\Omega$. For this reason, we need to be able to speak about the dimension of the boundary in a measure-theoretic way. However, at this level of generality of non-smoothness, there is no single natural dimension of a measure to consider; hence, it makes most sense to consider a notion of codimension of the boundary measure with respect to the ambient measure, as in the following definition. The condition in this definition is reminiscent of an Ahlfors regularity condition for measures, but here we relate two different measures to each other. Recall that the measure $\mu$ is supported on the domain $\Omega$.

DEFINITION 2.5. For $\Theta > 0$, we say that a Borel regular measure $\nu$ on $\partial\Omega$ is $\Theta$-*codimensional* to $\mu$ if there exists a constant $C > 0$ such that

$$C^{-1}\frac{\mu(B(x,r)\cap\Omega)}{r^\Theta} \le \nu(B(x,r)\cap\partial\Omega) \le C\frac{\mu(B(x,r)\cap\Omega)}{r^\Theta} \tag{2.6}$$

for all $x \in \partial\Omega$ and $0 < r < 2\,\mathrm{diam}(\partial\Omega)$.

Balls in $\partial\Omega$ are intersection of $\partial\Omega$ with balls in $\overline{\Omega}$, centered at points in $\partial\Omega$. The measure $\nu$ on $\partial\Omega$ can also be considered as a measure on $\overline{\Omega}$, but supported on $\partial\Omega$; while the measure $\mu$, which lives on $\Omega$, can be extended as a null measure to $\partial\Omega$. Therefore, in this note, we will consider both $\mu$ and $\nu$ as measures on $\overline{\Omega}$, but interpreted as above, and by balls we mean balls in $\overline{\Omega}$. Hence, (2.6) can also be written as

$$C^{-1}\frac{\mu(B(x,r))}{r^\Theta} \le \nu(B(x,r)) \le C\frac{\mu(B(x,r))}{r^\Theta}$$

for $x \in \partial\Omega$.

It follows from the above definition that if $\mu$ is doubling, then so is $\nu$. As the boundary of a bounded domain is itself bounded, if $\mu$ is doubling and $\Omega$ is bounded, then $\nu(\partial\Omega) < \infty$.

STANDING ASSUMPTIONS 2.7. Throughout this note, we fix $1 < p < \infty$, set $\Omega$ to be a bounded uniform domain equipped with a doubling measure $\mu$ such that the metric measure space $(\Omega, d, \mu)$ supports a $p$-Poincaré inequality, and assume that the boundary $\partial\Omega$ supports a measure $\nu$ that is $\Theta$-codimensional to $\mu$ for some $0 < \Theta < p$. Note that in this setting, $\nu(\partial\Omega) < \infty$.

**2.2. Function spaces.** In this section, we define the function spaces that will be considered in this paper. We consider functions defined on $\Omega$ that are in an appropriate Sobolev class. Within the context of metric measure spaces, there exist many possible approaches to Sobolev classes; here, we begin with the approach by upper gradients, see [**25, 36**].

DEFINITION 2.8. We say that a measurable function $u$ is in the *Dirichlet space* $D^{1,p}(\Omega)$ if

$$\|u\|^p_{D^{1,p}(\Omega)} := \inf_{g\in D_p(u)} \int_\Omega g^p\, d\mu < \infty.$$

By the Poincaré inequality, every function $u \in D^{1,p}(\Omega)$ is in $L^p_{\mathrm{loc}}(\Omega)$. Since $\Omega$ is bounded and the Poincaré inequality holds globally, we actually have that $D^{1,p}$-functions are in $L^p(\Omega)$, and so $D^{1,p}(\Omega) = N^{1,p}(\Omega)$ as sets; see the discussion following Definition 2.14 for similar phenomenon related to Besov spaces. Recall that $D_p(u)$ is the collection of all upper gradients of $u$ that are in $L^p(\Omega)$.

DEFINITION 2.9. We say that a measurable function $u$ is in $\widetilde{N}^{1,p}(\Omega)$ if

$$\|u\|_{N^{1,p}(\Omega)} := \|u\|_{L^p(\Omega)} + \|u\|_{D^{1,p}(\Omega)} < \infty.$$

This defines a seminorm; taking the quotient space modulo the null-space of the seminorm $\|\cdot\|_{N^{1,p}(\Omega)}$ yields the *Newton-Sobolev* space $N^{1,p}(\Omega)$.

Functions in $L^p(\Omega)$ are well defined up to sets of $\mu$-measure zero. However, changing a function from $N^{1,p}(\Omega)$ on arbitrary sets of $\mu$-measure zero can result in a function that no longer has a $p$-weak upper gradient in $L^p(\Omega)$; thus care needs to be taken if one wishes to modify functions in $N^{1,p}(\Omega)$. It turns out that if $u_1$ and $u_2$ are two functions in $\widetilde{N}^{1,p}(\Omega)$, then $\|u_1 - u_2\|_{N^{1,p}(\Omega)} = 0$ if and only if the set $\{x \in \Omega : u_1(x) \neq u_2(x)\}$ has $p$-capacity zero, as defined below. For more on this, we refer the reader to [**25**, **36**].

DEFINITION 2.10. For a set $E \subset \Omega$, we define its *$p$-capacity* as the number

$$\mathrm{Cap}_p(E) := \inf \|u\|_{N^{1,p}(\Omega)},$$

where the infimum is taken over all $u \in N^{1,p}(\Omega)$ satisfying $0 \leq u \leq 1$ on all of $\Omega$ and $u = 1$ on $E$.

As $p$ is fixed throughout this note, we will sometimes refer to the capacity of a set as opposed to the $p$-capacity of a set.

Another approach to Sobolev spaces of functions defined on metric measure spaces is due to a combination of a differential structure developed by Cheeger for Lipschitz functions, see [**8**], and the extension of such a structure to functions in $N^{1,p}$-classes, see [**14**, Theorems 9 and 10].

DEFINITION 2.11. A metric measure space $(\Omega, d, \mu)$ is said to support a *Cheeger differential structure* of dimension $N \in \mathbb{N}$ if there exists a collection of coordinate patches $\{(\Omega_\alpha, \varphi_\alpha)\}$ and a $\mu$-measurable inner product $\langle \cdot, \cdot \rangle_x$, $x \in \Omega_\alpha$, on $\mathbb{R}^N$ such that

(1) each $\Omega_\alpha$ is a measurable subset of $\Omega$ with positive measure and $\bigcup_\alpha \Omega_\alpha$ has full measure;
(2) each $\varphi_\alpha : \Omega_\alpha \to \mathbb{R}^N$ is Lipschitz;
(3) for every function $u \in D^{1,p}(\Omega)$, for $\mu$-a.e. $x \in \Omega_\alpha$ there is a vector $\nabla u(x) \in \mathbb{R}^N$ such that

$$\operatorname*{ess\,lim\,sup}_{\Omega_\alpha \ni y \to x} \frac{|u(y) - u(x) - \langle \nabla u(x), \varphi_\alpha(y) - \varphi_\alpha(x) \rangle_x|}{d(y,x)} = 0.$$

The vector $\nabla u$ is called the *Cheeger gradient* of $u$.

Note that there may be more than one possible Cheeger differential structure on a given space, and so we choose the one with the further property that $|\nabla u(x)|_x$ is comparable to the $p$-weak upper gradient of $u$ for $\mu$-a.e. $x \in \Omega$, where $|\nabla u(x)|_x^2 = \langle \nabla u(x), \nabla u(x) \rangle_x$, see [**10**, **25**]. This choice of differential structure can replace the upper gradient structure used to define $N^{1,p}(\Omega)$ to obtain an equivalent definition when $\Omega$ is a doubling metric measure space supporting a Poincaré inequality, provided that $1 < p < \infty$. In what follows, we suppress the dependence of the inner product and norm on the point $x$.

While Sobolev spaces form the natural space of functions on $\Omega$ to consider, the natural function-space on $\partial\Omega$ to consider is the Besov space.

DEFINITION 2.12. We say that a measurable function $f$ is in $\dot{B}^\theta_{p,p}(\partial\Omega)$ if

$$\|f\|^p_{HB^\theta_{p,p}(\partial\Omega)} := \int_{\partial\Omega} \int_{\partial\Omega} \frac{|f(y) - f(x)|^p}{d(y,x)^{\theta p} \nu(B(y, d(y,x)))} \, d\nu(x)\, d\nu(y) < \infty.$$

The quantity $\|f\|^p_{HB^\theta_{p,p}(\partial\Omega)}$ is called the *Besov energy* of $f$ in this note.

For constant functions $c$, we have $\|c\|_{HB^{\theta}_{p,p}(\partial\Omega)} = 0$. In fact, the collection of all functions with this seminorm zero is precisely the collection of all $\nu$-a.e. constant functions.

DEFINITION 2.13. We define the *homogeneous Besov space* $HB^{\theta}_{p,p}(\partial\Omega)$ to be $\dot{B}^{\theta}_{p,p}(\partial\Omega)/\mathbb{R}$. The natural norm on $HB^{\theta}_{p,p}(\partial\Omega)$ coming from being a quotient space modulo constants is given by

$$\inf_{c\in\mathbb{R}} \|f + c\|_{HB^{\theta}_{p,p}(\partial\Omega)}$$

whenever $f \in \dot{B}^{\theta}_{p,p}(\partial\Omega)$. However, $\|f + c\|_{HB^{\theta}_{p,p}(\partial\Omega)}$ trivially equals $\|f\|_{HB^{\theta}_{p,p}(\partial\Omega)}$ for every constant $c$, and so

$$\inf_{c\in\mathbb{R}} \|f + c\|_{HB^{\theta}_{p,p}(\partial\Omega)} = \|f\|_{HB^{\theta}_{p,p}(\partial\Omega)}.$$

DEFINITION 2.14. We set the *inhomogeneous Besov space* $B^{\theta}_{p,p}(\partial\Omega)$ to be the space $\dot{B}^{\theta}_{p,p}(\partial\Omega) \cap L^p(\partial\Omega)$. The natural norm on $B^{\theta}_{p,p}(\partial\Omega)$, coming from being an intersection of normed spaces, is given by

$$\max\{\|f\|_{HB^{\theta}_{p,p}(\partial\Omega)}, \|f\|_{L^p(\partial\Omega)}\} \approx \|f\|_{HB^{\theta}_{p,p}(\partial\Omega)} + \|f\|_{L^p(\partial\Omega)}.$$

It follows from [**6**, Lemma 2.2] that every function in the class $\dot{B}^{\theta}_{p,p}(\partial\Omega)$ is necessarily in $L^p(\partial\Omega)$ since $\nu(\partial\Omega)$ is finite. Hence, *as sets*, we have also that $B^{\theta}_{p,p}(\partial\Omega) = \dot{B}^{\theta}_{p,p}(\partial\Omega)$.

In the next section and in the rest of the paper, we exploit connections between Sobolev functions on $\Omega$ and Besov functions on $\partial\Omega$. We need corresponding connections between $HB^{\theta}_{p,p}(\partial\Omega)$ and a homogeneous space associated with Sobolev classes. In our setting, the *homogeneous Newton-Sobolev class* $HN^{1,p}(\Omega)$ is the natural partner. To define $HN^{1,p}(\Omega)$, we consider an equivalence relationship $\sim$ on $N^{1,p}(\Omega)$ as follows: given $f_1, f_2 \in N^{1,p}(\Omega)$, we say that $f_1 \sim f_2$ if

$$\inf_{g\in D_p(f_1-f_2)} \int_{\Omega} g^p \, d\mu = 0.$$

Then $HN^{1,p}(\Omega)$ is the collection of all equivalence classes from $N^{1,p}(\Omega)$; note that if $f \in HN^{1,p}(\Omega)$ and $f_1, f_2$ are two representatives of $f$, then by the Poincaré inequality we have that $f_1 - f_2$ is constant. Thus, equivalence classes consist of collections of Sobolev functions (that is, functions from $N^{1,p}(\Omega)$) that differ from each other only by a constant. The norm $\|f\|_{HN^{1,p}(\Omega)}$ on $HN^{1,p}(\Omega)$ is given by

$$\|f\|_{HN^{1,p}(\Omega)} = \|f_1\|_{D^{1,p}(\Omega)},$$

for any choice of representative from the class $f$. Note that this norm is independent of the choice of representative $f_1$ of $f$.

**2.3. Trace and extension operators.** Here, we present the known trace and extension results in the setting of bounded uniform domains due to Malý [**34**]. Let $\theta = 1 - \Theta/p$. As $0 < \Theta < p$, we have that $0 < \theta < 1$.

THEOREM 2.15. *Under the standing assumptions 2.7 above, there is a surjective bounded linear trace operator* $\mathrm{Tr} : N^{1,p}(\Omega) \to \dot{B}^{\theta}_{p,p}(\partial\Omega)$ *with*

$$\| \mathrm{Tr}(u)\|_{HB^{\theta}_{p,p}(\partial\Omega)} \leq \| \mathrm{Tr} \| \|\nabla u\|_{L^p(\Omega)}$$

*and there is a bounded linear extension operator* $E : \dot{B}^{\theta}_{p,p}(\partial\Omega) \to N^{1,p}(\Omega)$ *with*

$$\|\nabla E f\|_{L^p(\Omega)} \leq \|E\| \|f\|_{HB^{\theta}_{p,p}(\partial\Omega)}.$$

*Moreover,* $\mathrm{Tr} \circ E$ *is the identity operator on* $\dot{B}^{\theta}_{p,p}(\partial\Omega)$. *The trace* $\mathrm{Tr}(u)$ *of a function* $u$ *is given as follows. For each* $\xi \in \partial\Omega$, $\mathrm{Tr}(u)(\xi)$ *is the number (if it exists) satisfying*

$$\lim_{r\to 0^+} \fint_{B(\xi,r)} |u - \mathrm{Tr}(u)(\xi)|\, d\mu = 0.$$

REMARK 2.16. Since we are more concerned with $HB^{\theta}_{p,p}(\partial\Omega)$ rather than $\dot{B}^{\theta}_{p,p}(\partial\Omega)$, the operator Tr in the rest of this note is actually the composition of the trace operator from Theorem 2.15 above with the canonical quotient map from $\dot{B}^{\theta}_{p,p}(\partial\Omega)$ to $HB^{\theta}_{p,p}(\partial\Omega)$. In particular,

$$\mathrm{Tr} : HN^{1,p}(\Omega) \to HB^{\theta}_{p,p}(\partial\Omega)$$

is a bounded linear trace operator. A similar argument also gives us

$$E : HB^{\theta}_{p,p}(\partial\Omega) \to HN^{1,p}(\Omega)$$

as a bounded linear extension operator such that $\mathrm{Tr} \circ E$ is the identity map on $HN^{1,p}(\Omega)$.

The original theorems of Malý [**34**] are expressed in terms of the inhomogeneous Besov space. However, when $\nu(\partial\Omega) < \infty$, as we assume here, the sets $\dot{B}^{\theta}_{p,p}(\partial\Omega)$ and $B^{\theta}_{p,p}(\partial\Omega)$ are the same, and so this theorem can be interpreted in terms of the homogeneous Besov space since only the Besov energy is involved. Although [**34**] gives estimates for the Besov energy of traces in terms of upper gradient energy of the Sobolev functions, we can replace the upper gradient energy in [**34**] with the Cheeger energy as the two are comparable.

## 3. Boundary-value problems

The focus of this section is to establish the existence of solutions to the Neumann boundary-value problems where the Neumann data is given by an operator. Previously, existence of solutions to the Neumann boundary-value problems was established in [**35**] under the condition that the Neumann data is given by a function on $\partial\Omega$, and so our result in this section extends the discussion from [**35**] to a wider class of Neumann data.

We first discuss the necessary background regarding the Dirichlet boundary-value problem when the boundary data is in an appropriate Besov class. We then formulate the Neumann boundary-value problem for boundary data in the dual space of an appropriate Besov class and end this section by demonstrating the existence of solutions to this problem.

DEFINITION 3.1. For $f \in B^{\theta}_{p,p}(\partial\Omega)$, we say that a function $u_f \in N^{1,p}(\Omega)$ is a *solution to the Dirichlet problem* with boundary data $f$ if

$$\int_{\Omega} |\nabla u_f|^p \, d\mu \le \int_{\Omega} |\nabla(u_f + v)|^p \, d\mu$$

for all $v \in N^{1,p}_0(\Omega)$ and $\mathrm{Tr}(u_f) = f$ pointwise $\nu$–a.e. in $\partial\Omega$. From [**31**], we know that we can modify such a solution on a set of $\mu$-measure zero so that it is locally Hölder continuous on $\Omega$; henceforth, it is this Hölder continuous representative that we denote $u_f$.

In other words, $u_f$ is a minimizer for the functional

$$\frac{1}{p}\int_{\Omega} |\nabla u|^p \, d\mu$$

over the class of $u \in N^{1,p}(\Omega)$ satisfying $\mathrm{Tr}(u) = f$ pointwise $\nu$–a.e. in $\partial\Omega$. This variational perspective provides the following characterization for a function $u$ to be *p-harmonic* in $\Omega$:

$$\int_{\Omega} |\nabla u|^{p-2} \langle \nabla u, \nabla v\rangle \, d\mu = 0 \tag{3.2}$$

for every $v \in N^{1,p}_0(\Omega)$. Note that for each $f \in B^{\theta}_{p,p}(\partial\Omega)$, the solution $u_f$ to the Dirichlet problem with boundary data $f$ exists, see for instance [**37**].

DEFINITION 3.3. When $f \in HB^{\theta}_{p,p}(\partial\Omega)$, by a *solution to the Dirichlet problem* with boundary data $f$ we mean the *collection* of solutions to the Dirichlet problem with boundary data $\tilde{f}$ for each representative function $\tilde{f}$ of $f$; note that $f$ is itself a collection of functions, two of which differ only by a constant $\nu$-almost everywhere. We denote this collection also by the notation $u_f$.

REMARK 3.4. Note that if $f_1, f_2$ are both representatives of $f \in HB^{\theta}_{p,p}(\partial\Omega)$, then $f_2 = f_1 + c$ for some constant $c \in \mathbb{R}$; in this case, $u_{f_2} = u_{f_1} + c$ *everywhere* on $\Omega$, where $u_{f_1}$ and $u_{f_2}$ are the solutions from Definition 3.1. So the collection $u_f$ that forms a solution to the Dirichlet boundary value problem with boundary data $f \in HB^{\theta}_{p,p}(\partial\Omega)$ is itself made up of functions of the form $u_{f_1} + c$, $c \in \mathbb{R}$, and hence $u_f \in HN^{1,p}(\Omega)$.

Note that if we replace $v \in N^{1,p}_0(\Omega)$ with $v + c$ in (3.2) whenever $c \in \mathbb{R}$, the equation (3.2) continues to remain valid; that is, that equation requires only that $v$ has constant trace on $\partial\Omega$. Therefore the analog of the Euler-Lagrange equation (3.2) in this homogeneous setting is that whenever $u$ is a representative of $u_f$ and $v$ is a representative of $v_0 \in HN^{1,p}_0(\Omega)$, we have

$$\int_{\Omega} |\nabla u|^{p-2} \langle \nabla u, \nabla v\rangle \, d\mu = 0. \tag{3.5}$$

Here, the class $HN^{1,p}_0(\Omega)$ is the collection of all $v_0 \in HN^{1,p}(\Omega)$ for which each representative $v$ of $v_0$ has constant trace at $\partial\Omega$ in the sense of Theorem 2.15. Note that if $v, w$ are two representatives of $v_0 \in HN^{1,p}(\Omega)$, then $v - w$ is constant, and so $\nabla v = \nabla w$ $\mu$-a.e. in $\Omega$. Thus, we can denote by $\nabla v_0$ the Cheeger derivative $\nabla v$ of any (equivalently, every) representative $v$ of $v_0$.

REMARK 3.6. Given that the exposition of this note is about the homogeneous classes, we emphasize the distinction between elements of the inhomogeneous spaces and their corresponding homogeneous spaces. Two functions in the space $N^{1,p}(\Omega)$ are the same function if, as functions on the metric space $\Omega$, they differ only on a subset of $\Omega$ of capacity zero - which is a more stringent requirement than differing only on a set of $\mu$-measure zero. Two functions in $HN^{1,p}(\Omega)$ are the same if representatives of these functions, as pointwise defined functions on $\Omega$, differ by only a constant outside a set of zero capacity.

In this setting, it is known that a $p$-harmonic function $u_f \in N^{1,p}(\Omega)$ exists for any choice of $f \in B^\theta_{p,p}(\partial\Omega)$. Solutions to the Neumann problem in this setting are known for boundary data in an appropriate integrable function class. In the present note, we extend this to include boundary data in the form of an arbitrary bounded linear functional $L$ on $HB^\theta_{p,p}(\partial\Omega)$.

DEFINITION 3.7. Given an operator $L \in \left(HB^\theta_{p,p}(\partial\Omega)\right)^*$, we say that a function $u_L \in HN^{1,p}(\Omega)$ is a *solution to the Neumann problem* with boundary data $L$ if

$$\frac{1}{p}\int_\Omega |\nabla u_L|^p \, d\mu - L(\mathrm{Tr}(u_L)) \le \frac{1}{p}\int_\Omega |\nabla v|^p \, d\mu - L(\mathrm{Tr}(v))$$

for all $v \in HN^{1,p}(\Omega)$.

This is equivalent to saying that $u_L$ is a minimizer for the functional

$$I(u) := \frac{1}{p}\int_\Omega |\nabla u|^p \, d\mu - L(\mathrm{Tr}(u))$$

over the class of $u \in HN^{1,p}(\Omega)$. The following lemma gives us a characterization of solutions to the Neumann problem in the form of the Euler-Lagrange equation associated with the functional $I$.

LEMMA 3.8. *Let $L \in \left(HB^\theta_{p,p}(\partial\Omega)\right)^*$. Then $u$ is a minimizer of $I$ if and only if*

$$\int_\Omega |\nabla u|^{p-2}\langle \nabla u, \nabla v\rangle \, d\mu - L(\mathrm{Tr}(v)) = 0 \tag{3.9}$$

*for every $v \in HN^{1,p}(\Omega)$.*

Note that $L \in \left(HB^\theta_{p,p}(\partial\Omega)\right)^*$ necessarily maps constant functions to 0. As such, this lemma implies, taking $v \in HN^{1,p}_0(\Omega) \subset HN^{1,p}(\Omega)$, that a solution to the Neumann problem is necessarily $p$-harmonic; here $HN^{1,p}_0(\Omega)$ is as described in Remark 3.4.

PROOF. For $v \in HN^{1,p}(\Omega)$ and $h \in \mathbb{R}$, consider

$$I(u+hv) = \frac{1}{p}\int_\Omega |\nabla(u+hv)|^p \, d\mu - L(\mathrm{Tr}(u+hv)).$$

If $u$ is a minimizer of $I$, then the map $h \mapsto I(u+hv)$ has a minimum at $h=0$, whence $\frac{d}{dh}I(u+hv)\big|_{h=0} = 0$. By linearity,

$$L(\mathrm{Tr}(u+(h+\eta)v)) - L(\mathrm{Tr}(u+hv)) = \eta L(\mathrm{Tr}(v))$$

for any $\eta \in \mathbb{R}$, and so

$$\frac{d}{dh}L(\mathrm{Tr}(u+hv)) = L(\mathrm{Tr}(v)).$$

We also calculate

$$\frac{d}{dh}\int_\Omega |\nabla(u+hv)|^p\,d\mu = p\int_\Omega |\nabla(u+hv)|^{p-2}\langle\nabla(u+hv),\nabla v\rangle\,d\mu,$$

and so, setting $h=0$, we obtain

$$\frac{d}{dh}I(u+hv)\bigg|_{h=0} = \int_\Omega |\nabla u|^{p-2}\langle\nabla u,\nabla v\rangle\,d\mu - L(\mathrm{Tr}(v)).$$

Conversely, assume that $u$ satisfies (3.9) for every $v\in HN^{1,p}(\Omega)$. The energy $I$ is convex on $HB^\theta_{p,p}(\partial\Omega)$ Since $1<p<\infty$; therefore, a function $u$ minimizes $I$ if and only if for each $v\in HB^\theta_{p,p}(\partial\Omega)$ we have that

$$\frac{d}{dh}I(u+hv)\bigg|_{h=0} = 0,$$

which in turn is equivalent to the validity of (3.9), completing the proof. □

LEMMA 3.10. *Given $L\in\left(HB^\theta_{p,p}(\partial\Omega)\right)^*$ and the functional $I$ on $HN^{1,p}(\Omega)$ given by*

$$I(v) := \frac{1}{p}\int_\Omega |\nabla v|^p\,d\mu - L(\mathrm{Tr}(v)),$$

*the quantity $\alpha := \inf\{I(v) : v\in HN^{1,p}(\Omega)\}$ is finite.*

Recall that $L\in\left(HB^\theta_{p,p}(\partial\Omega)\right)^*$ necessarily maps constant functions to 0.

PROOF. The class of constant functions, $v$, is in $HN^{1,p}(\Omega)$ and $\mathrm{Tr}(v)$ is the class of constant functions on $\partial\Omega$, so $L(v)=0$. Hence, $I(v)=0$, and so $\alpha\le 0$. Thus, it suffices to show that the set $\{I(v) : v\in HN^{1,p}(\Omega)\}$ is bounded below.

Let $v\in HN^{1,p}(\Omega)$. Then, using the boundedness of $L$ and Tr, as well as Theorem 2.15, we have that

$$\begin{aligned} I(v) &\ge \frac{1}{p}\,\|\nabla v\|^p_{L^p(\Omega)} - \|L\|\,\|\mathrm{Tr}(v)\|_{HB^\theta_{p,p}(\Omega)} \\ &\ge \frac{1}{p}\,\|\nabla v\|^p_{L^p(\Omega)} - \|L\|\,\|\mathrm{Tr}\,\|\|\nabla v\|_{L^p(\Omega)}. \end{aligned} \tag{3.11}$$

For $C>0$, the map $x\mapsto \frac{1}{p}x^p - Cx$ for $x>0$ is bounded below with a minimum of $\frac{1-p}{p}C^{\frac{p}{p-1}}$. As such, it follows that

$$I(v) \ge \frac{1-p}{p}\left(\|L\|\|\,\mathrm{Tr}\,\|\right)^{\frac{p}{p-1}}. \tag{3.12}$$

This lower bound is independent of $v$, and so $\alpha$ is bounded below. □

THEOREM 3.13. *Given $L\in\left(HB^\theta_{p,p}(\partial\Omega)\right)^*$, there exists a solution $u_L$ on $\Omega$ to the Neumann problem with boundary data $L$.*

If $L$ is the zero functional, then $I$ is minimized by $u_L=0$. In what follows, we assume that $\|L\|>0$.

PROOF. Recall the definition of $\alpha$ from Lemma 3.10 above. Assuming that $\|L\|>0$, we know that $\alpha\ne 0$.

Let $\{u_k\}\subset HN^{1,p}(\Omega)$ be a sequence such that $I(u_k)\to\alpha$ as $k\to\infty$ and $I(u_k)\le 2|\alpha|$ for all $k$. Since $u_k\in HN^{1,p}(\Omega)$, for each $k$ we choose a representative of $u_k$, also denoted by $u_k\in N^{1,p}(\Omega)$, such that $(u_k)_\Omega = \frac{1}{\mu(\Omega)}\int_\Omega u_k\,d\mu = 0$. From this and the fact that $\Omega$ is bounded, by the $(p,p)$–Poincaré inequality applied to the ball that is $\Omega$, we have that

$$\int_\Omega |u_k|^p\,d\mu = \int_\Omega |u_k - (u_k)_\Omega|^p\,d\mu \le C\int_\Omega |\nabla u_k|^p\,d\mu \tag{3.14}$$

where $C$ is a constant depending on $\Omega$, but not $u_k$.

From (3.11), it follows that for each $k$,

$$I(u_k) \ge \|\nabla u_k\|_{L^p(\Omega)}\left[\frac{1}{p}\|\nabla u_k\|^{p-1}_{L^p(\Omega)} - \|L\|\|\operatorname{Tr}\|\right].$$

From this it follows that either $\|\nabla u_k\|^{p-1}_{L^p(\Omega)} \le 2p\|L\|\|\operatorname{Tr}\|$ or else

$$2|\alpha| \ge I(u_k) \ge \|\nabla u_k\|_{L^p(\Omega)}\left[2\|L\|\|\operatorname{Tr}\| - \|L\|\|\operatorname{Tr}\|\right] = \|L\|\|\operatorname{Tr}\|\|\nabla u_k\|_{L^p(\Omega)}.$$

In either case, we have that the sequence $\{\|\nabla u_k\|_{L^p(\Omega)}\}$ is bounded with

$$\|\nabla u_k\|_{L^p(\Omega)} \le (2p\|L\|\|\operatorname{Tr}\|)^{\frac{1}{p-1}} + \frac{2|\alpha|}{\|L\|\|\operatorname{Tr}\|} \tag{3.15}$$

for each $k$. Hence, (3.14) implies that the sequence $\{u_k\}$ is bounded in $N^{1,p}(\Omega)$. Since this is a reflexive space, there is a function $u \in N^{1,p}(\Omega)$ to which a subsequence of $\{u_k\}$, which we will continue to denote by $\{u_k\}$, converges weakly. We will now show that $u$ is a minimizer for $I$, that $u$ is $p$-harmonic, and that $L = L_{\operatorname{Tr}(u)}$. Here again, we denote the equivalence class in $HN^{1,p}(\Omega)$ corresponding to $u \in N^{1,p}(\Omega)$ also by $u$.

Mazur's lemma implies that there is a sequence of finite convex combinations $\{\widetilde{u}_n\} \subset N^{1,p}(\Omega)$ given by

$$\widetilde{u}_n = \sum_{k=n}^{N(n)} \lambda_{n,k}u_k$$

for $n = 1, 2, \ldots$, such that $\{\widetilde{u}_n\}$ converges strongly to $u$ in $N^{1,p}(\Omega)$. Continuing our abuse of notation, we also denote the equivalence classes in $HN^{1,p}(\Omega)$ with representative $\widetilde{u}_n \in N^{1,p}(\Omega)$ also by $\widetilde{u}_n$. Then the strong convergence of $\widetilde{u}_n$ to $u$ in the norm of $N^{1,p}(\Omega)$ implies the strong convergence of the sequence of classes $\widetilde{u}_n$ to the equivalence class $u$ in the norm of $HN^{1,p}(\Omega)$. From this strong convergence and the boundedness of the linear operators $L$ and Tr on $HN^{1,p}(\Omega)$, it follows that $L(\operatorname{Tr}(\widetilde{u}_n)) \to L(\operatorname{Tr}(u))$ as $n \to \infty$; moreover, $\|\nabla\widetilde{u}_n\|_{L^p(\Omega)} \to \|\nabla u\|_{L^p(\Omega)}$ as $n \to \infty$ since

$$\left|\|\nabla\widetilde{u}_n\|_{L^p(\Omega)} - \|\nabla u\|_{L^p(\Omega)}\right| \le \|\nabla\widetilde{u}_n - \nabla u\|_{L^p(\Omega)} \le \|\widetilde{u}_n - u\|_{N^{1,p}(\Omega)}.$$

Thus,

$$\begin{aligned} I(\widetilde{u}_n) &= \frac{1}{p}\int_\Omega |\nabla\widetilde{u}_n|^p\,d\mu - L(\operatorname{Tr}(\widetilde{u}_n)) \\ &\to \frac{1}{p}\int_\Omega |\nabla u|^p\,d\mu - L(\operatorname{Tr}(u)) = I(u). \end{aligned} \tag{3.16}$$

We aim to show that $u$ is a minimizer of $I$ by showing that $I(\widetilde{u}_n) \to \alpha$ as $n \to \infty$. To that end, notice that for each $n$,

$$\begin{aligned}(pI(\widetilde{u}_n) + pL(\mathrm{Tr}(\widetilde{u}_n)))^{1/p} = \|\nabla \widetilde{u}_n\|_{L^p(\Omega)} &\le \sum_{k=n}^{N(n)} \lambda_{n,k}\|\nabla u_k\|_{L^p(\Omega)} \\ &\le \left(\sum_{k=n}^{N(n)} \lambda_{n,k}\|\nabla u_k\|^p_{L^p(\Omega)}\right)^{\frac{1}{p}} \left(\sum_{k=n}^{N(n)} \lambda_{n,k}\right)^{\frac{p}{p-1}} \\ &= \left(\sum_{k=n}^{N(n)} \lambda_{n,k}\|\nabla u_k\|^p_{L^p(\Omega)}\right)^{\frac{1}{p}}\end{aligned}$$

by Hölder's inequality. Similarly, $\|\nabla u_k\|_{L^p(\Omega)} = (pI(u_k) + p\,L(\mathrm{Tr}(u_k)))^{1/p}$ for each $k$, and so from the linearity of $L$, we have that

$$\begin{aligned}(I(\widetilde{u}_n) + L(\mathrm{Tr}(\widetilde{u}_n)))^{1/p} &\le \left(\sum_{k=n}^{N(n)} \lambda_{n,k}(I(u_k) + L(\mathrm{Tr}(u_k)))\right)^{\frac{1}{p}} \\ &= \left(\sum_{k=n}^{N(n)} \lambda_{n,k}I(u_k) + L(\mathrm{Tr}(\widetilde{u}_n))\right)^{\frac{1}{p}}.\end{aligned}$$

At this point, we use that $\{u_k\}$ was chosen to be a minimizing sequence. Fix $\varepsilon > 0$ and select $N$ such that $\alpha \le I(u_k) < \alpha + \varepsilon$ for all $k \ge N$. Then, for $n \ge N$, we have that

$$\begin{aligned}(I(\widetilde{u}_n) + L(\mathrm{Tr}(\widetilde{u}_n)))^{1/p} &< \left(\sum_{k=n}^{N(n)} \lambda_{n,k}(\alpha + \varepsilon) + L(\mathrm{Tr}(\widetilde{u}_n))\right)^{\frac{1}{p}} \\ &= (\alpha + \varepsilon + L(\mathrm{Tr}(\widetilde{u}_n)))^{\frac{1}{p}},\end{aligned}$$

and so $\alpha \le I(\widetilde{u}_n) < \alpha + \varepsilon$ for such $n$. That is, $I(\widetilde{u}_n) \to \alpha$ as $n \to \infty$. From (3.16), we now know that $I(u) = \alpha$, that is, $u$ is a minimizer of the energy functional $I$. Hence we can set $u_L = u$. □

## 4. Betwixt Dirichlet and Neumann

In this section, we construct the Dirichlet-to-Neumann and Neumann-to-Dirichlet maps, and show that they are bounded in terms of the norms of the extension and trace operators, respectively, whose existence is guaranteed by Theorem 2.15.

We begin this section by showing the existence of the Dirichlet-to-Neumann map in this setting.

THEOREM 4.1. *Given $f \in HB^\theta_{p,p}(\partial\Omega)$ with solution $u_f$ to the Dirichlet problem with boundary data $f$, there exists a bounded linear functional $L_f$ in $\left(HB^\theta_{p,p}(\partial\Omega)\right)^*$ given by*

$$L_f(g) = \int_\Omega |\nabla u_f|^{p-2}\,\langle \nabla u_f, \nabla Eg\rangle\, d\mu$$

*such that $u_f$ is a solution to the Neumann problem with boundary data $L_f$.*

Note that linearity of $L_f$ follows from the linearity of the extension operator, Cheeger gradient, inner product, and integral.

PROOF. We begin by verifying that the functional $L_f$ is well defined: that it is independent of the choice of extension operator $E$ and of representative of $g$. To this end, suppose that $v_1, v_2 \in N^{1,p}(\Omega)$ such that $\mathrm{Tr}(v_1) = \mathrm{Tr}(v_2) + c$ $\nu$-almost everywhere for some constant $c \in \mathbb{R}$, whence $v_1 - v_2 \in HN_0^{1,p}(\Omega)$. Then,

$$\int_\Omega |\nabla u_f|^{p-2} \langle \nabla u_f, \nabla(v_1 - v_2)\rangle \, d\mu = 0$$

since $u_f$ is $p$-harmonic in $\Omega$, see for instance Remark 3.4. Suppose that $g_1, g_2$ are two representatives of some $g \in HB^\theta_{p,p}(\partial\Omega)$. Then $g_1 - g_2$ is a constant up to a set of $\nu$ measure zero, and so $E(g_1 - g_2)$ is a constant up to a set of $\mu$ measure zero. Then,

$$\int_\Omega |\nabla u_f|^{p-2} \langle \nabla u_f, \nabla E(g_1 - g_2)\rangle \, d\mu = 0,$$

from where it follows that the definition of $L_f(g)$ is independent of the choice of representative of $g$, and of the extension $Eg_1$ of the representative function $g_1$ of $g$.

Now we show that $L_f$ is indeed bounded. For $g \in HB^\theta_{p,p}(\partial\Omega)$, we have that

$$\begin{aligned} |L_f(g)| &\le \int_\Omega |\nabla u_f|^{p-1} |\nabla Eg| \, d\mu \le \left(\int_\Omega |\nabla u_f|^p \, d\mu\right)^{\frac{p-1}{p}} \left(\int_\Omega |\nabla Eg|^p \, d\mu\right)^{\frac{1}{p}} \\ &\le \|E\| \|\nabla u_f\|^{p-1}_{L^p(\Omega)} \|g\|_{HB^\theta_{p,p}(\partial\Omega)}. \end{aligned}$$

We used the Cauchy-Schwarz inequality and Hölder's inequality, followed by the extension Theorem 2.15. From this calculation it follows that the operator norm of $L_f$ satisfies $\|L_f\| \le \|E\| \|\nabla u_f\|^{p-1}_{L^p(\Omega)}$. Since $u_f$ is $p$-harmonic, it minimizes the Dirichlet $p$-energy, and so,

$$\|\nabla u_f\|_{L^p(\Omega)} \le \|\nabla Ef\|_{L^p(\Omega)} \le \|E\| \|f\|_{HB^\theta_{p,p}(\partial\Omega)}.$$

It follows that

$$\|L_f\| \le \|E\|^p \|f\|^{p-1}_{HB^\theta_{p,p}(\partial\Omega)}. \tag{4.2}$$

For $v \in N^{1,p}(\Omega)$, its trace $\mathrm{Tr}(v)$ is in $HB^\theta_{p,p}(\partial\Omega)$ and so

$$L_f(\mathrm{Tr}(v)) = \int_\Omega |\nabla u_f|^{p-2} \langle \nabla u_f, \nabla E\,\mathrm{Tr}(v)\rangle \, d\mu.$$

Since $\mathrm{Tr} \circ E$ is the identity, we know that $\mathrm{Tr}(E\,\mathrm{Tr}(v)) = \mathrm{Tr}(v)$, and therefore $E\,\mathrm{Tr}(v) - v \in N_0^{1,p}(\Omega)$. Since $u_f$ is $p$-harmonic, it follows that

$$\int_\Omega |\nabla u_f|^{p-2} \langle \nabla u_f, \nabla(E\,\mathrm{Tr}(v) - v)\rangle \, d\mu = 0,$$

and so

$$L_f(\mathrm{Tr}(v)) = \int_\Omega |\nabla u_f|^{p-2} \langle \nabla u_f, \nabla v\rangle \, d\mu.$$

This is, $u_f$ is a solution to the Neumann problem with boundary data $L_f$. □

COROLLARY 4.3. *The Dirichlet-to-Neumann map*

$$\mathrm{DtN} : HB^{\theta}_{p,p}(\partial\Omega) \to \left(HB^{\theta}_{p,p}(\partial\Omega)\right)^*$$

*given by* $\mathrm{DtN}(f) = L_f$ *is well-defined and bounded, with*

$$\|\,\mathrm{DtN}\,\| := \sup_{\|f\|_{HB^{\theta}_{p,p}(\partial\Omega)}=1} \|L_f\| \leq \|E\|^p.$$

PROOF. From Theorem 4.1, it follows that DtN is well defined and takes Dirichlet data to Neumann data with $u_f$ being also the solution to the Neumann problem with Neumann data $\mathrm{DtN}(f)$. Furthermore, from (4.2) in the proof of Theorem 4.1 it follows that DtN is bounded with

$$\sup_{\|f\|_{HB^{\theta}_{p,p}(\partial\Omega)}=1} \|L_f\| \leq \|E\|^p.$$

□

We now proceed by constructing the Neumann-to-Dirichlet map.

THEOREM 4.4. *Given* $L \in \left(HB^{\theta}_{p,p}(\partial\Omega)\right)^*$, *let* $u_L$ *be the solution to the Neumann problem with boundary data* $L$. *Then* $u_L$ *is the solution to the Dirichlet problem with boundary data* $\mathrm{Tr}(u_L)$. *Moreover, we have that* $L = L_{\mathrm{Tr}(u_L)}$.

PROOF. As $u_L$ is the solution to a Neumann problem, it is $p$-harmonic, and its trace is equal to $\mathrm{Tr}(u_L)$, and so $u_L$ is the solution to the Dirichlet problem with boundary data $\mathrm{Tr}(u_L)$.

For $v \in HN^{1,p}(\Omega)$, its trace $\mathrm{Tr}(v)$ is in $HB^{\theta}_{p,p}(\partial\Omega)$ and so by Lemma 3.8,

$$L(\mathrm{Tr}(v)) = \int_{\Omega} |\nabla u_L|^{p-2} \langle \nabla u_L, \nabla v \rangle \, d\mu.$$

Since $\mathrm{Tr} \circ E$ is the identity, we know that $\mathrm{Tr}(E \circ \mathrm{Tr}(v)) = \mathrm{Tr}(v)$, and therefore $E\,\mathrm{Tr}(v) - v \in HN^{1,p}_0(\Omega)$. Since $u_L$ is $p$-harmonic, it follows that

$$\int_{\Omega} |\nabla u_L|^{p-2} \langle \nabla u_L, \nabla (E \circ \mathrm{Tr}(v) - v) \rangle \, d\mu = 0,$$

and so

$$\begin{aligned} L(\mathrm{Tr}(v)) &= \int_{\Omega} |\nabla u_L|^{p-2} \langle \nabla u_L, \nabla v \rangle \, d\mu \\ &= \int_{\Omega} |\nabla u_L|^{p-2} \langle \nabla u_L, \nabla E(\mathrm{Tr}(v)) \rangle \, d\mu = L_{\mathrm{Tr}(u_L)}(\mathrm{Tr}(v)). \end{aligned}$$

The result follows from the fact that the trace operator is surjective. □

COROLLARY 4.5. *The Neumann-to-Dirichlet map*

$$\mathrm{NtD} : \left(HB^{\theta}_{p,p}(\partial\Omega)\right)^* \to HB^{\theta}_{p,p}(\partial\Omega)$$

*given by* $\mathrm{NtD}(L) := \mathrm{Tr}(u_L)$ *is well-defined and bounded, with*

$$\|\,\mathrm{NtD}\,\| := \sup_{\|L\|=1} \|\,\mathrm{Tr}(u_L)\|_{HB^{\theta}_{p,p}(\partial\Omega)} \leq \left[(2p)^{\frac{1}{p-1}} + \frac{2(p-1)}{p}\right] \|\,\mathrm{Tr}\,\|^{\frac{p}{p-1}}.$$

PROOF. From Theorem 4.4, it follows that NtD is well defined and takes Neumann data to Dirichlet data with solution being equal to the solution of the Neumann problem. Furthermore, by (3.12) and then taking the infimum over all $v \in HN^{1,p}(\Omega)$, we know that

$$0 \geq \alpha \geq \frac{1-p}{p} \left( \|L\| \, \| \operatorname{Tr} \| \right)^{\frac{p}{p-1}}.$$

So, by (3.15), when $L \in (HB^{\theta}_{p,p}(\partial\Omega))^*$ and $u_L$ is the corresponding minimizer constructed in Theorem 3.13, we have

$$\begin{aligned} \|\nabla u_L\|_{L^p(\Omega)} &\leq \left(2p \, \|L\| \, \| \operatorname{Tr} \|\right)^{\frac{1}{p-1}} + \frac{p-1}{p} \frac{2}{\|L\| \, \| \operatorname{Tr} \|} \left( \|L\| \, \| \operatorname{Tr} \| \right)^{\frac{p}{p-1}} \\ &= \left[ (2p)^{\frac{1}{p-1}} + \frac{2(p-1)}{p} \right] \left( \|L\| \, \| \operatorname{Tr} \| \right)^{\frac{1}{p-1}}. \end{aligned} \tag{4.6}$$

Technically, the inequality (3.15) is known to hold for the minimizing sequence $\nabla u_k$, but as this inequality is respected by convex combinations of $\nabla u_k$, and the corresponding convex combination converges strongly to $\nabla u_L$, this inequality holds also for $\nabla u_L$.

The above inequality, along with Theorem 2.15, implies that NtD is bounded with

$$\begin{aligned} \sup_{\|L\|=1} \| \operatorname{Tr}(u_L) \|_{HB^{\theta}_{p,p}(\partial\Omega)} &\leq \sup_{\|L\|=1} \| \operatorname{Tr} \| \|\nabla u_L\|_{L^p(\Omega)} \\ &\leq \left[ (2p)^{\frac{1}{p-1}} + \frac{2(p-1)}{p} \right] \| \operatorname{Tr} \|^{\frac{p}{p-1}}. \end{aligned}$$

□

THEOREM 4.7. *The operators* DtN *and* NtD *are inverses of each other. Moreover,*

$$c_p \, \| \operatorname{Tr} \|^{\frac{p}{1-p}} \leq \|DtN\| \leq \|E\|^p,$$

*where*

$$c_p := \left[ (2p)^{\frac{1}{p-1}} + \frac{2(p-1)}{p} \right]^{-1}.$$

PROOF. For every $L \in \left(HB^{\theta}_{p,p}(\partial\Omega)\right)^*$, we have

$$\operatorname{DtN} \circ \operatorname{NtD}(L) = \operatorname{DtN}(\operatorname{Tr}(u_L)) = L_{\operatorname{Tr}(u_L)} = L$$

by Theorem 4.4.

For every $f \in HB^{\theta}_{p,p}(\partial\Omega)$, we have

$$\operatorname{NtD} \circ \operatorname{DtN}(f) = \operatorname{NtD}(L_f) = \operatorname{Tr}(u_{L_f}),$$

where $u_{L_f}$ be the solution to the Neumann problem with the boundary data $L_f$. In order to show that $\operatorname{Tr}(u_{L_f}) = f$, it suffices to show that $\operatorname{Tr}(u_f) = \operatorname{Tr}(u_{L_f})$ in $HB^{\theta}_{p,p}(\partial\Omega)$, where $u_f \in HN^{1,p}(\Omega)$ is the solution to the Dirichlet problem with boundary data $f$.

For any $v \in HN^{1,p}(\Omega)$, since $u_f$ is $p$-harmonic and $\operatorname{Tr}(E(\operatorname{Tr}(v))) = \operatorname{Tr}(v)$, we have that

$$L_f(\operatorname{Tr}(v)) = \int_{\Omega} |\nabla u_f|^{p-2} \langle \nabla u_f, \nabla E \circ \operatorname{Tr}(v) \rangle \, d\mu = \int_{\Omega} |\nabla u_f|^{p-2} \langle \nabla u_f, \nabla v \rangle \, d\mu.$$

On the other hand, since $u_{L_f}$ is a solution to the Neumann problem with boundary data $L_f$, we have

$$L_f(\mathrm{Tr}(v)) = \int_\Omega |\nabla u_{L_f}|^{p-2} \langle \nabla u_{L_f}, \nabla v\rangle \, d\mu.$$

Hence, setting $v = u_{L_f} - u_f$ and using linearity of the Cheeger gradient,

$$\int_\Omega |\nabla u_{L_f}|^{p-2} \langle \nabla u_{L_f}, \nabla u_{L_f} - \nabla u_f\rangle \, d\mu = \int_\Omega |\nabla u_f|^{p-2} \langle \nabla u_f, \nabla u_{L_f} - \nabla u_f\rangle \, d\mu.$$

From here, it follows that

$$\int_\Omega \big\langle |\nabla u_{L_f}|^{p-2}\nabla u_{L_f} - |\nabla u_f|^{p-2}\nabla u_f, \nabla u_{L_f} - \nabla u_f\big\rangle \, d\mu = 0,$$

and so $\nabla u_{L_f} - \nabla u_f = 0$ $\mu$-a.e. in $\Omega$, that is, $u_{L_f} - u_f \in HN_0^{1,p}(\Omega)$. Therefore, $\mathrm{Tr}\, u_{L_f} = \mathrm{Tr}\, u_f$ as members of the space $HB^\theta_{p,p}(\partial\Omega)$. The claim about the norm of the Dirichlet-to-Neumann transform operator DtN follows now from combining Corollary 4.3 with Corollary 4.5. $\square$

## References


[1] W. Arendt and A. F. M. ter Elst, *The Dirichlet-to-Neumann operator on rough domains*, J. Differential Equations **251** (2011), no. 8, 2100–2124, DOI 10.1016/j.jde.2011.06.017. MR2823661

[2] J. Björn and N. Shanmugalingam, *Poincaré inequalities, uniform domains and extension properties for Newton-Sobolev functions in metric spaces*, J. Math. Anal. Appl. **332** (2007), no. 1, 190–208, DOI 10.1016/j.jmaa.2006.09.064. MR2319654

[3] A. Björn, J. Björn, and N. Shanmugalingam, *Extension and trace results for doubling metric measure spaces and their hyperbolic fillings* (English, with English and French summaries), J. Math. Pures Appl. (9) **159** (2022), 196–249, DOI 10.1016/j.matpur.2021.12.003. MR4377995

[4] M. Bonk, J. Heinonen, and P. Koskela, *Uniformizing Gromov hyperbolic spaces*, Astérisque **270** (2001), viii+99. MR1829896

[5] A. Butaev and G. Dafni, *Locally uniform domains and extension of bmo functions* (English, with English and Finnish summaries), Ann. Fenn. Math. **48** (2023), no. 2, 567–594. MR4628611

[6] L. Capogna, J. Kline, R. Korte, N. Shanmugalingam, and M. Snipes, *Neumann problems for p-harmonic functions, and induced nonlocal operators in metric measure spaces*, preprint, (2022) arXiv:2204.00571, to appear in Amer. J. Math.

[7] F. Cardoso and R. Mendoza, *On the hyperbolic Dirichlet to Neumann functional*, Comm. Partial Differential Equations **21** (1996), no. 7-8, 1235–1252, DOI 10.1080/03605309608821224. MR1399197

[8] J. Cheeger, *Differentiability of Lipschitz functions on metric measure spaces*, Geom. Funct. Anal. **9** (1999), no. 3, 428–517, DOI 10.1007/s000390050094. MR1708448

[9] B. Colbois, A. Girouard, C. Gordon, and D. Sher, *Some recent developments on the Steklov eigenvalue problem*, Rev. Mat. Complut. **37** (2024), no. 1, 1–161, DOI 10.1007/s13163-023-00480-3. MR4695859

[10] J. Cheeger, *Differentiability of Lipschitz functions on metric measure spaces*, Geom. Funct. Anal. **9** (1999), no. 3, 428–517, DOI 10.1007/s000390050094. MR1708448

[11] R. R. Coifman and G. Weiss, *Analyse harmonique non-commutative sur certains espaces homogènes* (French), Lecture Notes in Mathematics, Vol. 242, Springer-Verlag, Berlin-New York, 1971. Étude de certaines intégrales singulières. MR499948

[12] E. Durand-Cartagena, J. A. Jaramillo, and N. Shanmugalingam, *Geometric characterizations of p-Poincaré inequalities in the metric setting*, Publ. Mat. **60** (2016), no. 1, 81–111. MR3447735

[13] A. Feizmohammadi, K. Krupchyk, L. Oksanen, and G. Uhlmann, *Reconstruction in the Calderón problem on conformally transversally anisotropic manifolds*, J. Funct. Anal. **281** (2021), no. 9, Paper No. 109191, 25, DOI 10.1016/j.jfa.2021.109191. MR4291510

[14] B. Franchi, P. Hajłasz, and P. Koskela, *Definitions of Sobolev classes on metric spaces*, Ann. Inst. Fourier (Grenoble) **49** (1999), no. 6, 1903–1924, DOI 10.5802/aif.1742. MR1738070

[15] F. W. Gehring and B. G. Osgood, *Uniform domains and the quasihyperbolic metric*, J. Analyse Math. **36** (1979), 50–74 (1980), DOI 10.1007/BF02798768. MR581801

[16] R. Gibara, R. Korte, and N. Shanmugalingam, *Solving a Dirichlet problem on unbounded domains via a conformal transformation*, Math. Ann. **389** (2024), no. 3, 2857–2901, DOI 10.1007/s00208-023-02705-8. MR4753076

[17] R. Gibara and N. Shanmugalingam, *Trace and extension theorems for homogeneous Sobolev and Besov spaces for unbounded uniform domains in metric measure spaces* (Russian, with Russian summary), Tr. Mat. Inst. Steklova **323** (2023), no. Teoriya Funktsiĭ Mnogikh Deĭstvitel'nykh Peremennykh i ee Prilozheniya, 107–126, DOI 10.4213/tm4362; English transl., Proc. Steklov Inst. Math. **323** (2023), no. 1, 101–119. MR4716517

[18] A. Girouard, M. Karpukhin, M. Levitin, and I. Polterovich, *The Dirichlet-to-Neumann map, the boundary Laplacian, and Hörmander's rediscovered manuscript*, J. Spectr. Theory **12** (2022), no. 1, 195–225, DOI 10.4171/jst/399. MR4404812

[19] A. Gogatishvili, P. Koskela, and N. Shanmugalingam, *Interpolation properties of Besov spaces defined on metric spaces*, Math. Nachr. **283** (2010), no. 2, 215–231, DOI 10.1002/mana.200810242. MR2604119

[20] P. Hajłasz, *Sobolev spaces on metric-measure spaces*, Heat kernels and analysis on manifolds, graphs, and metric spaces (Paris, 2002), Contemp. Math., vol. 338, Amer. Math. Soc., Providence, RI, 2003, pp. 173–218, DOI 10.1090/conm/338/06074. MR2039955

[21] P. Hajłasz and P. Koskela, *Sobolev met Poincaré*, Mem. Amer. Math. Soc. **145** (2000), no. 688, x+101, DOI 10.1090/memo/0688. MR1683160

[22] D. Hauer, *The p-Dirichlet-to-Neumann operator with applications to elliptic and parabolic problems*, J. Differential Equations **259** (2015), no. 8, 3615–3655, DOI 10.1016/j.jde.2015.04.030. MR3369257

[23] J. Heinonen, *Lectures on analysis on metric spaces*, Universitext, Springer-Verlag, New York, 2001, DOI 10.1007/978-1-4613-0131-8. MR1800917

[24] J. Heinonen and P. Koskela, *Quasiconformal maps in metric spaces with controlled geometry*, Acta Math. **181** (1998), no. 1, 1–61, DOI 10.1007/BF02392747. MR1654771

[25] J. Heinonen, P. Koskela, N. Shanmugalingam, and J. T. Tyson, *Sobolev spaces on metric measure spaces*, New Mathematical Monographs, vol. 27, Cambridge University Press, Cambridge, 2015. An approach based on upper gradients, DOI 10.1017/CBO9781316135914. MR3363168

[26] D. A. Herron and P. Koskela, *Uniform and Sobolev extension domains*, Proc. Amer. Math. Soc. **114** (1992), no. 2, 483–489, DOI 10.2307/2159672. MR1075947

[27] V. Isakov, *Inverse problems for partial differential equations*, Applied Mathematical Sciences, vol. 127, Springer-Verlag, New York, 1998, DOI 10.1007/978-1-4899-0030-2. MR1482521

[28] P. W. Jones, *Extension theorems for BMO*, Indiana Univ. Math. J. **29** (1980), no. 1, 41–66, DOI 10.1512/iumj.1980.29.29005. MR554817

[29] P. W. Jones, *Quasiconformal mappings and extendability of functions in Sobolev spaces*, Acta Math. **147** (1981), no. 1-2, 71–88, DOI 10.1007/BF02392869. MR631089

[30] C. E. Kenig and T. Toro, *Free boundary regularity for harmonic measures and Poisson kernels*, Ann. of Math. (2) **150** (1999), no. 2, 369–454, DOI 10.2307/121086. MR1726699

[31] J. Kinnunen and N. Shanmugalingam, *Regularity of quasi-minimizers on metric spaces*, Manuscripta Math. **105** (2001), no. 3, 401–423, DOI 10.1007/s002290100193. MR1856619

[32] P. Koskela and P. MacManus, *Quasiconformal mappings and Sobolev spaces*, Studia Math. **131** (1998), no. 1, 1–17. MR1628655

[33] Y. Li, S. Ponnusamy, and Q. Zhou, *Sphericalization and flattening preserve uniform domains in nonlocally compact metric spaces*, J. Aust. Math. Soc. **112** (2022), no. 1, 68–89, DOI 10.1017/S1446788719000582. MR4362399

[34] L. Malý, *Trace and extension theorems for Sobolev-type functions in metric spaces*, preprint (2017), `arXiv:1704.06344`

[35] O. Martio and J. Sarvas, *Injectivity theorems in plane and space*, Ann. Acad. Sci. Fenn. Ser. A I Math. **4** (1979), no. 2, 383–401, DOI 10.5186/aasfm.1978-79.0413. MR565886

[36] N. Shanmugalingam, *Newtonian spaces: an extension of Sobolev spaces to metric measure spaces*, Rev. Mat. Iberoamericana **16** (2000), no. 2, 243–279, DOI 10.4171/RMI/275. MR1809341

[37] N. Shanmugalingam, *Harmonic functions on metric spaces*, Illinois J. Math. **45** (2001), no. 3, 1021–1050. MR1879250

[38] M. E. Taylor, *Partial differential equations. II*, Applied Mathematical Sciences, vol. 116, Springer-Verlag, New York, 1996. Qualitative studies of linear equations, DOI 10.1007/978-1-4757-4187-2. MR1395149

[39] G. Uhlmann, *Inverse problems: seeing the unseen*, Bull. Math. Sci. **4** (2014), no. 2, 209–279, DOI 10.1007/s13373-014-0051-9. MR3228575

[40] Q. Zhou and S. Ponnusamy, *Gromov hyperbolicity in the free quasiworld. I*, Studia Math. **268** (2023), no. 1, 23–49, DOI 10.4064/sm210825-7-3. MR4504439

[41] Q. Zhou, Y. He, S. Ponnusamy, and Q. Luo, *Egg-yolk principle for uniformizing Gromov hyperbolic domains*, Bull. Sci. Math. **188** (2023), Paper No. 103333, 23, DOI 10.1016/j.bulsci.2023.103333. MR4642172

Department of Mathematical Sciences, P.O. Box 210025, University of Cincinnati, Cincinnati, OH 45221–0025, U.S.A.
*Email address*: gibararn@ucmail.uc.edu

Department of Mathematical Sciences, P.O. Box 210025, University of Cincinnati, Cincinnati, OH 45221–0025, U.S.A.
*Email address*: shanmun@uc.edu

Contemporary Mathematics
Volume **820**, 2025
https://doi.org/10.1090/conm/820/16420

# Becker–Pommerenke type inequalities and domains of univalence

O. S. Kudryavtseva and A. P. Solodov

Abstract. We present the estimate of common value of holomorphic self-map of a unit disc at two distinct points in terms of the angular derivative at repulsive boundary fixed point and depending on the localization of the attractive fixed point. The result refines Becker–Pommerenke estimate in the case of one fixed point and allows to solve the problem of finding domains of univalence for functions with two fixed points.

## 1. Introdaction

Let $\mathscr{B}$ be the set of holomorphic maps of the unit disc $\mathbb{D} = \{z \in \mathbb{D}\colon |z| < 1\}$ into itself. It is well known that fixed points of a function $f \in \mathscr{B}$ have a decisive influence on its geometric and analytic properties. We recall that a point $\zeta \in \overline{\mathbb{D}}$ is a fixed point of $f \in \mathscr{B}$ if $\lim_{r\to 1^-} f(r\zeta) = \zeta$. In particular, if $\zeta \in \partial\mathbb{D}$, then the above definition is equivalent to the existence of the angular limit $\angle\lim_{z\to\zeta} f(z) = \zeta$. In case of boundary fixed point, there is always a finite or infinite angular limit $\angle\lim_{z\to\zeta}(f(z) - \zeta)/(z - \zeta)$. This limit is called the angular derivative of function $f$ at point $\zeta$ and is denoted by $f'(\zeta)$. The finite value of the angular derivative is positive. The set of all fixed points of $f \in \mathscr{B}$ is not empty. Moreover, if $f$ differs from a linear fractional transformation, then there is a unique fixed point that has the property of attraction (see [**1**]). In the literature this point is called a Denjoy–Wolff point of $f$. A Denjoy–Wolff point $q$ can be located both inside the unit disc $\mathbb{D}$ (then $|f'(q)| \leqslant 1$) and on its boundary $\partial\mathbb{D}$ (then $0 < f'(q) \leqslant 1$). The remaining fixed points, if any, are located on the boundary $\partial\mathbb{D}$ and are repulsive.

We denote by $\mathscr{B}\{1\}$ the class of functions $f$ in $\mathscr{B}$ with the boundary fixed point $\zeta = 1$:

$$\mathscr{B}\{1\} = \bigl\{f \in \mathscr{B}\colon \angle\lim_{z\to 1} f(z) = 1\bigr\}.$$

On the class $\mathscr{B}\{1\}$ the following inequality holds (see [**2**]).

Theorem 1.1 (Becker and Pommerenke). *Let $f \in \mathscr{B}\{1\}$, and let $a, b \in \mathbb{D}$ be distinct points such that $f(a) = f(b) = c$. Then*

$$f'(1)\,\frac{1 - |c|^2}{|1 - c|^2} \geqslant \frac{1 - |a|^2}{|1 - a|^2} + \frac{1 - |b|^2}{|1 - b|^2}. \tag{1.1}$$

---

2020 *Mathematics Subject Classification.* Primary 30C55; Secondary 30C75.

The work of the second author was supported by the Theoretical Physics and Mathematics Advancement Foundation “BASIS”.

We refine Theorem 1.1 in the case of two fixed points, one of which is attractive.

THEOREM 1.2. *Let $f \in \mathscr{B}\{1\}$, and let $q$, $q \neq 1$, be a Denjoy–Wolff point of $f$ and $a, b \in \mathbb{D}$ be distinct points such that $f(a) = f(b) = c$.*

(i) *If $q = 0$, then*

$$f'(1)\frac{1-|c|^2}{|1-c|^2} \geqslant \frac{1-|a|^2}{|1-a|^2} + \frac{1-|b|^2}{|1-b|^2} + \left|1 + \frac{c}{ab}\frac{1-\bar{c}}{1-c}\frac{1-a}{1-\bar{a}}\frac{1-b}{1-\bar{b}}\right|^2 \left(1 - \frac{|c|^2}{|a|^2|b|^2}\right)^{-1}. \tag{1.2}$$

*(For $z = 1$ we set $|1-z|^2/(1-|z|^2) = 0$).*

(ii) *If $q = -1$, then*

$$4\left(f'(1)\frac{1-|c|^2}{|1-c|^2} - \frac{1-|a|^2}{|1-a|^2} - \frac{1-|b|^2}{|1-b|^2}\right)\left(\frac{1-|c|^2}{|1+c|^2} - \frac{1-|a|^2}{|1+a|^2} - \frac{1-|b|^2}{|1+b|^2}\right) \geqslant \left|1 + \frac{1-\bar{c}}{1-c}\frac{1+c}{1+\bar{c}}\frac{1-a}{1-\bar{a}}\frac{1+\bar{a}}{1+a}\frac{1-b}{1-\bar{b}}\frac{1+\bar{b}}{1+b}\right|^2. \tag{1.3}$$

## 2. Proof of Becker–Pommerenke type inequalities

LEMMA 2.1. *Let $f \in \mathscr{B}$ and $\angle\lim_{z\to\pm1} f(z) = \pm1$. Then $f'(-1)f'(1) \geqslant 1$.*

PROOF. If $f$ is a linear fractional transformation of the disc $\mathbb{D}$ into itself, i. e.

$$f(z) = \frac{z-u}{1-uz}, \quad u \in (-1,1),$$

then it is easy to check the equality $f'(-1)f'(1) = 1$. If $f$ differs from a linear fractional transformation, then one of the fixed points $\zeta = \pm1$ is a repulsive. For certainty, we will assume that $f'(1) > 1$. Consider the function

$$g(z) = \frac{f'(1)(1-z)f(z) - z(1-f(z))}{f'(1)(1-z) - (1-f(z))}$$

and show that $g$ belongs to the class $\mathscr{B}$. We estimate the real part of the composition $h \circ g(z)$, where $h(z) = (1+z)/(1-z)$ is a linear fractional transformation of the unit disc $\mathbb{D}$ into the right half-plane $\mathbb{H} = \{w \in \mathbb{C}\colon \operatorname{Re} w > 0\}$,

$$\begin{aligned}\operatorname{Re}(h\circ g(z)) = \operatorname{Re}\frac{1+g(z)}{1-g(z)} &= \frac{1}{f'(1)-1}\operatorname{Re}\left(f'(1)\frac{1+f(z)}{1-f(z)} - \frac{1+z}{1-z}\right)\\ &= \frac{1}{f'(1)-1}\left(f'(1)\frac{1-|f(z)|^2}{|1-f(z)|^2} - \frac{1-|z|^2}{|1-z|^2}\right).\end{aligned}$$

By the Julia–Carathéodory theorem (see [**3**]), the second factor is positive. Thus, we obtain $\operatorname{Re}(h\circ g(z)) > 0$ for all $z \in \mathbb{D}$. It means that $g \in \mathscr{B}$. Moreover, it is easy to see that the function $g$ preserves the property of the point $\zeta = -1$ to be fixed and

$$g'(-1) = \frac{f'(-1)f'(1)-1}{f'(1)-1}.$$

Hence, $g'(-1) > 0$ and $f'(1) > 1$ imply $f'(-1)f'(1) > 1$. □

PROOF THEOREM 1.2. Consider the function

$$g(z) = \frac{1-\bar{c}}{1-c}\frac{f(z)-c}{1-\bar{c}f(z)}.$$

It is clear that $g$ belongs to the class $\mathscr{B}\{1\}$ and $g(a) = g(b) = 0$. By the Schwarz–Pick lemma (see [**4**]), $g$ admits a representation

$$g(z) = \frac{1-\bar{a}}{1-a}\frac{z-a}{1-\bar{a}z}\frac{1-\bar{b}}{1-b}\frac{z-b}{1-\bar{b}z}h(z),$$

where either $h \in \mathscr{B}\{1\}$ or $h(z) \equiv 1$. If $h(z) \equiv 1$, then direct calculations show that equality holds in both (1.2) and (1.3).

Now let's consider a nontrivial case. If $h \in \mathscr{B}\{1\}$, then

$$h(z) = \frac{1-a}{1-\bar{a}}\frac{1-\bar{a}z}{z-a}\frac{1-b}{1-\bar{b}}\frac{1-\bar{b}z}{z-b}\frac{1-\bar{c}}{1-c}\frac{f(z)-c}{1-\bar{c}f(z)}$$

and its angular derivative at $z = 1$ has the following form:

$$h'(1) = f'(1)\frac{1-|c|^2}{|1-c|^2} - \frac{1-|a|^2}{|1-a|^2} - \frac{1-|b|^2}{|1-b|^2}. \tag{2.1}$$

(i) To prove the inequality (1.2), we use the Julia–Carathéodory theorem:

$$h'(1) \geqslant \frac{|1-h(0)|^2}{1-|h(0)|^2}. \tag{2.2}$$

It is easy to calculate that

$$h(0) = -\frac{c}{ab}\frac{1-\bar{c}}{1-c}\frac{1-a}{1-\bar{a}}\frac{1-b}{1-\bar{b}}. \tag{2.3}$$

Combining (2.1)–(2.3) yields the inequality (1.2).

(ii) It remains to prove the inequality (1.3). Note that

$$\begin{aligned} h(-1) &= -\varkappa, \\ h'(-1) &= \varkappa\left(\frac{1-|c|^2}{|1+c|^2} - \frac{1-|a|^2}{|1+a|^2} - \frac{1-|b|^2}{|1+b|^2}\right), \end{aligned}$$

where

$$\varkappa = \frac{1-\bar{c}}{1-c}\frac{1+c}{1+\bar{c}}\frac{1-a}{1-\bar{a}}\frac{1+\bar{a}}{1+a}\frac{1-b}{1-\bar{b}}\frac{1+\bar{b}}{1+b}.$$

Consider the linear fractional transformation

$$S(z) = \frac{(3+\varkappa)z + \varkappa - 1}{(1-\varkappa)z + 3\varkappa + 1}.$$

It takes the unit disc $\mathbb{D}$ into itself and satisfies the conditions $S(1) = 1$, $S(-1) = -\bar{\varkappa}$. Then the composition $u(z) = S \circ h(z)$ belongs to the class $\mathscr{B}$ and preserves the property of the points $z = \pm 1$ to be fixed. By Lemma 2.1, we obtain $u'(-1)u'(1) \geqslant 1$. Since

$$\begin{aligned} u'(1) &= f'(1)\frac{1-|c|^2}{|1-c|^2} - \frac{1-|a|^2}{|1-a|^2} - \frac{1-|b|^2}{|1-b|^2}, \\ u'(-1) &= \frac{4}{|1+\varkappa|^2}\left(\frac{1-|c|^2}{|1+c|^2} - \frac{1-|a|^2}{|1+a|^2} - \frac{1-|b|^2}{|1+b|^2}\right), \end{aligned}$$

the last inequality is equivalent to (1.3).

□

## 3. Applications

Becker and Pommerenke used inequality (1.1) to find the domain of univalence of $f$ in $\mathscr{B}\{1\}$.

THEOREM 3.1 (Becker and Pommerenke). *Let $f \in \mathscr{B}\{1\}$ and let $f'(1) < \infty$. Then $f$ is univalent in the domain*

$$\mathscr{P} = \left\{ z \in \mathbb{D}\colon f'(1)\, \frac{1-|f(z)|^2}{|1-f(z)|^2}\, \frac{|1-z|^2}{1-|z|^2} < 2 \right\}. \tag{3.1}$$

PROOF. Note that the domain $\mathscr{P}$ is not empty because by the Julia–Carathéodory theorem the left-hand side of (3.1) is always greater than 1 and certainly less than 2 in a Stolz angle of sufficiently small radius, with opening arbitrarily close to $\pi$. If $f$ is not univalent in $\mathscr{P}$, then there exist distinct points $a, b \in \mathscr{P}$ such that $f(a) = f(b) = c$. Since the inequality in (3.1) holds at both $a$ and $b$, namely,

$$f'(1)\, \frac{1-|c|^2}{|1-c|^2} < 2\, \frac{1-|a|^2}{|1-a|^2} \quad \text{and} \quad f'(1)\, \frac{1-|c|^2}{|1-c|^2} < 2\, \frac{1-|b|^2}{|1-b|^2},$$

we see that

$$f'(1)\, \frac{1-|c|^2}{|1-c|^2} < \frac{1-|a|^2}{|1-a|^2} + \frac{1-|b|^2}{|1-b|^2}.$$

This contradicts (1.1). □

Consider the problem of domain of univalence on class of functions. For any $\alpha > 1$ in $\mathscr{B}\{1\}$ we single out the subclass $\mathscr{B}_\alpha\{1\}$ of functions satisfying a constraint on the value of the angular derivative at the fixed point:

$$\mathscr{B}_\alpha\{1\} = \{f \in \mathscr{B}\{1\}\colon f'(1) = \alpha\}.$$

It was shown in [**5**] that for any $\alpha > 1$ there are no non-empty domains of univalence in $\mathscr{B}_\alpha\{1\}$. In other words, the class $\mathscr{B}_\alpha\{1\}$ is too wide and we need to consider smaller classes. The most natural restriction of $\mathscr{B}_\alpha\{1\}$ is its subclass consisting of functions having, in addition to the repulsive fixed point $\zeta = 1$, an attractive fixed point. Such point necessarily exists (it is a Denjoy–Wolff point) and the additional condition only fixes its location (inside $\mathbb{D}$ or at its boundary). So, let's represent the class $\mathscr{B}_\alpha\{1\}$ as a union of disjoint subclasses distinguished by the condition of the location of attractive fixed point $q$:

$$\mathscr{B}_\alpha\{1\} = \bigcup_{q \in \overline{\mathbb{D}}} \mathscr{B}_\alpha[q, 1],$$

where $\mathscr{B}_\alpha[q, 1] = \mathscr{B}_\alpha\{1\} \cap \mathscr{B}[q]$. Here $\mathscr{B}[q]$, $q \in \overline{\mathbb{D}}$, denotes the class of functions $f$ in $\mathscr{B}$ with a Denjoy–Wolff point $q$.

The existence of non-empty domains of univalence on the indicated subclasses for some values of $\alpha$ was established in [**6**]. Further, for simplicity of formulations, we will assume $q = 0$ (the case of the interior Denjoy–Wolff point) and $q = -1$ (the case of the boundary Denjoy–Wolff point).

Theorem 1.2 allows to find a sharp domain of univalence on both class $\mathscr{B}_\alpha[0, 1]$ and class $\mathscr{B}_\alpha[-1, 1]$.

THEOREM 3.2. *Let $f \in \mathscr{B}_\alpha[0,1]$, $\alpha \in (1,4)$. Then $f$ is univalence in the domain*

$$\mathscr{D}(\alpha) = \left\{ z \in \mathbb{D} \colon \frac{\left|1 - 2z + |z|^2\right|}{1 - |z|^2} < \frac{1}{\sqrt{\alpha - 1}} \right\}.$$

*Moreover, for any domain $\mathscr{V}$ such that $\mathscr{D}(\alpha) \subset \mathscr{V} \subset \mathbb{D}$ and $\mathscr{V} \neq \mathscr{D}(\alpha)$, there exists a function $f \in \mathscr{B}_\alpha[0,1]$ that is not univalent in $\mathscr{V}$.*

THEOREM 3.3. *Let $f \in \mathscr{B}_\alpha[-1,1]$, $\alpha \in (1,4)$. Then $f$ is univalence in the domain*

$$\mathscr{U}(\alpha) = \left\{ z \in \mathbb{D} \colon \frac{|1 - z^2|}{1 - |z|^2} < \sqrt{\frac{\alpha}{\alpha - 1}} \right\}.$$

*Moreover, for any domain $\mathscr{V}$ such that $\mathscr{U}(\alpha) \subset \mathscr{V} \subset \mathbb{D}$ and $\mathscr{V} \neq \mathscr{U}(\alpha)$, there exists a function $f \in \mathscr{B}_\alpha[-1,1]$ that is not univalent in $\mathscr{V}$.*

Theorem 3.2 was proven in [**7**]. Theorem 3.3 was announced in [**8**].

## References


[1] G. Valiron, *Fonctions analytiques* (French), Presses Universitaires de France, Paris, 1954. MR61658

[2] J. Becker and C. Pommerenke, *Angular derivatives for holomorphic self-maps of the disk*, Comput. Methods Funct. Theory **17** (2017), no. 3, 487–497, DOI 10.1007/s40315-017-0199-x. MR3686894

[3] L. V. Ahlfors, *Conformal invariants: topics in geometric function theory*, McGraw-Hill Series in Higher Mathematics, McGraw-Hill Book Co., New York-Düsseldorf-Johannesburg, 1973. MR357743

[4] G. M. Goluzin, *Geometric theory of functions of a complex variable*, Translations of Mathematical Monographs, Vol. 26, American Mathematical Society, Providence, RI, 1969. MR247039

[5] O. S. Kudryavtseva and A. P. Solodov, *Two-sided estimates for domains of univalence for classes of holomorphic mappings of the disk into itself with two fixed points* (Russian, with Russian summary), Mat. Sb. **210** (2019), no. 7, 120–144, DOI 10.4213/sm9095; English transl., Sb. Math. **210** (2019), no. 7, 1019–1042. MR3985724

[6] V. V. Goryaĭnov, *Holomorphic mappings of the unit disk into itself with two fixed points* (Russian, with Russian summary), Mat. Sb. **208** (2017), no. 3, 54–71, DOI 10.4213/sm8802; English transl., Sb. Math. **208** (2017), no. 3-4, 360–376. MR3629076

[7] A. P. Solodov, *The exact domain of univalence on the class of holomorphic maps of a disc into itself with an interior and a boundary fixed points* (Russian, with Russian summary), Izv. Ross. Akad. Nauk Ser. Mat. **85** (2021), no. 5, 190–218, DOI 10.4213/im9053; English transl., Izv. Math. **85** (2021), no. 5, 1008–1035. MR4324012

[8] V. V. Goryainov, O. S. Kudryavtseva, and A. P. Solodov, *Estimate for the domain of univalence in the class of holomorphic self-maps of a disk with two boundary fixed points*, Dokl. Math. **108** (2023), no. 1, 326–330, DOI 10.1134/s1064562423700874. Translation of Dokl. Akad. Nauk **512** (2023), 96–101. MR4673791



LOMONOSOV MOSCOW STATE UNIVERSITY, MOSCOW CENTER OF FUNDAMENTAL AND APPLIED MATHEMATICS, 119991 MOSCOW, RUSSIA

VOLGOGRAD STATE TECHNICAL UNIVERSITY, 400005 VOLGOGRAD, RUSSIA
*Email address*: `kudryavceva_os@mail.ru`

LOMONOSOV MOSCOW STATE UNIVERSITY, MOSCOW CENTER OF FUNDAMENTAL AND APPLIED MATHEMATICS, 119991 MOSCOW, RUSSIA
*Email address*: `apsolodov@mail.ru`

Contemporary Mathematics
Volume **820**, 2025
https://doi.org/10.1090/conm/820/16421

# Quasiconformal distortion of hyperbolic distances

Toshiyuki Sugawa

Abstract. In the present paper, we first give a quasiconformal Schwarz lemma with uniqueness assertion. Then sharp estimates are given for hyperbolic distances between two points or two hyperbolic geodesics in the hyperbolic plane (or, more generally, hyperbolic Riemann surfaces) under quasiconformal mappings in terms of the Hersch-Pfluger distortion function. This unifies theorems of Basmajian in 2000 and Shiga in 2005.

## 1. Introduction

A complex-valued continuous function $f$ on a domain $D \subset \mathbb{C}$ is called $K$-quasiregular for a number $K \geq 1$ if $f$ has locally square integrable partial derivatives on $D$ (in the sense of distributions) such that $|f_{\bar{z}}| \leq k|f_z|$ almost everywhere on $D$, where $k = (K-1)/(K+1) \in [0,1)$. If, in addition, $f$ is a homeomorphism of $D$ onto another domain $D' \subset \mathbb{C}$, then $f$ is called $K$-quasiconformal. It is well known that the composition of a $K_1$-quasiregular map with a $K_2$-quasiregular map is $K_1K_2$-quasiregular map. Since a conformal mapping is 1-quasiconformal, $K$-quasiregularity is invariant under pre- and post-composition with conformal mappings. In particular, $K$-quasiregularity and $K$-quasiconformality can be defined for continous functions on Riemann surfaces or homeomorphisms between Riemann surfaces, respectively. One of the important issues in Geometric Function Theory is to estimate the change of magnitude for function-theoretical quantities under $K$-quasiconformal mappings. For example, the hyperbolic length of a closed geodesic is quasi-invariant under quasiconformal mappings. More precisely, for a $K$-quasiconformal map $f : R \to R'$ between hyperbolic Riemann surfaces, the sharp inequalities $\ell_R([\gamma])/K \leq \ell_{R'}([f(\gamma)]) \leq K\ell_R([\gamma])$ holds for every homotopically non-trivial closed curve $\gamma$ in $R$, where $[\gamma]$ denotes the free homotopy class of a closed curve $\gamma$ in $R$ and $\ell_R([\gamma])$ denotes the hyperbolic length of the closed geodesic in the homotopy class $[\gamma]$ in $R$. This is known as Wolpert's theorem [**12**]. Another well-known result is the following theorem due to A. Mori [**10**]. Throughout the paper, we will denote the open unit disk $\{z \in \mathbb{C} : |z| < 1\}$ by $\mathbb{D}$.

---

2020 *Mathematics Subject Classification.* Primary 30F45; Secondary 30C62, 33E05.
*Key words and phrases.* Hyperbolic distance, quasiconformal mapping,
This research is partly supported by JSPS KAKENHI Grant no. JP23H01078 (conducted by Prof. Katsuhiko Matsuzaki).

THEOREM 1.1 (Mori). *Let $f : \mathbb{D} \to \mathbb{D}$ be a $K$-quasiconformal homeomorphism with $f(0) = 0$. Then*

$$|f(z) - f(w)| < 16|z - w|^{1/K}, \quad z, w \in \mathbb{D} \text{ with } z \neq w.$$

Note that the uniform Hölder continuity above for $f$ and $f^{-1}$ guarantees a homeomorphic extension of $f$ to the closure of $\mathbb{D}$. It is conjectured that the best inequality of this type is $|f(z) - f(w)| \leq 16^{1-1/K}|z - w|^{1/K}$ but it is still an open problem. See [**1**] for more information. When $w = 0$, Mori's theorem implies $|f(z)| < 16|z|^{1/K}$ for $0 < |z| < 1$. This may be regarded as a variant of Schwarz's lemma. Indeed, we have a sharp form of Schwarz's lemma for $K$-quasiregular maps. The following result was shown by Hersch and Pfluger [**5**] (see also [**8**, Chap.II §3]) except for the uniqueness part. For convenience of the reader, we will give a proof below as well as some expressions of the extremal mapping.

THEOREM 1.2. *Let $f$ be a $K$-quasiregular mapping of $\mathbb{D}$ into $\mathbb{D}$ with $f(0) = 0$. Then*

$$|f(z)| \leq \varphi_K(|z|), \quad z \in \mathbb{D}.$$

*The inequality is sharp for each point $z \neq 0$ and the extremal mapping is unique up to rotation.*

Here, $\varphi_K : (0,1) \to (0,1)$, $0 < K < +\infty$, is the Hersch-Pfluger distortion function (see the next section for details). As a consequence, we obtain explicit estimates of the change of the hyperbolic distance between two points in the disk under a quasiregular map. Recall that the hyperbolic distance induced by the hyperbolic metric $|dz|/(1 - |z|^2)$ (of Gaussian curvature $-4$) is given by

$$d_{\mathbb{D}}(z_1, z_2) = \operatorname{arctanh}\left|\frac{z_1 - z_2}{1 - \bar{z}_1 z_2}\right|, \quad z_1, z_2 \in \mathbb{D},$$

where $\operatorname{arctanh} x = \frac{1}{2}\log\frac{1+x}{1-x}$ for $x \geq 0$.

For convenience, for $0 < K < +\infty$, we define a function by

$$\psi_K(x) = \operatorname{arctanh}[\varphi_K(\tanh x)], \quad x \in (0, +\infty). \tag{1.1}$$

In other words, $\psi_K = \tanh^{-1} \circ \varphi_K \circ \tanh$. Since $\varphi_K \circ \varphi_L = \varphi_{KL}$ and $\varphi_1 = \mathrm{id}$, we have $\psi_K \circ \psi_L = \psi_{KL}$ and $\psi_K^{-1} = \psi_{1/K}$ for $K, L \in (0, +\infty)$. This is a distortion function for hyperbolic distances.

THEOREM 1.3. *Let $f : \mathbb{D} \to \mathbb{D}$ be a $K$-quasiregular map from the unit disk $\mathbb{D}$ into itself. Then*

(1.2)
$$d_{\mathbb{D}}(f(z_1), f(z_2)) \leq \operatorname{arctanh}[\varphi_K(\tanh d_{\mathbb{D}}(z_1, z_2))] = \psi_K(d_{\mathbb{D}}(z_1, z_2)), \quad z_1, z_2 \in \mathbb{D}.$$

*If $f : \mathbb{D} \to \mathbb{D}$ is $K$-quasiconformal, then*

$$\psi_{1/K}(d_{\mathbb{D}}(z_1, z_2)) \leq d_{\mathbb{D}}(f(z_1), f(z_2)) \leq \psi_K(d_{\mathbb{D}}(z_1, z_2)), \quad z_1, z_2 \in \mathbb{D}. \tag{1.3}$$

The following estimates give useful information about the distortion function $\psi_K$.

LEMMA 1.4. *Let* $0 < K < +\infty$ *and* $x > 0$.

(1) *For a fixed* $\sigma > 0$,

$$\psi_K(\sigma)\min\{x/\sigma, (x/\sigma)^{1/K}\} \le \psi_K(x) \le \psi_K(\sigma)\max\{x/\sigma, (x/\sigma)^{1/K}\}.$$

(2) *If* $K \ge 1$, $\psi_K(x) \le Kx + (K-1)\log 2$ *and* $\psi_{1/K}(x) \ge x/K - (1-1/K)\log 2$.

(3) *If* $K \ge 1$, $\psi_{1/K}(x) \ge \operatorname{arctanh}(4^{1-K}\tanh^K x) \ge 4^{1-K}\tanh^K x$.

(4) $\lim_{\sigma\to+\infty}\big(\psi_K(\sigma) - K\sigma\big) = (K-1)\log 2$.

The inequality in (1.2) appears as (3.13) in [**8**, p.66]. Put $d = d_{\mathbb{D}}(z_1, z_2), d^* = d_{\mathbb{D}}(f(z_1), f(z_2))$ for simplicity. Letting $\sigma = 1/2$ in Lemma 1.4 (1), we obtain the estimate

$$2d^* \le 2\psi_K(1/2)\max\{2d, (2d)^{1/K}\},$$

which is exactly the estimate given by Hariri-Klén-Vuorinen [**4**, Theorem 16.39]. By Lemma 1.4 (2), we obtain $2\psi_K(1/2) \le K + (K-1)\log 4$. They give the slightly better estimate $2\psi_K(1/2) \le K + (K-1)\log[2(1+\sqrt{1-e^{-2}})]$. We remark that a weaker estimate was found by Kiernan [**7**].

The next topic is on the hyperbolic distance between two hyperbolic lines in the hyperbolic plane. We take the upper half-plane $\mathbb{H} = \{z \in \mathbb{C} : \operatorname{Im} z > 0\}$ as a model of the hyperbolic plane here. The hyperbolic metric $|dz|/(2\operatorname{Im} z)$ on the upper half-plane $\mathbb{H}$ induces the hyperbolic distance

$$d_{\mathbb{H}}(z_1, z_2) = \operatorname{arctanh}\left|\frac{z_1 - z_2}{z_1 - \bar{z}_2}\right|, \qquad z_1, z_2 \in \mathbb{H}.$$

We define the hyperbolic distance between two hyperbolic lines $L_1$ and $L_2$ in $\mathbb{H}$ to be $\inf\{d_{\mathbb{H}}(z_1, z_2) : z_1 \in L_1, z_2 \in L_2\}$. When $L_1$ and $L_2$ do not intersect and are not asymptotic, it is the hyperbolic distance of the intersection points of $L_1$ and $L_2$ with the unique hyperbolic line $L$ perpendicular to both $L_1$ and $L_2$. For a quasiconformal homeomorphism $f : \mathbb{H} \to \mathbb{H}$, the image $f(L)$ of a hyperbolic line $L$ is not a hyperbolic line in general. We denote by $[f(L)]$ the hyperbolic line sharing endpoints with $f(L)$. Note that $f$ extends homeomorphically to the closure $\overline{\mathbb{H}} = \mathbb{H} \cup \mathbb{R} \cup \{\infty\}$ by Mori's theorem above. We will denote the extended homeomorphism of $\overline{\mathbb{H}}$ by the same symbol $f$. In the following theorem, (1.4) is due to Basmajian [**3**] and (1.5) is due to Shiga [**11**]. We remark that the hyperbolic distance in [**3**] is the same as here but that in [**11**] is $2d_{\mathbb{H}}(z, w)$ so that the Gaussian curvature $-1$.

THEOREM 1.5 (Basmajian and Shiga). *Let* $f : \mathbb{H} \to \mathbb{H}$ *be a* $K$*-quasiconformal map. Let* $\delta$ *be the hyperbolic distance between two hyperbolic lines* $L_1, L_2$ *in* $\mathbb{H}$ *and let* $\delta^*$ *be the hyperbolic distance between* $L_1^* = [f(L_1)]$ *and* $L_2^* = [f(L_2)]$. *Let* $M > 0$ *be a given number. Then*

$$\delta/K - C_K \le \delta^* \le K\delta + C_K, \tag{1.4}$$

*where* $C_K = \frac{1}{2}\operatorname{arccosh}\big(2^{-(K-1)^2}e^{6(K+1)^2\sqrt{K-1}}\big)$, *and if* $\delta \le M$ *in addition, then*

$$\delta^K/A \le \delta^* \le A\delta^{1/K}, \tag{1.5}$$

*where* $A > 1$ *is a constant depending only on* $K$ *and* $M$.

In the present paper, we give the sharp estimate for the distance of two hyperbolic lines. Unexpectedly, the same distortion function as above works for this case.

THEOREM 1.6. *Let $f : \mathbb{H} \to \mathbb{H}$ be a $K$-quasiconformal map. Let $\delta$ be the hyperbolic distance between two hyperbolic lines $L_1, L_2$ in $\mathbb{D}$ and let $\delta^*$ be the hyperbolic distance between $L_1^* = [f(L_1)]$ and $L_2^* = [f(L_2)]$. Then,*

$$\psi_{1/K}(\delta) \le \delta^* \le \psi_K(\delta). \tag{1.6}$$

*Both inequalities are sharp.*

As a corollary, by choosing $\sigma = 1$ in Lemma 1.4 (1), we obtain the following consequences, which contain the above theorems of Basmajian and Shiga.

COROLLARY 1.7. *Under the same assumptions as in Theorem 1.6, the following inequalities hold for a fixed $\sigma > 0$ :*

$$\psi_{1/K}(\sigma) \min\{\delta/\sigma, (\delta/\sigma)^{1/K}\} \le \delta^* \le \psi_K(\sigma) \max\{\delta/\sigma, (\delta/\sigma)^{1/K}\}. \tag{1.7}$$

*In particular,*

$$\frac{\delta}{K} - \left(1 - \frac{1}{K}\right) \log 2 \le \delta^* \le K\delta + (K-1)\log 2. \tag{1.8}$$

The organization of the present paper is as follows. In Section 2, we give definitions of special functions that are required for the exposition. Also, a proof of Theorem 1.2 will be given. Section 3 will be devoted to the proof of Theorem 1.6. In Section 4, we will give proofs of Theorem 1.3 and Lemma 1.4. In the last section, a short account will be given for applications to hyperbolic Riemann surfaces.

## 2. Preliminaries and a proof of quasiconformal Schwarz's lemma

We summarize basic facts about special functions which are needed in the following. For more details, the reader can consult monographs [**2**] and [**4**]. Let $A$ be a ring domain, that is, a doubly connected domain in $\mathbb{C}$. We will exclude the case when $A$ is a punctured plane. Then $A$ is conformally equivalent to the genuine annulus of the form $A(M) = \{z \in \mathbb{C} : 1 < |z| < M\}$ for some $1 < M \le +\infty$. The quantity $m = \log M$ is called the *modulus* of $A$ and denoted by $\operatorname{mod} A$. This quantity is also quasi-invariant under quasiconformal mappings.

LEMMA 2.1. *Let $f : D \to D'$ be a $K$-quasiconformal map. Then*

$$\frac{\operatorname{mod} A}{K} \le \operatorname{mod} f(A) \le K \operatorname{mod} A$$

*for all ring domains $A$ contained in $D$.*

It is also known that this property characterizes quasiconformal maps. See [**8**].

The ring $G_c = \mathbb{D} \setminus [0, c]$ is called the *Grötzsch ring* for $c \in \mathbb{C}$ with $0 < |c| < 1$ and it has the following extremal property (see [**8**, p. 54] except for the equality assertion, which is contained in the main result of Jenkins [**6**]).

LEMMA 2.2. *Let $0 < |c| < 1$. Suppose that a ring domain $A$ in $\mathbb{D}$ separates the two points $0$ and $c$ from the unit circle $\partial\mathbb{D}$. Then $\operatorname{mod} A \le \operatorname{mod} G_c = \operatorname{mod} G_{|c|}$. Equality holds only when $A = G_c$.*

It is therefore important to describe and estimate the quantity

$$\mu(r) = \operatorname{mod} G_r, \quad 0 < r < 1.$$

We recall that

$$\mathcal{K}(r) = \int_0^1 \frac{dx}{\sqrt{(1-x^2)(1-r^2x^2)}}$$

is the complete elliptic integral of the first kind. Then $\mu(r)$ is expressed as follows:

$$\mu(r) = \frac{\pi}{2}\frac{\mathcal{K}(r')}{\mathcal{K}(r)}, \quad r' = \sqrt{1-r^2},\ 0<r<1.$$

It is known that $\mu(r)$ (strictly) decreases from $+\infty$ to 0 as $r$ moves from 0 to 1 (see [**2**] for instance). For $0 < K < +\infty$, the function

$$\varphi_K(r) = \mu^{-1}(\mu(r)/K), \quad 0<r<1,$$

is known as the *Hersch-Pfluger distortion function.* By definition, we have $\varphi_K \circ \varphi_L = \varphi_{KL}$ and $\varphi_K^{-1} = \varphi_{1/K}$. By the definition of the modulus, there is a unique conformal homeomorphism $\phi_r : G_r \to A(e^{\mu(r)})$ normalized so that the boundary point $r \in \partial G_r$ corresponds to $1 \in \partial A(e^{\mu(r)})$.

The following result due to Li, Liu and Zeng [**9**, Lemma 4] will be required in the proof of uniquness part of Theorem 1.2. Recall that $A(M) = \{z \in \mathbb{C} : 1 < |z| < M\}$.

LEMMA 2.3. *Let $K \in [1, +\infty)$ and $M > 1$. If a homeomorphism $F : A(M) \to A(M^{1/K})$ is $K$-quasiconformal, then there is a unimodular constant $\lambda$ such that either $F(z) = \lambda z|z|^{1/K-1}$ or $F(z) = \lambda M^{1/K} z|z|^{-1/K-1}$.*

We are now ready to prove Theorem 1.2.

PROOF OF THEOREM 1.2. Let $f : \mathbb{D} \to \mathbb{D}$ be a $K$-quasiregular map with $f(0) = 0$. Using the Stoïlow factorization, we decompose $f$ into the form $h \circ g$, where $g : \mathbb{D} \to \mathbb{D}$ is a $K$-quasiconformal homeomorphism with $g(0) = 0$ and $h : \mathbb{D} \to \mathbb{D}$ is a holomorphic map with $h(0) = 0$. The (classical) Schwarz Lemma implies $|h(w)| \le |w|$ for $w \in \mathbb{D}$ and that equality holds at a non-zero point only when $h(w) \equiv e^{i\theta}w$ for a real constant $\theta$. Next, take a non-zero $z_0 \in \mathbb{D}$ and set $z_1 = g(z_0)$, $r_0 = |z_0|$ and $r_1 = |z_1|$. Then we consider the ring domain $A_1 = g(G_{z_0})$. Since $A_1$ separates 0 and $z_1$ from $\partial\mathbb{D}$, the Grötzsch ring theorem (Lemma 2.2) implies that $\operatorname{mod} A_1 \le \mu(r_1)$. On the other hand, by quasi-invariance of the modulus, we have $\operatorname{mod} A_1 \ge (\operatorname{mod} G_{z_0})/K = \mu(r_0)/K$. Hence, $\mu(r_0)/K \le \mu(r_1)$ and, equivalently, $r_1 \le \varphi_K(r_0)$. We now conclude that $|g(z_0)| \le \varphi_K(|z_0|)$. Hence,

$$|f(z_0)| = |h(g(z_0))| \le |g(z_0)| \le \varphi_K(|z_0|). \tag{2.1}$$

We now consider the equality case. Assume that the equality $|f(z_0)| = \varphi_K(|z_0|)$ holds under the above situation. Then the equalities $|h(g(z_0))| = |g(z_0)|$ and $|g(z_0)| = \varphi_K(|z_0|)$ must hold in (2.1). In particular, $r_1 = \varphi_K(r_0)$, and, equivalently, $\mu(r_1) = \mu(r_0)/K$. By the classical Schwarz lemma, $h(w) \equiv e^{i\theta}w$ for a real constant $\theta$. Also, in view of the above proof, we have the equality $\operatorname{mod} A_1 = \mu(r_1)$. Then the equality assertion in Lemma 2.2 implies that $A_1 = G_{z_1}$. Let $\theta_0 = \arg z_0$, $\theta_1 = \arg z_1$ and consider the map

$$F(z) = (\phi \circ g)(e^{i\theta_0}\phi_{r_0}^{-1}(z)), \quad z \in A(e^{\mu(r_0)}).$$

where $\phi(w) = \phi_{r_1}(e^{-i\theta_1}w)$ is a conformal homeomorphism from $g(G_{z_0}) = G_{z_1}$ onto the annulus $A(e^{\mu(r_1)}) = A(e^{\mu(r_0)/K})$ with $\phi(z_1) = 1$. Since $F$ is $K$-quasiconformal homeomorphism from $A(M)$ onto $A(M^{1/K})$ with $F(1) = 1$, where $M = e^{\mu(r_0)}$,

Lemma 2.3 now implies that $F(z) = z|z|^{1/K-1}$. Thus the map $g$ is uniquely determined and expressed by $g(z) = e^{i\theta_1}(\phi_{r_1}^{-1} \circ F \circ \phi_{r_0})(e^{-i\theta_0}z)$. Finally, we obtain a form of $f$ as follows:

$$f(z) = \lambda(\phi_{r_1}^{-1} \circ F \circ \phi_{r_0})(e^{-i\theta_0}z), \quad z \in \mathbb{D},$$

where $\lambda$ is a unimodular constant, $z_0 = r_0 e^{i\theta_0}, r_1 = \varphi_K(r_0), F(z) = z|z|^{1/K-1}$ and $\phi_r : G_r \to A(e^{\mu(r)}) = \{1 < |w| < e^{\mu(r)}\}$ is a conformal homeomorphism with $\phi(r) = 1$. □

The above expression appears in [**8**, p. 63] but discussions for uniquness is missing there.

## 3. Estimates of hyperbolic distances between hyperbolic lines

In this section, we prove Theorem 1.6. Let $L_1$ and $L_2$ be two hyperbolic lines in $\mathbb{H}$ which do not intersect and are not asymptotic (otherwise, the hyperbolic distance is 0). We note that the hyperbolic distance is invariant under the action of $\mathrm{Aut}(\mathbb{H}) = \mathrm{PSL}(2,\mathbb{R})$. Therefore, we may assume that the endpoints of $L_1$ are $x_1 = -1/r$ and $x_2 = -1$ and the endpoints of $L_2$ are $x_3 = 1$ and $x_4 = 1/r$ for some $0 < r < 1$. Let us compute the hyperbolic distance $\delta$ between $L_1$ and $L_2$. It is easy to see that $L_1$ is the intersection of the circle $|z + m| = \rho$ with $\mathbb{H}$, where $m = (1 + 1/r)/2$ and $\rho = (-1 + 1/r)/2$. Similarly, $L_2$ is the intersection of the circle $|z - m| = \rho$ with $\mathbb{H}$. By the symmetry in the imaginary axis, we detect that a circle $C$ of the form $|z| = R$ for some $R > 0$ is perpendicular to $L_1$ and $L_2$. Assume that $Re^{i\theta}$ is the intersection point of $C$ with $L_2$ for some $\theta \in (0, \pi/2)$. Then, by a simple fact in elementary geometry, we have $\sin\theta = \rho/m = (1-r)/(1+r)$. We are now ready to calculate the hyperbolic distance as follows:

$$\delta = 2d_{\mathbb{H}}(Ri, Re^{i\theta}) = 2\,\mathrm{arctanh}\left|\frac{e^{i\theta} - i}{e^{i\theta} + i}\right| = 2\,\mathrm{arctanh}\,\frac{\cos\theta}{1 + \sin\theta}.$$

Since $\cos\theta = \sqrt{1 - \sin^2\theta} = 2\sqrt{r}/(1 + r)$, we finally have $\delta = 2\,\mathrm{arctanh}\sqrt{r}$, which is equivalent to $r = \mathrm{arctanh}^2(\delta/2)$. Then the function

$$F_r(z) = \int_0^z \frac{dt}{\sqrt{(1-t^2)(1-r^2t^2)}}$$

maps $\mathbb{H}$ conformally onto the rectangle $Q = (-\mathcal{K}(r), \mathcal{K}(r)) \times (0, \mathcal{K}(r'))$, where $r' = \sqrt{1-r^2}$, in such a way that $x_1, x_2, x_3, x_4$ are mapped to $-\mathcal{K}(r) + i\mathcal{K}(r')$, $-\mathcal{K}(r)$, $\mathcal{K}(r)$, $\mathcal{K}(r) + i\mathcal{K}(r')$, respectively. Recall that the modulus of the rectangle is defined by $\mathrm{mod}\, Q = 2\mathcal{K}(r)/\mathcal{K}(r') = \pi/\mu(r)$ (see [**8**, I.§4.1]).

Let $f : \mathbb{H} \to \mathbb{H}$ be a $K$-quasiconformal mapping. As before, we may assume that the endpoints of $L_1^* = [f(L_1)]$ are $x_1^* = -1/r^*, x_2^* = -1$ and the endpoints of $L_2^* = [f(L_2)]$ are $x_3^* = 1, x_4^* = 1/r^*$ for some $r^* \in (0, 1)$. Since the endpoints of $L_j$ are mapped to those of $f(L_j)$, we indeed have $f(x_j) = x_j^*$ for $j = 1, 2, 3, 4$. Then $F_{r^*}$ maps $\mathbb{H}$ conformally onto the rectangle $Q^*$, which is similarly defined as above. Since the moduli of rectangles are quasi-invariant under quasiconformal mappings, we have $K^{-1}\,\mathrm{mod}\, Q \le \mathrm{mod}\, Q^* \le K\,\mathrm{mod}\, Q$, and, equivalently, $\mu(r)/K \le \mu(r^*) \le K\mu(r)$, which in turn implies

$$\varphi_{1/K}(r) \le r^* \le \varphi_K(r).$$

By using the relations $r = \operatorname{arctanh}^2(\delta/2)$ and $r^* = \operatorname{arctanh}^2(\delta^*/2)$, we obtain the double inequality

$$\Psi_{1/K}(\delta) \le \delta^* \le \Psi_K(\delta), \tag{3.1}$$

where

$$\Psi_K(x) = 2\operatorname{arctanh}\sqrt{\varphi_K(\tanh^2(x/2))}, \quad K, x \in (0, +\infty). \tag{3.2}$$

We next show that the inequalities in (3.1) are both sharp. We choose hyperbolic lines $L_1$ and $L_2$ as above with $r = \operatorname{arctanh}^2(\delta/2)$. First we choose $r^* = \varphi_K(r)$ so that $\mu(r^*) = \mu(r)/K$. Note that the affine transformation

$$\alpha(x + yi) = \frac{\mathcal{K}(r^*)}{\mathcal{K}(r)}x + \frac{\mathcal{K}((r^*)')}{\mathcal{K}(r')}yi$$

maps the rectangle $Q$ onto $Q^*$. In general, the affine transformation $x+yi \mapsto ax+byi$ is $\max\{a/b, b/a\}$-quasiconformal for positive constants $a, b$. Therefore, the above $\alpha$ is $K$-quasiconformal. Now the $K$-quasiconformal homeomorphism $f = F_{r^*}^{-1} \circ \alpha \circ F_r$ of $\mathbb{H}$ maps $-1/r, -1, 1, 1/r$ to $-1/r^*, -1, 1, 1/r^*$, respectively. Thus the right-hand inequality turns to equality for the map $f$. We can construct a $K$-quasiconformal mapping by choosing $r^* = \varphi_{1/K}(r)$ to obtain an equality in the left-hand side in (3.1). The proof of Theorem 1.6 will be completed by showing the next lemma.

LEMMA 3.1. *The functions $\psi_K$ and $\Psi_K$ are identically equal for each $K > 0$.*

PROOF. First we note the relation [**2**, Theorem 10.5 (3)]

$$\varphi_K\left(\frac{2\sqrt{r}}{1+r}\right) = \frac{2\sqrt{\varphi_K(r)}}{1+\varphi_K(r)}.$$

This follows from $\varphi_K \circ \varphi_2 = \varphi_2 \circ \varphi_K = \varphi_{2K}$ and the formula $\varphi_2(r) = 2\sqrt{r}/(1+r)$ which is a consequence of the identity $\mu(r) = 2\mu(\sqrt{r}/(1+r))$ (see (2.3) in [**8**, p. 60]). By the double-angle formula $\tanh 2x = (2\tanh x)/(1+\tanh^2 x)$, we have $\tanh(2\operatorname{arctanh} t) = 2t/(1+t^2)$. We now let $r = \tanh^2(x/2)$. Then $2\sqrt{r}/(1+r) = \tanh x$ and

$$\begin{aligned}\tanh[\Psi_K(x)] &= \tanh\left[2\operatorname{arctanh}\sqrt{\varphi_K(r)}\right] = \frac{2\sqrt{\varphi_K(r)}}{1+\varphi_K(r)} = \varphi_K\left(\frac{2\sqrt{r}}{1+r}\right)\\ &= \varphi_K(\tanh x) = \tanh[\psi_K(x)].\end{aligned}$$

Thus we have proved $\Psi_K(x) = \psi_K(x)$ as required. □

## 4. Explicit estimates

Theorems 1.2 and 1.6 are sharp but not very explicit. We need some estimates of $\varphi_K$ and related functions to obtain explicit estimates. The following result is contained in Theorem 5.43 (1) and (2) of [**2**] (see also [**4**, p.167]). We remark that the second inequality appears already in [**8**, p. 65].

LEMMA 4.1. *For $1 \le K < +\infty, 0 < r < 1$*

$$4^{1-K}r^K \le \varphi_{1/K}(r) \quad \textit{and} \quad \varphi_K(r) \le 4^{1-1/K}r^{1/K}.$$

We remark that the number 4 cannot be replaced by a smaller number in the above lemma (see [**8**, p. 65]). In conjunction with Theorem 1.2, we have a version of the quasiconformal Schwarz Lemma:

$$|f(z)| \le 4^{1-1/K}|z|^{1/K}, \quad z \in \mathbb{D},$$

for a $K$-quasiregular map $f : \mathbb{D} \to \mathbb{D}$ with $f(0) = 0$. This even contains the classical Schwarz Lemma when $K = 1$.

We next prove Theorem 1.3. Let $f : \mathbb{D} \to \mathbb{D}$ be a $K$-quasiregular map and $z_1, z_2 \in \mathbb{D}$ and $w_j = f(z_j)$ for $j = 1, 2$. Then the mappings $L(z) = (z - z_1)/(1 - \bar{z}_1 z)$ and $M(w) = (w - w_1)/(1 - \bar{w}_1 w)$ are analytic automorphisms of $\mathbb{D}$ with $L(z_1) = 0$ and $M(w_1) = 0$. Let $g = M \circ f \circ L^{-1} : \mathbb{D} \to \mathbb{D}$ and put $z_0 = L(z_2)$ and $w_0 = M(w_2)$. Then $g$ is a $K$-quasiregular map with $g(z_0) = w_0$ and $d_{\mathbb{D}}(z_1, z_2) = d_{\mathbb{D}}(0, z_0)$. Theorem 1.2 now implies

$$\begin{aligned} d_{\mathbb{D}}(w_1, w_2) &= d_{\mathbb{D}}(0, w_0) \\ &= \operatorname{arctanh}|w_0| \le \operatorname{arctanh}\varphi_K(|z_0|) \\ &= \operatorname{arctanh}[\varphi_K(\tanh d_{\mathbb{D}}(0, z_0))], \end{aligned}$$

which confirms validity of the inequality (1.2).

We now prove Lemma 1.4. To this end, we first recall the following results.

LEMMA 4.2. *Let* $K \in (0, +\infty)$.

(i) *The function* $g(r) = \operatorname{arctanh}(\varphi_K(r))/\operatorname{arctanh} r$ *is strictly decreasing in* $0 < r < 1$ *for* $K > 1$ *and strictly increasing in* $0 < r < 1$ *for* $K < 1$.
(ii) *The function* $h(r) = \operatorname{arctanh}(\varphi_K(r))/(\operatorname{arctanh} r)^{1/K}$ *is strictly increasing in* $0 < r < 1$ *for* $K > 1$ *and strictly decreasing in* $0 < r < 1$ *for* $K < 1$.

Assertion (i) is found at Theorem 10.9 (3) in [**2**] whereas assertion (ii) is contained in Lemma 16.44 in [**4**].

PROOF OF LEMMA 1.4. Recall that $\psi_K(x) = \operatorname{arctanh}[\varphi_K(\tanh x)]$. In order to show part (1), we follow the proof of Theorem 16.39 in [**4**]. We first assume that $K > 1$. Then $f(x) := \psi_K(x)/x = g(\tanh x)$ is decreasing in $0 < x < +\infty$ by Lemma 4.2 (i). In particular, $\psi_K(x)/x = f(x) \le f(\sigma) = \psi_K(\sigma)/\sigma$ for $\sigma \le x < +\infty$. We next see that $f(x) := \psi_K(x)/x^{1/K} = h(\tanh x)$ is increasing in $0 < x < +\infty$ by Lemma 4.2 (ii). In particular, $\psi_K(x)/x^{1/K} = f(x) \le f(\sigma) = \psi_K(\sigma)/\sigma^{1/K}$ for $0 < x \le \sigma$. Now the assertion of the lemma has been shown for $K > 1$. For $K < 1$, the inequality signs should be reversed in the above and the assertion is shown also in this case.

We next show part (2). Recall the formula

$$\frac{1 - \varphi_K(r)}{1 + \varphi_K(r)} = \varphi_{1/K}\left(\frac{1 - r}{1 + r}\right), \quad 0 < r < 1,$$

from Theorem 10.5 (2) in [**2**]. Put $r = \tanh x$ and $s = (1 - r)/(1 + r) = e^{-2x}$ for convenience. Then, by the first inequality in Lemma 4.1, we have

$$\begin{aligned} \psi_K(x) &= \frac{1}{2}\log\frac{1 + \varphi_K(r)}{1 - \varphi_K(r)} = \frac{1}{2}\log\frac{1}{\varphi_{1/K}(s)} \\ &\le \frac{1}{2}\log(4^{K-1}s^{-K}) = (K - 1)\log 2 + Kx. \end{aligned}$$

Thus the first inequality in (2) has been shown. To show the second inequality, we substitute $x = \psi_{1/K}(y)$ into the first inequality and make use of the relation $y = \psi_K(x)$. After rearrangement of the terms, we obtain the second inequality.

In order to show part (3), we only need to use Lemma 4.1 and the elementary inequality $\operatorname{arctanh} t > t$ for $0 < t < 1$.

Finally, we show (4). By Lemma 10.8 in [**2**], we know that $r^{-1/K}\varphi_K(r) \to 4^{1-1/K}$ as $r \to 0$ for each $K \in (0, +\infty)$. Therefore, for the above $s = (1-r)/(1+r)$, we have $\varphi_{1/K}(s) = 4^{1-K}s^K(1 + o(1))$ as $s \to 0$. In particular, for $r = \tanh \sigma$ with $s = e^{-2\sigma}$,

$$\psi_K(\sigma) = -\frac{1}{2}\log \varphi_{1/K}(s) = -\frac{1}{2}\log\left(4^{1-K}s^K(1+o(1))\right) = (K-1)\log 2 + K\sigma + o(1)$$

as $\sigma \to +\infty$. From this, (4) follows. □

The author learned the above simple proof of part (2) from Matti Vuorinen.

## 5. Applications to Riemann surfaces

Let $R$ be a hyperbolic Riemann surface. Let $p : \mathbb{H} \to R$ be a holomorphic universal covering. Then the hyperbolic metric $\lambda_R(w)|dw|$ of $R$ is defined by the formula

$$\lambda_{\mathbb{H}}(z)|dz| = \lambda_R(p(z))|p'(z)||dz|, \quad z \in \mathbb{H},$$

where $\lambda_{\mathbb{H}}(z) = 1/(2\operatorname{Im} z)$. It induces the complete distance $d_R(w_1, w_2)$ on $R$ by

$$d_R(w_1, w_2) = \inf_{\alpha} \ell_R(\alpha), \quad \text{where } \ell_R(\alpha) = \int_{\alpha} \lambda_R(w)|dw|,$$

and $\alpha$ runs over all curves $\alpha$ joining $w_1$ and $w_2$ in $R$. $d_R$ is called the hyperbolic distance of $R$. It is known that there is a curve $\alpha$ attaining the infimum, which is called the *hyperbolic geodesic* joining $w_1$ and $w_2$. A lift of the hyperbolic geodesic to $\mathbb{H}$ with respect to $p$ is indeed a hyperbolic line segment in $\mathbb{H}$.

Let $\alpha$ be a closed curve in $R$ which is not homotopic to a point or a puncture in $R$. In the free homotopy class $[\alpha]$ of $\alpha$, there is the unique curve $\alpha_0$ which minimizes the hyperbolic length among $[\alpha]$. It is called the (hyperbolic) *closed geodesic*. The curve $\alpha_0$ is indeed the projection under $p$ of a hyperbolic line, which is the axis of a hyperbolic element of the covering transformation group with respect to $p$ (a Fuchsian group uniformizing $R$). We have

$$\ell_R[\alpha] = \inf_{\beta \in [\alpha]} \ell_R(\beta) = \ell_R(\alpha_0).$$

The quasiconformal distortion of hyperbolic distances is obtained immediately from Theorem 1.3.

THEOREM 5.1. *Suppose that $f$ is a $K$-quasiregular map from a hyperbolic Riemann surface $R$ to another hyperbolic Riemann surface $S$. Let $w_1$ and $w_2$ be points in $R$. Then*

$$d_S(f(w_1), f(w_2)) \le \operatorname{arctanh}[\varphi_K(\tanh d_R(w_1, w_2))] = \psi_K(d_R(w_1, w_2)). \tag{5.1}$$

*If $f : R \to S$ is a $K$-quasiconformal homeomorphism, then*

$$\psi_{1/K}(d_R(w_1, w_2)) \le d_S(f(w_1), f(w_2)) \le \psi_K(d_R(w_1, w_2)). \tag{5.2}$$

*In particular,*

(5.3)

$$\frac{1}{K}d_R(w_1,w_2)-\left(1-\frac{1}{K}\right)\log 2\le d_S(f(w_1),f(w_2))\le Kd_R(w_1,w_2)+(K-1)\log 2.$$

PROOF. We take analytic universal covering maps $p:\mathbb{D}\to R$ and $q:\mathbb{D}\to S$ such that $p(0)=w_1$ and $q(0)=f(w_1)$. Then we can take a lift $\tilde f:\mathbb{D}\to\mathbb{D}$ of $f$ so that $q\circ\tilde f=f\circ p$ and $\tilde f(0)=0$. Now we lift the hyperbolic geodesic $\alpha$ joining $w_1$ and $w_2$ in $R$ to $\tilde\alpha$ starting at 0 and ending at $z_0$ with $p(z_0)=w_2$. Then $d_R(w_1,w_2)=d_{\mathbb{D}}(0,z_0)$. Since $\tilde f:\mathbb{D}\to\mathbb{D}$ is $K$-quasiregular, by Theorem 1.3 we have $d_{\mathbb{D}}(0,\tilde f(z_0))\le\psi_K(d_{\mathbb{D}}(0,z_0))$. Since $q(\tilde f(z_0))=f(p(z_0))=f(w_2)$, we have $d_S(f(w_1),f(w_2))\le d_{\mathbb{D}}(0,\tilde f(z_0))$. We now combine these observations to get (5.1). When $f$ is $K$-quasiconformal, we apply (5.1) to the inverse map $f^{-1}:S\to R$ to obtain $d_R(w_1,w_2)\le\psi_K(d_S(f(w_1),f(w_2)))$. Since $\psi_K$ is monotone increasing and $\psi_K^{-1}=\psi_{1/K}$, we obtain the first inequality in (5.2). The second inequality in (5.3) follows from Lemma 1.4. By interchanging the roles of $\delta$ and $\delta^*$ in the second inequality in (5.3) to obtain the first inequality in (5.3). □

We obtain a quasiconformal distortion theorem for hyperbolic distances between two closed geodesics in a similar mannar by Theorem 1.6.

THEOREM 5.2. *Let $f:R\to S$ be a $K$-quasiconformal homeomorphism between hyperbolic Riemann surfaces $R$ and $S$ for some $K\ge 1$. Let $\alpha_1$ and $\alpha_2$ be closed hyperbolic geodesics in $R$ and $\alpha_1^*$ and $\alpha_2^*$ be the closed hyperbolic geodesics in $S$ representing the free homotopy classes $[f(\alpha_1)]$ and $[f(\alpha_2)]$ in $S$. The hyperbolic distances $\delta$ between $\alpha_1$ and $\alpha_2$ in $R$ and $\delta^*$ between $\alpha_1^*$ and $\alpha_2^*$ in $S$ are related in the following sharp inequalities:*

$$\psi_{1/K}(\delta)=\operatorname{arctanh}[\varphi_{1/K}(\tanh\delta)]\le\delta^*\le\operatorname{arctanh}[\varphi_K(\tanh\delta)]=\psi_K(\delta). \tag{5.4}$$

*and thus, for a fixed $\sigma>0$,*

$$\psi_{1/K}(\sigma)\min\{\delta/\sigma,(\delta/\sigma)^{1/K}\}\le\delta^*\le\psi_K(\sigma)\max\{\delta/\sigma,(\delta/\sigma)^{1/K}\}.$$

*In particular,*

$$\frac{\delta}{K}-\left(1-\frac{1}{K}\right)\log 2\le\delta^*\le K\delta+(K-1)\log 2.$$

PROOF. It is enough to show the assertion when $\alpha_1\ne\alpha_2$. We first assume that $\delta>0$. Let $\alpha$ be the shortest geodesic segment joining a point $w_1\in\alpha_1$ and a point $w_2\in\alpha_2$ in $R$ so that $\delta=\ell_R(\alpha)$. We take analytic universal covering maps $p:\mathbb{H}\to R$ and $q:\mathbb{H}\to S$ such that $p(i)=w_1$ and $q(i)=f(w_1)$ and let $\tilde f:\mathbb{H}\to\mathbb{H}$ be the lift of $f$ with $\tilde f(i)=i$ and $q\circ\tilde f=f\circ p$. Note that $\tilde f$ is a $K$-quasiconformal homeomorphism of $\mathbb{H}$ onto itself. Let $\Gamma$ and $\Gamma^*$ be covering groups of $p$ and $q$, respectively. Note that $\Gamma$ and $\Gamma^*$ are Fuchsian groups contained in $\mathrm{Aut}(\mathbb{H})=\mathrm{PSL}(2,\mathbb{R})$ and that the mapping $\chi:\gamma\mapsto\tilde f\circ\gamma\circ\tilde f^{-1}$ gives a group isomorphism of $\Gamma$ onto $\Gamma^*$. Let $\tilde\alpha_1$ be the lift of $\alpha_1$ with respect to $p$ with initial point at $i$. Its terminal point can be expressed as $\gamma_1(i)$ for a hyperbolic element $\gamma_1\in\Gamma$. Let $L_1$ be the axis of $\gamma_1$. Then $L_1$ is a hyperbolic line in $\mathbb{H}$. Let $\tilde\alpha$ be the lift of $\alpha$ with respect to $p:\mathbb{H}\to R$ starting at $i$. We denote by $z_0$ the terminal point of $\tilde\alpha$. Note that $p(z_0)=w_2$. Let $\tilde\alpha_2$ be the lift of $\alpha_2$ with respect to $p$ with initial point at $z_0$. Its terminal point can be expressed as $\gamma_2(z_0)$ for a hyperbolic element $\gamma_2 \in \Gamma$. Note that $\gamma_1 \neq \gamma_2$ because $\alpha_1 \neq \alpha_2$. Let $L_2$ be the axis of $\gamma_2$. Then the hyperbolic distance between $L_1$ and $L_2$ is given by $\delta = \ell_{\mathbb{H}}(\tilde{\alpha}) = d_{\mathbb{H}}(i, z_0)$. It is easy to see that $L_j^* = [\tilde{f}(L_j)]$ is the axis of the hyperbolic element $\gamma_j^* = \chi(\gamma_j)$ for $j = 1, 2$. Therefore, $\alpha_j^* = q(L_j^*)$ for $j = 1, 2$. Here we note that the two endpoints of $L_1^*$ do not separate the endpoints of $L_2^*$ in $\partial\mathbb{H} \approx S^1$ and therefore $L_1^*$ and $L_2^*$ do not intersect and are not asymptotic. Let $\tilde{\beta}$ be the shortest hyperbolic geodesic segment joining $L_1^*$ and $L_2^*$. Then $\beta = q(\tilde{\beta})$ is a curve joining $\alpha_1^*$ and $\alpha_2^*$ in $S$. By Theorem 1.6, the hyperbolic distance $\delta_0^*$ between $L_1^*$ and $L_2^*$ satisfies the inequalities

$$\delta^* \leq \ell_S(\beta) = \ell_{\mathbb{H}}(\tilde{\beta}) = \delta_0^* \leq \psi_K(\delta).$$

Thus we have shown the second inequality in (5.4). Here, we shall show that $\alpha_1^*$ and $\alpha_2^*$ do not intersect under the assumption $\delta > 0$. Indeed, otherwise, we can take connected components $M_1, M_2$ of $q^{-1}(\alpha_1^*), q^{-1}(\alpha_2^*)$, respectively, so that $M_1$ intersects $M_2$. Since the hyperbolic lines $M_1$ and $M_2$ intersect, the endpoints of $M_1$ separate those of $M_2$ in $\partial\mathbb{H}$. Hence $L_1' = [\tilde{f}^{-1}(M_1)]$ and $L_2' = [\tilde{f}^{-1}(M_2)]$ intersect. Since $p(L_j') = \alpha_j$ for $j = 1, 2$, this implies that $\alpha_1$ and $\alpha_2$ intersect, which would violate the assumption $\delta > 0$. Thus we conclude that the hyperbolic distance $\delta^*$ between $\alpha_1^*$ and $\alpha_2^*$ is positive if $\delta > 0$. In particular, we can now interchange the roles of $\delta$ and $\delta^*$ in the above to obtain $\delta \leq \psi_K(\delta^*)$, which is equivalent to $\psi_K^{-1}(\delta) = \psi_{1/K}(\delta) \leq \delta^*$. Thus the first inequality in (5.4) has been shown under the assumption that $\delta > 0$.

Next we assume that $\delta = 0$, that is, $\alpha_1$ and $\alpha_2$ intersect. Then, as was shown above, $\alpha_1^*$ and $\alpha_2^*$ intersect, too. Thus (5.4) clearly holds in this case.

The latter part of the theorem follows from Lemma 1.4. □

## Acknowledgment

The author is grateful to Professor Lixin Liu for sending his joint paper [**9**] to the author. He would also like to thank Professors Matti Vuorinen and Hiroshige Shiga for valuable suggestions. Finally, he thanks an anonymous referree for careful checking of the mansucript and helpful comments.

## References

[1] G. D. Anderson, M. K. Vamanamurthy, and M. Vuorinen, *Topics in special functions. II*, Conform. Geom. Dyn. **11** (2007), 250–270, DOI 10.1090/S1088-4173-07-00168-3. MR2354098

[2] G. D. Anderson, M. K. Vamanamurthy, and M. K. Vuorinen, *Conformal invariants, inequalities, and quasiconformal maps*, Canadian Mathematical Society Series of Monographs and Advanced Texts, John Wiley & Sons, Inc., New York, 1997. With 1 IBM-PC floppy disk (3.5 inch; HD); A Wiley-Interscience Publication. MR1462077

[3] A. Basmajian, *Quasiconformal mappings and geodesics in the hyperbolic plane*, In the tradition of Ahlfors and Bers (Stony Brook, NY, 1998), Contemp. Math., vol. 256, Amer. Math. Soc., Providence, RI, 2000, pp. 1–4, DOI 10.1090/conm/256/03992. MR1759665

[4] P. Hariri, R. Klén, and M. Vuorinen, *Conformally invariant metrics and quasiconformal mappings*, Springer Monographs in Mathematics, Springer, Cham, 2020, DOI 10.1007/978-3-030-32068-3. MR4179585

[5] J. Hersch and A. Pfluger, *Généralisation du lemme de Schwarz et du principe de la mesure harmonique pour les fonctions pseudo-analytiques* (French), C. R. Acad. Sci. Paris **234** (1952), 43–45. MR46452

[6] J. A. Jenkins, *Some uniqueness results in the theory of symmetrization. II*, Ann. of Math. (2) **75** (1962), 223–230, DOI 10.2307/1970170. MR133475

[7] P. J. Kiernan, *Quasiconformal mappings and Schwarz's lemma*, Trans. Amer. Math. Soc. **148** (1970), 185–197, DOI 10.2307/1995046. MR255807

[8] O. Lehto and K. I. Virtanen, *Quasiconformal mappings in the plane*, 2nd ed., Die Grundlehren der mathematischen Wissenschaften, Band 126, Springer-Verlag, New York-Heidelberg, 1973. Translated from the German by K. W. Lucas. MR344463

[9] S. Li, L. Liu, and C. Zeng, *The Schwarz' theorem for quasiconformal mappings on the condition of area distortion (Chinese)*, Chinese Ann. of Math. A **27** (2006), 27–38.

[10] A. Mori, *On quasi-conformality and pseudo-analyticity*, Trans. Amer. Math. Soc. **84** (1957), 56–77, DOI 10.2307/1992891. MR83024

[11] H. Shiga, *On the hyperbolic length and quasiconformal mappings*, Complex Var. Theory Appl. **50** (2005), no. 2, 123–130, DOI 10.1080/02781070412331328206. MR2122749

[12] S. Wolpert, *The length spectra as moduli for compact Riemann surfaces*, Ann. of Math. (2) **109** (1979), no. 2, 323–351, DOI 10.2307/1971114. MR528966



Graduate School of Information Sciences, Tohoku University, Aoba-ku, Sendai 980-8579, Japan

*Email address*: `sugawa@math.is.tohoku.ac.jp`

Contemporary Mathematics
Volume **820**, 2025
https://doi.org/10.1090/conm/820/16422

# Harmonic quasiconformal mappings, Ahlfors regular and Bishop-Jones domains

Vesna Todorčević

ABSTRACT. We give an overview of several conditions characterizing chord-arc and Ahlfors regular curves, Bishop-Jones domains and Bishop-Jones quasicircles in terms of Riemann mapping from the unit disk, welding map and, in case of quasicircles, quasiconformal Beltrami differential of an extension. We also consider generalizations of these results to harmonic quasiconformal mappings of unit disk to domain of corresponding type and pose several open problems.

## 1. Introduction

The geometry of domains has been an important topic in Geometric Function Theory over the last forty years. Martio and Sarvas [**27**] were first to consider and introduce the so-called *uniform domains* in the late 1970s. We recall that these are domains in which every two points can be joined by an arc of length bounded by a constant times the Euclidean distance between them such that each point on that arc divides the arc into two parts, at least one of them of length smaller than a constant times the distance of dividing point to the boundary of the domain (i.e. the arc does not come too close to the boundary of the domain).

Around the same time, Jones [**20**] characterized domains which have the property that bounded mean oscillation (BMO) functions can be extended from the domain to the whole space. Gehring and Osgood in [**16**] have proved that Jones domains are precisely uniform domains.

M. Zinsmeister studied Lavrentiev domains that are also called chord-arc, while G. David reinvented in [**13**] Ahlfors curves. In the late 80' Bishop and Jones [**10**] found a geometric characterization (which later became known as Bishop-Jones domains) in terms of analytic properties of the corresponding Riemannn map.

About that time several papers by Semmes, Astala, Zinsmeister and others ([**1**], [**2**], [**7**], [**9**], [**11**], [**12**], [**19**], [**21**], [**33**], [**34**], [**35**], [**41**], [**42**]) along with classical earlier results of Pommerenke gave a range of characterizations of various types of domains in terms of properties of corresponding Riemann map, welding map, Teichmüller spaces and Carleson measures.

2020 *Mathematics Subject Classification.* Primary 30C62.
*Key words and phrases.* Harmonic quasiconformal mappings, chord-arc curves and Ahlfors regular curves.

## 2. Geometry of domains and harmonic analysis

In this paper we study mappings $f : \mathbb{D} \to \Omega$, where $\mathbb{D} = \{z \in \mathbb{C} : |z| < 1\}$ and $\Omega$ is a simply connected domain in the complex plane. The reader is invited to consult [**5**], [**8**], [**15**], [**31**] and [**39**] for a proper bacground.

We recall that an *Ahlfors regular domain* $\Omega$ is a domain $\Omega \subset \mathbb{C}$ which is bounded by Ahlfors regular curve (see fig. 1), i.e. a rectifiable curve $\gamma$ such that any small disk of radius $r$ covers part of $\gamma$ of length at most $Cr$ for some $C > 0$. In other words,

$$(\forall z_0 \in \mathbb{C},\, 0 < r < \infty)\; l(\gamma \cap D(z_0, r)) \leqslant Cr,$$

FIGURE 1. Ahlfors regular curve

Recall also the following well known theorem.

THEOREM 2.1. *Suppose $f : \mathbb{D} \to \Omega$ is conformal and $\Omega$ is an arbitrary simply connected domain. Then*

$$\log f'(z) \in \mathcal{B},$$

*where $\mathcal{B}$ denotes the Bloch space, consisting of analytic functions $g : \mathbb{D} \to \mathbb{C}$ with*

$$\|g\|_{\mathcal{B}} := \sup_{|z|<1} (1 - |z|^2)|g'(z)| < \infty.$$

We note that in this context it is known that $\mathcal{B} \equiv \{g \in BMO(\mathbb{D}) :\; g \;\text{ analytic}\}$, see e.g. [**32**].

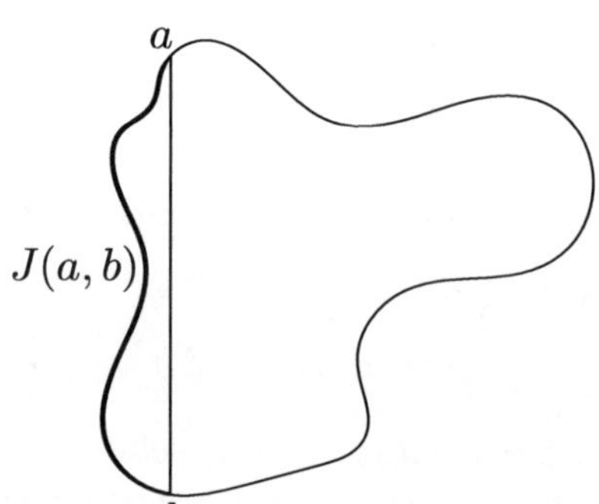


FIGURE 2. Chord-arc curve

Suppose that $\Omega$ is a *chord-arc domain*. This means that $J := \partial\Omega$ is a chord-arc curve (see fig. 2), i.e. $J$ is rectifiable Jordan curve and there is a constant $C < \infty$ such that

$$(\forall a, b \in J)\ \mathcal{H}^1[J(a,b)] \leq C|a-b|.$$

Here $J(a,b)$ denotes the shorter of the arcs of $J$ connecting $a$ and $b$.

Chord-arc domains are also called *Lavrentiev domains*, and their boundaries are precisely the images of a circle or a line under a bilipschitz mapping of $\mathbb{C}$. The constant $C$ is called a norm of chord-arc curve.

Then by a result of Pommerenke, see e.g. his book [**32**, pp. 155], we have the following:

THEOREM 2.2. *Suppose $f : \mathbb{D} \to \Omega$ is conformal with $\Omega$ a chord-arc domain, then*

$$\log f'(z) \in BMOA. \tag{2.1}$$

*Here* $BMOA := \{f\big|_{\partial\mathbb{D}} \in BMO(\partial\mathbb{D}) : f \ \ \text{analytic} \ \ \text{in} \ \ \mathbb{D}\}$.

Many further results in this direction have been established. In particular, (2.1) is shown by proving that $|f'| \in A_p$ for some $p$; here $A_p$ is the Muckenhoupt-class on the boundary $\partial\mathbb{D}$. Here we give the definition of Muckenhoupt classes.

DEFINITION 2.1. Let $w$ be a locally integrable nonnegative function in $\mathbb{R}^n$ such that $0 < w < \infty$ almost everywhere. Then $w$ belongs to the *Muckenhoupt class $A_p$* for $1 < p < \infty$, or that is an *$A_p$-weight* whenever there is a constant $c_{p,w}$ such that

$$\frac{1}{|B|}\int_B w\ dx \leq c_{p,w}\Big(\frac{1}{|B|}\int_B w^{\frac{1}{1-p}}\ dx\Big)^{1-p}$$

for all balls $B$ in $\mathbb{R}^n$. We let $w$ belong to $A_1$, or that $w$ is an *$A_1$-weight* if there is a constant $c_{1,w} \geq 1$ such that

$$\frac{1}{|B|}\int_B w\ dx \leq c_{1,w} \text{ess}\inf_B w$$

for all balls $B$ in $\mathbb{R}^n$. The union of all Muckenhoupt classes $A_p$ is denoted by $A_\infty$, or more precisely,

$$A_\infty = \bigcup_{p>1} A_p.$$

We shall say that $w$ is an *$A_\infty$-weight* if it belongs to $A_\infty$, i.e., if $w$ is an $A_p$-weight for some $p > 1$.

By Kellogs theorem, see e.g. [**15**, p. 62] we have the following:

THEOREM 2.3. *Suppose $f : \mathbb{D} \to \Omega$ is conformal, with $\partial\Omega$ a $C^{1,\alpha}$-Jordan curve. Then, $f \in C^1(\overline{\mathbb{D}})$ with $f'(z) \neq 0$ in $\overline{\mathbb{D}}$. In particular,*

$$\log f'(z) \in H^\infty,$$

*so that $f$ is bilipschitz.*

This result is also true for more general domains.

Suppose $f : \mathbb{D} \longrightarrow \Omega$ is conformal and $\Omega$ is domain with Jordan boundary. The following theorem was proven by Bishop and Jones.

THEOREM 2.4. (Bishop-Jones) *If $f$ is a conformal mapping in $\mathbb{D}$ then* $\text{Log} f' \in$ $BMOA(\mathbb{D})$ *if and only if the domain $\Omega$ satisfies:*

***(Bishop-Jones)*** *For any $z \in \Omega$, there exists a Lavrentiev domain $\Omega_z \subseteq \Omega$ containing $z$, of "norm" $\leqslant k(\Omega)$ and of diameter uniformly comparable to $d(z, \partial\Omega)$, such that $l(\partial\Omega \cap \partial\Omega_z) \geqslant c(\Omega) d(z, \partial\Omega)$.*

The domains which satisfy the condition (**BJ**) are called Bishop-Jones domains. In fact Bishop-Jones domains are generalization of Ahlfors domains. However, Ahlfors curves are not necessary quasicircles. Chord-arc curves are precisely Ahlfors curves which are also quasicircles.

Quasicircles are images of $\partial\mathbb{D} = \mathbf{T}$ under quasiconformal homeomorphism mappings of the complex plane $\mathbb{C}$.

We have the following geometric characterization of quasicircles: A Jordan curve $\Gamma$ is qusicircle if and only if there is a constant $C > 0$ such that

$$\text{diam}(J(a,b)) \leqslant C|a-b|, \quad \text{for } a, b \in \Gamma,$$

where $J(a,b)$ is the smaller of the two subarcs of $\Gamma$ between $a$ and $b$. Namely, we have the following inclusion relations:

The set of chord-arc curve is a subset of the set of Bishop-Jones quasicircles, which is a subset of the set of quasicircles.

Note that quasicircles typically are not rectifiable. One of most famous examples of a nonrectifiable quasicircle is the snowflake curve, also known as von Koch curve (see fig. 3). It is obtained as a limit of iterative procedure and it's Hausdorff dimension is between 1 and 2 ($\log_3(4)$).

Let us briefly recall some facts about Beltrami differentials of quasiconformal mappings.

For every Beltrami differential $\mu \in L^\infty(\mathbb{C})$ and $\|\mu\|_\infty < 1$ there is a quasiconformal solution to the Beltrami equation $f_{\bar{z}} = \mu f_z$ for $f : \overline{\mathbb{C}} \longrightarrow \overline{\mathbb{C}}$, unique up to postcomposing with a conformal mapping of $\overline{\mathbb{C}}$.

A positive measure $\lambda$ defined in a simply connected domain $\Omega$ is called a Carleson measure if

$$\|\lambda\|_c^2 = \sup\left\{\frac{\lambda(\Omega \cap D(Z,r))}{r} : z \in \partial\Omega,\ 0 < r < \text{diam}(\partial\Omega)\right\} < \infty$$

where $D(z,r)$ is the disk with centre $z$ and radius $r$.

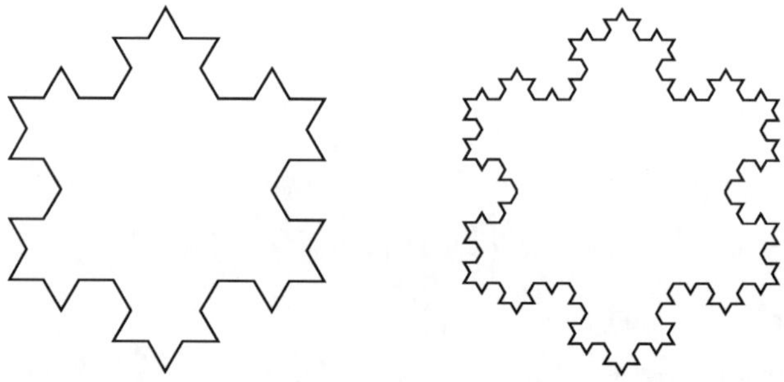

FIGURE 3. Two consecutive steps in construstion of Koch curve

## 3. Conformal welding

Suppose that $\Omega$ is a domain where $\partial\Omega$ is a Jordan curve and that $f : \mathbb{D} \longrightarrow \Omega$ and $g : \mathbb{C} \setminus \mathbb{D} \longrightarrow \mathbb{C} \setminus \Omega$ are conformal mappings.

By Caratheodory theorem both $f$ and $g$ have continuos extensions to $\mathbf{T} = \partial\mathbb{D}$. The mapping $h = g^{-1} \circ f$ is a homeomorphism of $\mathbf{T}$. This map is called a conformal welding (see fig. 4) with respect to $\Gamma = \partial\Omega = \partial(\mathbb{C} \setminus \Omega)$.

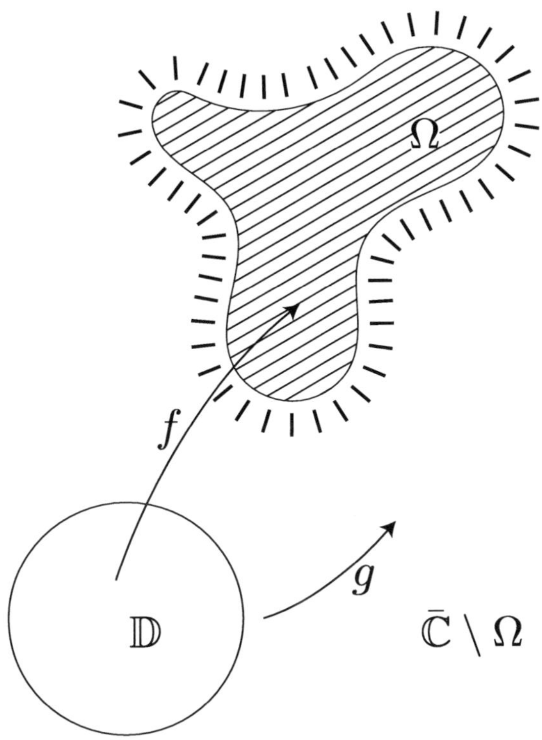


FIGURE 4. Welding mapping

The following theorem characterizes quasicircles in terms of welding mapping $h$.

THEOREM 3.1. *The following statements are equivalent:*

(1) $f$ *admits a quasiconformal extension to* $\bar{\mathbb{C}}$*;*
(2) $\Gamma$ *is quasicircle;*
(3) $h$ *is quasisymmetric i.e., it satisfies the following condition for some constant* $M > 0$

$$\frac{1}{M} \leqslant \frac{|h(e^{i(\theta+t)} - h(e^{i\theta}))|}{|h(e^{i\theta}) - h(e^{i(\theta-t)})|} \leqslant M, \quad \text{where } \theta \in [0, 2\pi),\ t > 0.$$

*The condition above is called the* $M$*-condition.*

Next, we recall theorem of Astala and Zinsmeister. Here, equivalent criteria are given for Lavrentiev curve of small norm in terms of welding map along with some other conditions.

THEOREM 3.2. [**7**] *The following are equivalent:*

(1) $\operatorname{Log} f' \in BMOA(\mathbb{D})$ *with a small norm,*
(2) $f(\mathbf{T})$ *is a Lavrentiev curve with a small norm,*
(3) $\operatorname{Log} h' \in BMO(\mathbf{T})$ *with a small norm,*
(4) $f$ *has a q.c. extension with dilatation* $\mu$ *such that* $|\mu|^2/(|z|^2-1)$ *is a Carleson measure with small norm.*

## 4. General Quasiconformal maps and the Universal Teichmüller space

The next natural step is a generalization of Pommerenke and Kellog's theorems 2.2 and 2.3 to quasiconformal mappings.

By a Theorem of Reimann, see e.g. [**5**, p. 347], we have the following¿

THEOREM 4.1. *Suppose that* $f : \mathbb{D} \to \Omega$ *is quasiconformal, then*

$$\log J(z,f) \in BMO(\mathbb{D}).$$

Further, if $f$ is conformal, then $2\Re e \log f'(z) \equiv 2\log|f'(z)| \equiv \log J(z,f)$. Hence Reimann's result is the natural counterpart to 2.1.

Some further analogies of $\log f'(z)$ and $\log J(z,f)$ are studied in [**4**]. However, the Theorems 2.2–2.3 do not seem to have natural counterparts in case of general quasiconformal mappings since quasiconformal maps $f : \mathbb{D} \to \mathbb{D}$ can be singular on the boundary $\partial\mathbb{D}$.

For every $\mu \in L^\infty(\mathbb{C}\setminus\mathbb{D})$ there exists a unique qc-mapping $f^\mu : (\mathbb{C}\setminus\mathbb{D}) \longrightarrow (\mathbb{C}\setminus\mathbb{D})$, such that

$$f^\mu_{\bar z}/f^\mu_z = \mu, \quad f^\mu(1) = 1, \quad f^\mu(i) = i, \quad f^\mu(-1) = -1.$$

Also for every $\mu \in L^\infty(\mathbb{C}\setminus\mathbb{D})$ there is a unique qc-mapping $f_\mu : \bar{\mathbb{C}} \longrightarrow \bar{\mathbb{C}}$ such that $(f_\mu)_{\bar z}/(f_\mu)_z = \mu$ for $z \in \mathbb{C}\setminus\mathbb{D}$ (equal to zero for $z \in \mathbb{D}$) normalized by $f_\mu(0) = 0$, $f_\mu'(0) = 1$ and $f_\mu''(0) = 0$.

It is known that

$$f^\mu|_{\mathbf{T}} = f^\nu|_{\mathbf{T}} \quad \Leftrightarrow \quad f_\mu|_{\mathbb{D}} = f_\nu|_{\mathbb{D}}. \tag{4.1}$$

Note that $f_\mu|_{\mathbb{D}}$ is a conformal mapping. If (4.1) is satisfied then Beltrami differentials are equivalent and equivalence classes form the Universal Teichmüller space ([**40**], [**36**], [**28**]). Theorem 3.2 gives a characterization of Lavrentiev curves with small norm in terms of welding mapping $h$ and Beltrami differential $\mu$, of a representative of corresponding Teichmüller class.

It should be noted, however, that there are simmilar characterizations of Bishop-Jones quasicircles and some other type of domains.

THEOREM 4.2. *The following are equivalent:*

(1) $|\mu(z)|^2/(|z|^2-1) \in CM(\mathbb{C}\setminus\mathbb{D})$,
(2) $\Gamma$ *is a Bishop-Jones quasicircle.*

For readers interested in exploring these and related topics on the universal Teichmüller space we suggest the references [**14**], [**24**], [**29**] and [**37**].

## 5. Harmonic Quasiconformal mappings

For harmonic quasiconformal mappings introduced by Martio [**26**], we can combine the above facts with the theorem of Todorčević [**25**] which shows that $\log J(z,f)$ is superharmonic for every harmonic homeomorphism $f$.

THEOREM 5.1. *Let $f : \Omega \to \mathbb{C}$ be a harmonic map whose Jacobian determinant $J = |f_z|^2 - |f_{\bar{z}}|^2$ is positive everywhere $\Omega$. Then $\log J$ is a superharmonic function.*

This theorem has an application in establishing the minimum principle for the Jacobian determinant which is a novelty for the new analytic proof of the Radó-Kneser-Choquet theorem given by T. Iwaniec and J. Onninen [**18**].

Hence in many ways, for harmonic quasiconformal mappings, the quantity $\log J(z, f)$ seems the natural counterpart for $\log f'(z)$. In particular, the question arises if the counterparts of Theorems 2.1–2.3 hold for HQC-maps and $\log J(z, f)$. The first one holds, and already follows from Theorem 4.1.

There are many studies of bilipschicity of maps ([**6**],[**30**][**17**],[**23**]), but properties of harmonic quasiconformal maps from a disk to an Ahlfors domain seems to be very little studied. In fact M. Zinsmeister generalized Pommerenke Theorem 2.2 to domains with Ahlfors regular boundary [**41**]. From other side D. Kalaj considered generalization of Pommerenke Theorem 2.2 to harmonic quasiconformal maps and proved several equivalent characterizations of the Muckenhoupt property of the derivative on the boundary [**22**].

THEOREM 5.2 ([**22**]). *Assume that $f$ is $K$-quasiconformal harmonic mapping of the unit disk onto a domain $\Omega$ by a chord-arc Jordan courve. Then*

(1) *there is a positive number $\kappa > 0$ such that*

$$\int_{\mathbf{T}} |\partial_t f(e^{it})|^{-\kappa}\, dt < \infty,$$

(2) *there is a number $\lambda > 1$ such that*

$$\int_{\mathbf{T}} |\partial_t f(e^{it})|^{\lambda}\, dt < \infty,$$

(3) $\log |\partial_t f(e^{it})| \in BMO(\mathbf{T})$,

(4) $\log f_z \in BMOA$.

Bounded chord-arc domains are exactly the bilipschitz images of the unit disk. However there is a more natural and general class of domains, the class of domains having an Ahlfors regular boundary. As mentioned before, $\Omega$ is a chord-arc domain if and only if it is both a quasidisk and has Ahlfors regular boundary. Our goal is to give a full generalization of Zinsmeister theorem for harmonic quasiconformal mappings so we pose the following problem (see [**38**]).

PROBLEM 1. If $f : \mathbb{D} \longrightarrow \Omega$ is harmonic and quasiconformal and if $\Omega$ is a Ahlfors regular domain, is $f$ absolutely continuous on the boundary?

Another goal is to extend the result of Kalaj [**22**] as follows.

PROBLEM 2. Is the corresponding boundary measure induced by mapping in the Muckenhoupt class $A_p$ for some $p$?

Note that the Kalaj's theorem 5.2 shows that if $f : \mathbb{D} \longrightarrow \Omega$ where $f$ is harmonic $K$-quasiconformal and $\Omega$ is chord-arc domain then $\log f_z \in BMOA$.

We can ask if the converse is true for $\log f_z \in BMOA$ and for the case when this $BMOA$ norm is small analogous to conditions in Bishop-Jones theorem and Astala-Zinstmeister's theorem (Theorems 2.4 and 3.2).

PROBLEM 3. Suppose that $f : \mathbb{D} \longrightarrow \Omega$ is a harmonic quasiconformal map and $\log f_z \in BMOA$. What can we say about domain $\Omega$? Is it true that $\log f_z \in BMOA$ is equivalent to the statement that $\Omega$ is Bishop-Jones domain?

We can also ask the following question:

PROBLEM 4. Suppose that $f : \mathbb{D} \longrightarrow \Omega$ is harmonic quasiconformal mapping such that $\|\log f_z\|_{BMOA} < \varepsilon$. Is it true that for sufficiently small $\varepsilon$ it holds that $\Omega$ is chord-arc?

## References


[1] L. V. Ahlfors, *Lectures on quasiconformal mappings*, 2nd ed., University Lecture Series, vol. 38, American Mathematical Society, Providence, RI, 2006. With supplemental chapters by C. J. Earle, I. Kra, M. Shishikura and J. H. Hubbard, DOI 10.1090/ulect/038. MR2241787

[2] K. Astala, *A remark on quasiconformal mappings and BMO-functions*, Michigan Math. J. **30** (1983), no. 2, 209–212, DOI 10.1307/mmj/1029002851. MR718266

[3] K. Astala and F. W. Gehring, *Quasiconformal analogues of theorems of Koebe and Hardy-Littlewood*, Michigan Math. J. **32** (1985), no. 1, 99–107, DOI 10.1307/mmj/1029003136. MR777305

[4] K. Astala and F. W. Gehring, *Injectivity, the BMO norm and the universal Teichmüller space*, J. Analyse Math. **46** (1986), 16–57, DOI 10.1007/BF02796572. MR861687

[5] K. Astala, T. Iwaniec, and G. Martin, *Elliptic partial differential equations and quasiconformal mappings in the plane*, Princeton Mathematical Series, vol. 48, Princeton University Press, Princeton, NJ, 2009. MR2472875

[6] K. Astala and V. Manojlović, *On Pavlovic's theorem in space*, Potential Anal. **43** (2015), no. 3, 361–370, DOI 10.1007/s11118-015-9475-4. MR3430458

[7] K. Astala and M. Zinsmeister, *Teichmüller spaces and BMOA*, Math. Ann. **289** (1991), no. 4, 613–625, DOI 10.1007/BF01446592. MR1103039

[8] S. Axler, P. Bourdon, and W. Ramey, *Harmonic function theory*, Graduate Texts in Mathematics, vol. 137, Springer-Verlag, New York, 1992, DOI 10.1007/b97238. MR1184139

[9] C. J. Bishop, *A counterexample in conformal welding concerning Hausdorff dimension*, Michigan Math. J. **35** (1988), no. 1, 151–159, DOI 10.1307/mmj/1029003689. MR931947

[10] C. J. Bishop and P. W. Jones, *Harmonic measure and arclength*, Ann. of Math. (2) **132** (1990), no. 3, 511–547, DOI 10.2307/1971428. MR1078268

[11] C. J. Bishop and P. W. Jones, *Harmonic measure, $L^2$ estimates and the Schwarzian derivative*, J. Anal. Math. **62** (1994), 77–113, DOI 10.1007/BF02835949. MR1269200

[12] R. R. Coifman and Y. Meyer, *Lavrentiev's curves and conformal mappings*, Report No. 5, Institute Mittag-Leffler, 1983.

[13] G. David, *Opérateurs intégraux singuliers sur certaines courbes du plan complexe* (French), Ann. Sci. École Norm. Sup. (4) **17** (1984), no. 1, 157–189. MR744071

[14] F. P. Gardiner and W. J. Harvey, *Universal Teichmüller space*, Handbook of complex analysis: geometric function theory, Vol. 1, North-Holland, Amsterdam, 2002, pp. 457–492, DOI 10.1016/S1874-5709(02)80016-6. MR1966201

[15] J. B. Garnett and D. E. Marshall, *Harmonic measure*, New Mathematical Monographs, vol. 2, Cambridge University Press, Cambridge, 2005, DOI 10.1017/CBO9780511546617. MR2150803

[16] F. W. Gehring and B. G. Osgood, *Uniform domains and the quasihyperbolic metric*, J. Analyse Math. **36** (1979), 50–74 (1980), DOI 10.1007/BF02798768. MR581801

[17] S. Hencl and P. Koskela, *Lectures on mappings of finite distortion*, Lecture Notes in Mathematics, vol. 2096, Springer, Cham, 2014, DOI 10.1007/978-3-319-03173-6. MR3184742

[18] T. Iwaniec and J. Onninen, *Radó-Kneser-Choquet theorem*, Bull. Lond. Math. Soc. **46** (2014), no. 6, 1283–1291, DOI 10.1112/blms/bdu084. MR3291265

[19] D. S. Jerison and C. E. Kenig, *Hardy spaces, $A_\infty$, and singular integrals on chord-arc domains*, Math. Scand. **50** (1982), no. 2, 221–247, DOI 10.7146/math.scand.a-11956. MR672926

[20] P. W. Jones, *Extension theorems for BMO*, Indiana Univ. Math. J. **29** (1980), no. 1, 41–66, DOI 10.1512/iumj.1980.29.29005. MR554817

[21] P. W. Jones, *Homeomorphisms of the line which preserve BMO*, Ark. Mat. **21** (1983), no. 2, 229–231, DOI 10.1007/BF02384312. MR727346

[22] D. Kalaj, *Muckenhoupt weights and Lindelöf theorem for harmonic mappings*, Adv. Math. **280** (2015), 301–321, DOI 10.1016/j.aim.2015.04.025. MR3350221

[23] P. Koskela, P. Lammi, and V. Manojlović, *Gromov hyperbolicity and quasihyperbolic geodesics* (English, with English and French summaries), Ann. Sci. Éc. Norm. Supér. (4) **47** (2014), no. 5, 975–990, DOI 10.24033/asens.2231. MR3294621

[24] O. Lehto, *Univalent functions and Teichmüller spaces*, Graduate Texts in Mathematics, vol. 109, Springer-Verlag, New York, 1987, DOI 10.1007/978-1-4613-8652-0. MR867407

[25] V. Manojlović, *Bi-Lipschicity of quasiconformal harmonic mappings in the plane*, Filomat **23** (2009), no. 1, 85–89, DOI 10.2298/FIL0901085M. MR2568617

[26] O. Martio, *On harmonic quasiconformal mappings*, Ann. Acad. Sci. Fenn. Ser. A I **425** (1968), 10. MR236382

[27] O. Martio and J. Sarvas, *Injectivity theorems in plane and space*, Ann. Acad. Sci. Fenn. Ser. A I Math. **4** (1979), no. 2, 383–401, DOI 10.5186/aasfm.1978-79.0413. MR565886

[28] K. Matsuzaki and H. Wei, *Space of chord-arc curves and BMO/VMO Teichmüller space* (English, with English and Finnish summaries), Ann. Fenn. Math. **48** (2023), no. 1, 27–42, DOI 10.1088/0253-6102/48/1/007. MR4499587

[29] S. Nag, *On the tangent space to the universal Teichmüller space*, Ann. Acad. Sci. Fenn. Ser. A I Math. **18** (1993), no. 2, 377–393. MR1234741

[30] M. Pavlović, *Boundary correspondence under harmonic quasiconformal homeomorphisms of the unit disk*, Ann. Acad. Sci. Fenn. Math. **27** (2002), no. 2, 365–372. MR1922194

[31] M. Pavlović, *Function classes on the unit disc—an introduction*, 2nd ed., De Gruyter Studies in Mathematics, vol. 52, De Gruyter, Berlin, 2019. MR4321142

[32] Ch. Pommerenke, *Boundary behaviour of conformsal maps*, Springer, 1992.

[33] S. W. Semmes, *The Cauchy integral, chord-arc curves, and quasiconformal mappings*, The Bieberbach conjecture (West Lafayette, Ind., 1985), Math. Surveys Monogr., vol. 21, Amer. Math. Soc., Providence, RI, 1986, pp. 167–183, DOI 10.1090/surv/021/14. MR875240

[34] S. Semmes, *A counterexample in conformal welding concerning chord-arc curves*, Ark. Mat. **24** (1986), no. 1, 141–158, DOI 10.1007/BF02384395. MR852832

[35] S. W. Semmes, *Quasiconformal mappings and chord-arc curves*, Trans. Amer. Math. Soc. **306** (1988), no. 1, 233–263, DOI 10.2307/2000836. MR927689

[36] Y. Shen and H. Wei, *Universal Teichmüller space and BMO*, Adv. Math. **234** (2013), 129–148, DOI 10.1016/j.aim.2012.10.017. MR3003927

[37] T. Sugawa, *The universal Teichmüller space and related topics*, Quasiconformal mappings and their applications, Narosa, New Delhi, 2007, pp. 261–289. MR2492507

[38] V. Todorčević, *Harmonic quasiconformal mappings and hyperbolic type metrics*, Springer, Cham, 2019, DOI 10.1007/978-3-030-22591-9. MR3967686

[39] M. Vuorinen, *Conformal geometry and quasiregular mappings*, Lecture Notes in Mathematics, vol. 1319, Springer-Verlag, Berlin, 1988, DOI 10.1007/BFb0077904. MR950174

[40] H. Wei and K. Matsuzaki, *BMO embeddings, chord-arc curves, and Riemann mapping parametrization*, Adv. Math. **417** (2023), Paper No. 108933, 34, DOI 10.1016/j.aim.2023.108933. MR4551880

[41] M. Zinsmeister, *Représentation conforme et courbes presque lipschitziennes* (French), Ann. Inst. Fourier (Grenoble) **34** (1984), no. 2, 29–44, DOI 10.5802/aif.961. MR746492

[42] M. Zinsmeister, *Domaines de Lavrent'iev* (French), Publications Mathématiques d'Orsay [Mathematical Publications of Orsay], vol. 85, Université de Paris-Sud, Département de Mathématiques, Orsay, 1985. MR818223

Mathematical Institute of Serbian Academy of Sciences and Arts, Belgrade, Serbia
*Email address*: vesna.todorcevic@fon.bg.ac.rs

Contemporary Mathematics
Volume **820**, 2025
https://doi.org/10.1090/conm/820/16423

# Dynamical behavior of meromorphic solutions for certain complex difference equations

Molla Basir Ahamed and Vasudevarao Allu

ABSTRACT. This paper aims to study meromorphic solutions of complex difference equations with their dynamics. Using Nevanlinna theory, we find solutions to certain complex difference equations. Furthermore, we show that finite order solutions do not have any wandering domains, considering certain dynamical behavior.

## 1. Introduction

Problems concerning wandering domains tend to be difficult. The question whether polynomials (and rational maps) can have wandering domains remained open for the best part of the twentieth century, until Sullivan [**36**] has given his celebrated negative answer using quasiconformal deformation theory. This proof generalizes to the class of transcendental entire functions for which the set of singular values is finite. Functions, with infinitely many singular values, may have wandering domains. For example, $f_1(z) = z - 1 + e^{-z} + 2\pi i$, $f_2(z) = z + 2\pi \sin z$ and $f_3(z) = z + \lambda \sin(2\pi z) + 1$ have wandering domains for suitable $\lambda$ (see [**7**] for detailed information). It is pertinent to mention that, not much is known about the existence of wandering domains of the meromorphic solutions to difference equations.

In this paper, utilizing Nevanlinna theory, we aim to study the existence of wandering domains for meromorphic solutions of some difference equations. More precisely, we find the exact form of the solutions of the difference equation, which will help us to ensure the existence of wandering domains.

We assume that the reader is familiar to the basic value distribution theory of Nevanlinna, and its standard notations such as the *proximity function* $m(r, f)$, *counting function* $N(r, f)$, and *characteristic function* $T(r, f)$, and $S(r, f)$ denotes any quantity satisfying $S(r, f) = o(T(r, f))$ as $r \to \infty$ outside some exceptional set of finite measure. For a detailed study on this topic, we refer to [**23, 30, 40**] and references therein.

---

2020 *Mathematics Subject Classification.* Primary 30D05, 30D35.

*Key words and phrases.* Transcendental meromorphic function, Fatou and Julia sets, Complex Dynamics, complex difference equation, finite order, wandering domains, iteration theory.

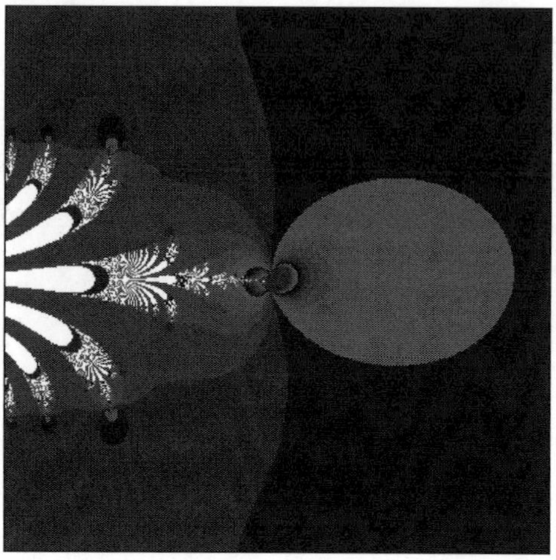

$f_1(z) = z - 1 + e^{-z} + 2\pi i$

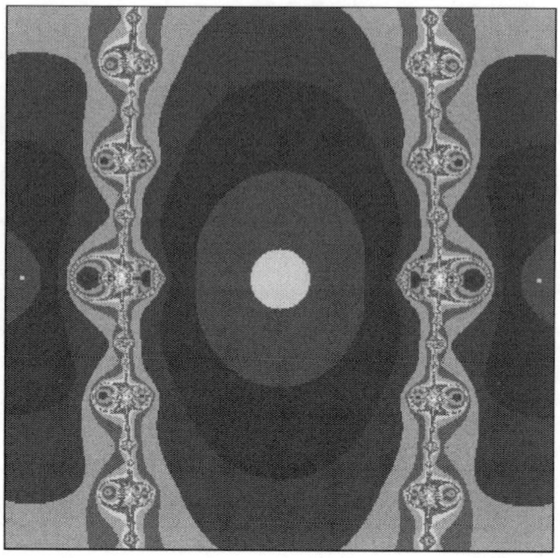

$f_2(z) = z + 2\pi \sin z$

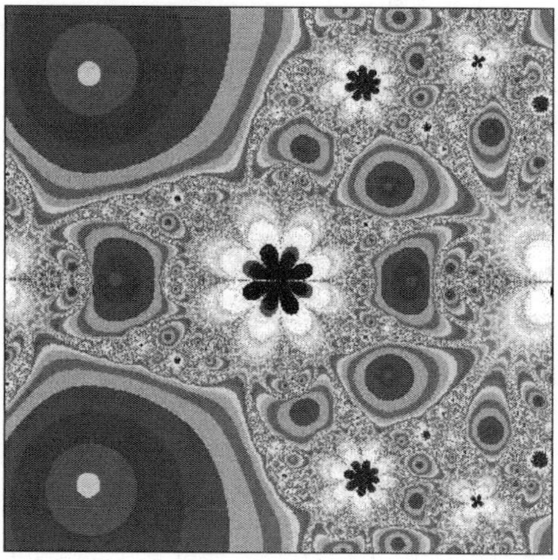

$f_3(z) = z + \sin z + 2\pi$

FIGURE 1. The dynamics of the functions $f_1, f_2, f_3$ have been illustrated in the figures.

For an introduction to iteration theory, we refer to the articles of Beardon [**6**], Carleson and Gamelin [**11**], and Steinmetz [**35**] as well as the lecture notes of Milnor for rational functions, Eremenko and Lyubich [**14**] for rational and entire functions, Bergweiler [**7**] for transcendental meromorphic functions. The classical references are Fatou [**16**] and Julia [**29**] for rationals, and Fatou [**17**] for transcendental entire functions.

Sullivan [**36**] has shown that rational functions do not have wandering domains. However, transcendental meromorphic functions, may have wandering domains (see [**2**–**4**, **13**, **36**] and references therein), whereas various classes of meromorphic functions without wandering domains are known (see [**2**, **5**, **8**, **9**, **20**] for more details). Bergweiler [**8**], Bergweiler and Terglane [**9**] have proved the non-existence of wandering domains for meromorphic function of finite order which satisfies the first order differential equation $f'(z) = q(z)e^{p(z)}(f(z) - z)$, where $q(z)$ is rational and $p(z)$ is a polynomial, and also for meromorphic solutions of one of the following differential equations

$$\begin{aligned} f'(z) &= r(z)(f(z) - z)^2 \\ f'(z) &= r(z)(f(z) - z)(f(z) - \tau) \\ f'(z)^2 &= r(z)(f(z) - z)^2(f(z) - \tau) \\ f'(z)^2 &= r(z)(f(z) - z)^2(f(z) - \tau)(f(z) - \delta), \end{aligned}$$

where $r(z)$ is a rational function, and $\tau, \delta \in \mathbb{C}$. The no-wandering domain theorem is a result on dynamical systems, established by Dennis Sullivan [**36**] in 1985 which states that a rational map $f : \hat{\mathbb{C}} \to \hat{\mathbb{C}}$ with $\deg(f) \geq 2$ does not have a wandering domain, where $\hat{\mathbb{C}}$ denotes the Riemann sphere. This theorem does not hold for arbitrary maps, for example, the transcendental map $f(z) = z + 2\pi \sin z$ has wandering domains, and $f(z) = z + \sin z + 2\pi$ has a bounded simply connected wandering domain. In fact, the result can be generalized to many situations where the functions naturally belong to a finite-dimensional parameter space, most notably to transcendental entire and meromorphic functions with a finite number of singular values. Baker [**1**, **2**] has given examples of multiply connected wandering domains, which are now known as Baker wandering domains.

In 1999, Wang [**37**] proved that if $f$ is a meromorphic solution of the differential equations

$$(f'(z))^n = q(z)p(f(z))(f'(z) - a)^t(f(z) - z)^m, \tag{1.1}$$

$$(f'(z))^n = q(z)p(f(z))e^{\mathcal{Q}(z)}(f(z) - z)^m, \tag{1.2}$$

where $q(z)$ is a rational function, $p(z)$ and $\mathcal{Q}(z)$ be two polynomials, and $m, n, \in \mathbb{N}$, then $f$ does not have wandering domains.

In 2019, Korhonen *et al.* [**32**] established some results on wandering domains concerning some solution of differential equations. More precisely, the authors have proved that every non-zero solution $f$ of the differential equation

$$f''(z) + A(z)f(z) = H(z),$$

along with its $f^{(k)}$, $k \in \mathbb{Z}$, have no Baker wandering domains, where $H$ is an entire function with $\sigma(H) < \sigma(A)$, $A \in \mathcal{A}$, here $\mathcal{A}$ is the set of all functions of exponential growth. Halburd and Korhonen [**21**, **22**], Chiang and Feng [**12**] have established independently the difference analogue lemma of the logarithmic derivative lemma.

Since then various properties of meromorphic solutions of difference equations similar to those of differential equations have been extensively studied. In what follows, $c$ always means a non-zero constant. We define the *shift* $I_c$ and *difference operator* $\Delta_c f$ (see [**21, 26**] ) of a meromorphic function $f$, respectively, by $I_c(f) = f(z+c)$ and $\Delta_c(f) = f(z+c) - f(z) = (I_c - 1)f(z)$ for all $z \in \mathbb{C}$.

Regarding periodicity of meromorphic functions, Heittokangas *et al.* [**24, 25**] have considered the problem of value sharing for shifts of meromorphic functions and proved a result to find the periodicity of meromorphic functions. Zhang [**41**] have shown that replacing the derivative $f'(z)$ of a meromorphic function $f$ in the equations (1.1) and (1.2) by its *shift* $I_c(f)$, a similar kind of results can be established showing that the meromorphic solutions do not have wandering domains.

We see that the wandering domain of the solutions of some differential equations (see e.g. [**28, 32, 37, 38**] and references therein) has been studied extensively but it is not known much about the existence of wandering domains to the solutions of difference equations. Since the derivative $f'$ can be treated as a difference counterpart of a meromorphic function $f$, one of our main aims is to study the similar dynamical behavior of solutions to the difference equations which are analogous to the differential equations (1.1) and (1.2).

The main aim of this paper is look for a possible answers to the following question.

QUESTION 1.3. What can be said about the wandering domains of $f$ when derivatives $f'$ is replaced by difference operators $\Delta_c(f)$ in (1.1) and (1.2)?

A general form of the difference equation that is analogous to the differential equation (1.1) can be written as

$$(1.4) \qquad (\Delta_c(f(z)))^n = P(f(z), z)\,(\Delta_c(f) - a)^t\,,$$

where $a$ is a non-zero constant, $P(f(z), z)$ is a polynomial in $f$ with degree $p$, the coefficients of $P(f(z), z)$ are small functions of $f$.

The following question is inevitable in our study.

QUESTION 1.5. Can we find the precise form of the solutions for the difference equation (1.4)?

Our aim of this paper is to answer Questions 1.3 and 1.5 affirmatively.

## 2. Main results

We obtain Theorem 2.1 to Theorem 2.7 as the main results of this paper, in order to answer Questions 1.3 and 1.5 affirmatively. We first state the following result with regard to (1.4) estimating between $n, t$ and $p$.

THEOREM 2.1. *Let $f$ be a finite-order meromorphic solution of the equation* (1.4)*. Then*

$$\max\{t, n\} \geq \frac{p}{2} \geq \frac{n}{2} - t.$$

The study of non-existence of wandering domains for non-rational functions has been extensively studied. But for the rational functions in view of the theory of Sullivan's non-existence of wandering domains for finite-order solution of the equation(1.4), we obtain the following dynamical property for the finite-order solution of (1.4) which is similar to the dynamical property of solutions of the equations (1.1) and (1.2).

COROLLARY 2.2. *The finite order meromorphic solution of the difference equation* $(\Delta_c(z))^n = P(f(z), z)\,(\Delta_c(f) - a)^t$ *do not have wandering domains.*

We now consider a difference equation which is analogous to the differential equation $(f')^n = q(z)P(f(z), z)e^{\mathcal{Q}(z)}\,(f(z) - z)^m$. We see that, a general setting of this equation is

$$(\Delta_c(f))^n = q(z)e^{\mathcal{Q}(z)}P(f(z), z), \tag{2.3}$$

where $q(z)$, $\mathcal{Q}(z)$ are non-constant polynomials, and $P(f(z), z)$ is a polynomial in $f$ with polynomials as coefficients. Replacing $f(z+c)$ by $f(z)$ in (2.3), we can write (2.3) as $(-\Delta_{-c}(f))^n = q(z)e^{\mathcal{Q}(z)}P(f(z-c), z)$. Without loss of generality, we consider (2.5) in the following result.

THEOREM 2.4. *Let* $q(z)$, $\mathcal{Q}(z)$ *be non-constant polynomials, and* $P(z, f)$ *be a polynomial in* $f$ *with degree* $p$, *the coefficients of* $P(z, f)$ *are also polynomials. If* $n > p$, *and* $f$ *is a finite order entire solution of the difference equation*

$$(-\Delta_c(f))^n = q(z)e^{\mathcal{Q}(z)}P(f(z+c), z), \tag{2.5}$$

*then* $\sigma(f) = \deg(\mathcal{Q}(z))$.

In the next result, we consider the difference equation

$$(-\Delta_c(f))^n = q(z)e^{\mathcal{Q}(z)}f(z+c)^p, \tag{2.6}$$

and prove a result showing that the precise form of the solutions of (2.6) can be obtained and this will eventually help us to draw a conclusion of the existence of wandering domains.

THEOREM 2.7. *Let* $p, n \geq 2$ *be integers, and* $n > p$, $p|n$. *Let* $c \in \mathbb{C} \setminus \{0\}$, *and* $q(z)$, $\mathcal{Q}(z)$ *be polynomials such that* $\mathcal{Q}(z)$ *is not a constant, and* $q(z) \not\equiv 0$. *If* $f$ *is a finite order entire solution of the difference equation* (2.6), *then* $f$ *must be of the form* $f(z) = ce^{\beta(z)}$, *where* $\beta(z) = c_1 z + c_2$, $c_1, c \in \mathbb{C} \setminus \{0\}$ *and* $c_2 \in \mathbb{C}$.

REMARK 2.8. There are finite order entire functions that satisfy the difference equation (2.6). The following example illustrate this fact.

EXAMPLE 2.9. Let $f(z) = Ce^{z+3}$, where $C$ is a non-zero constant. If $n = 6$, $p = 2$, then $f(z)$ is a finite order entire solution of the equation (2.6) with $q(z) = \big(C^4(1 - e^c)^6\big)/e^{pc}$, $\mathcal{Q}(z) = 4z + 12$, and $c \neq 2k\pi i$, $k \in \mathbb{Z}$.

REMARK 2.10. If $\deg(\beta(z)) > 1$, then the function $f(z) = Ce^{\beta(z)}$ cannot be a solution of the difference equation (2.6) in Theorem 2.7.

EXAMPLE 2.11. Let $f(z) = e^{z^2}$ and $n = 2$, $p = 1$. It is clear that $\beta(z) = z^2$ with $\deg(\beta(z)) > 1$. A simple computation shows that $f(z)$ is a solution of (2.6) only if

$$q(z) = \frac{\left(1 - 2e^{2z+1} + e^{4z+2}\right)}{e^{\mathcal{Q}(z) - z^2 + 2z + 1}},$$

where $\mathcal{Q}(z)$ is any non-constant polynomial. But we note that $q(z)$ is not a polynomial in $z$.

We recall here a result of Baker **[3]** which is about the non-existence of wandering domains for a class of entire functions.

THEOREM C. [**3**] *Let $P(z)$ and $Q(z)$ be polynomials with $Q(z)$ non-constant. Then*

$$f(z) = \int_0^x P(t)e^{Q(t)}dt \tag{2.12}$$

*has no wandering domains. Particularly, the form $f(z) = P_1(z)e^{Q_1(z)}$, where $P_1(z)$ and $Q_1(z)$ have the special case of (2.12).*

Combining Theorems C and 2.7, we obtain the following corollary of Theorem 2.7 which is our desired goal of this paper. As a matter of fact, Question 1.3 is answered affirmatively.

COROLLARY 2.13. *Under the hypothesis of Theorem 2.7, every finite order solutions of (2.6) have no wandering domains.*

## 3. Key Lemmas and proof of the main results

In this section, we present some necessary lemmas which will play a key role in proving the main results.

LEMMA A. [**19**] *Meromorphic solutions of first order algebraic differential equations are of finite order.*

LEMMA B. [**31**] *Let $g(z)$ be a transcendental meromorphic solution of finite order of the difference equation $P(z, g) = 0$, where $P(z, g)$ is a difference polynomial in $g(z)$, and its shift. If $P(z, a) \neq 0$ for a slowly moving target function $a$, that is, $T(r, a) = S(r, g)$, then*

$$m\left(r, \frac{1}{g-a}\right) = S(r, g).$$

LEMMA C. [**12**] *Let $f(z)$ be a meromorphic function with order $\sigma \equiv \sigma(f)$,$\sigma < \infty$, and let $c$ be a fixed non-zero complex number. Then for each $\epsilon > 0$, we have*

$$T(r, f(z + c)) = T(r, f) + O\left(r^{\sigma-1+\epsilon}\right) + O(\log r).$$

LEMMA 3.1. *Let $f(z)$ be a transcendental meromorphic function of finite order. Then $S(r, \Delta_c(f))$ can be replaced by $S(r, f)$.*

PROOF. Since $f$ is transcendental, it is evident that $O(\log r) = S(r, f)$. Thus, in view of Lemma C, we obtain

$$\begin{aligned} T\left(r, \Delta_c(f)\right) &\leq T(r, f(z + c)) + T(r, f(z)) + O(1) \\ &\leq 2\,T(r, f) + O\left(r^{\sigma-1+\epsilon}\right) + O(\log r) \\ &\leq 2\,T(r, f) + O\left(r^{\sigma-1+\epsilon}\right) + S(r, f). \end{aligned}$$

In view of the above estimate, it is evident that

$$S(r, \Delta_c(f)) = o(T(r, \Delta_c(f))) = o(T(r, f)) = S(r, f)$$

as $r \to \infty$ outside some exceptional set of finite measure. With these estimates the lemma follows. □

The next lemma is due to the result of Mohonko and Mohonko [**33**] (see also [**30**, p. 29]).

LEMMA D. [**33**] *Let $f(z)$ be a meromorphic function. Then, for all irreducible rational functions in $f$,*

$$\mathcal{R}(f(z),z)=\left(\sum_{j=0}^{p}a_j(z)f(z)^j\right)\Big/\left(\sum_{k=0}^{q}b_k(z)f(z)^i\right)$$

*with meromorphic coefficients $a_j(z)$, $b_k(z)$, the characteristic function of $\mathcal{R}(f(z),z)$ satisfies $T\left(r,\mathcal{R}(f(z),z)\right)=\max\{p,q\}\,T(r,f)+O(\Psi(z))$, where*

$$\Psi(r)=\max_{k,j}\left\{T(r,a_j),T(r,b_k)\right\}.$$

*In the particular case when*

$$T(r,a_j)=S(r,f),\quad j=0,1,\dots,p,\ T(r,b_k)=S(r,f),\quad k=0,1,\dots,q,$$

*we have $T(r,\mathcal{R}(f(z),z))=\max\{p,q\}\,T(r,f)+S(r,f)$.*

We now discuss the proof of the main results.

PROOF OF THEOREM 2.1. By the assumption of $\sigma(f)<+\infty$, we obtain

$$\begin{aligned}T\left(r,P(f(z),z)\right)&\leq T\left(r,\frac{\Delta_c(f)^n}{\left(\Delta_c(f)-a\right)^t}\right)\\&\leq\max\{n,t\}\,T\left(r,\Delta_c(f)\right)+S\left(r,\Delta_c(f)\right).\end{aligned}$$

In view of the Lemmas C, 3.1 and D, a simple computation shows that

$$\begin{aligned}p\,T(r,f)&\leq\max\{n,t\}\,T(r,f(z+c))+\max\{n,t\}\,T(r,f)+S(r,f)\\&\leq 2\max\{n,t\}\,T(r,f)+S(r,f).\end{aligned}$$

Hence, it is easy to see that

$$\max\{n,t\}\geq\frac{p}{2}.\tag{3.2}$$

A simple computation shows that

$$\begin{aligned}n\,T\left(r,\Delta_c(f)\right)&\leq T\left(r,\Delta_c(f)^n\right)+O(1)\\&=T\left(r,\left(\Delta_c(f)-a\right)^t P(f(z),z)\right)\\&\leq t\,T\left(r,\Delta_c(f)\right)+p\,T(r,f)+S(r,f)\\&\leq(2t+p)\,T(r,f)+S(r,f),\end{aligned}$$

and this implies

$$p\geq n-2t.\tag{3.3}$$

Therefore, it follows from (3.2) and (3.3) that

$$\max\{n,t\}\geq\frac{p}{2}\geq\frac{n}{2}-t.$$

This completes the proof. □

Proof of Theorem 2.4. Suppose that $f$ is an entire solution of (2.5) with $\sigma(f)\equiv\sigma<+\infty$. By Lemma C and in view of (2.5), a simple computation shows that

$$\begin{aligned}2n\,T(r,f)&\leq T\left(r,(-\Delta_c(f))^n\right)\\&=T\left(r,q(z)e^{\mathcal{Q}(z)}P(f(z+c),z)\right)\\&\leq T\left(r,e^{\mathcal{Q}(z)}\right)+T\left(r,P(f(z+c),z)\right)+O(\log r)\\&\leq T\left(r,e^{\mathcal{Q}(z)}\right)+p\,T(r,f)+O\left(r^{\sigma-1+\epsilon}\right)+O(\log r).\end{aligned}$$

Thus we have

$$(n-p)\,T(r,f)\leq T\left(r,e^{\mathcal{Q}(z)}\right)+O\left(r^{\sigma-1+\epsilon}\right)+O(\log r).$$

Since $n>p$, it is easy to see that $\sigma(f)\leq\deg(\mathcal{Q})$. By Lemma C, we obtain

$$\begin{aligned}T\left(r,e^{\mathcal{Q}(z)}\right)&\leq T\left(r,\frac{(-\Delta_c(f))^n}{q(z)P((f(z+c)),z)}\right)+O(1)\\&\leq T\left(r,(\Delta_c(f))^n\right)+T\left(r,q(z)P((f(z+c)),z)\right)+O(1)\\&\leq n\,T(r,f)+p\,T(r,f)+O\left(r^{\sigma-1+\epsilon}\right)+O(\log r)\\&\leq(n+p)\,T(r,f)+O\left(r^{\sigma-1+\epsilon}\right)+O(\log r),\end{aligned}$$

which shows that

$$\deg(\mathcal{Q}(z))\leq\sigma(f). \tag{3.4}$$

In view of (3.4), we obtain $\sigma(f)=\deg(\mathcal{Q}(z))$. This completes the proof. □

Proof of Theorem 2.7. By the Hadamard Factorization Theorem, we write $f(z)=T(z)e^{\beta(z)}$, where $T$ and $f$ have the same set of zeros and $\beta$ is a polynomial. Substituting this expression of $f$ in the equation $(-\Delta_cf)^n=q(z)e^{\mathcal{Q}(z)}\left(f(z+c)\right)^p$ we obtain

$$\frac{\left(T(z)-T(z+c)e^{\beta(z+c)-\beta(z)}\right)^n}{q(z)\left(T(z+c)\right)^p}=e^{\mathcal{Q}(z)+p\beta(z+c)-n\beta(z)}. \tag{3.5}$$

Set

$$G(z):=T(z)-T(z+c)e^{\beta(z+c)-\beta(z)}.$$

Thus, (3.5) can be written as

$$\frac{(G(z))^n}{q(z)\left(T(z+c)\right)^p}=e^{\mathcal{Q}(z)+p\beta(z+c)-n\beta(z)}. \tag{3.6}$$

Suppose that $G$ has infinitely many zeros. Since $q(z)$ is a polynomial, it can only have a finite number of zeros. Therefore, there exists a zero $z_0$ of $G$ for which none of the numbers $z_m=z_0+mc$, where $m\in\mathbb{N}\cup\{0\}$, are zeros of $q(z)$. However, these numbers $z_m$, where $m\in\mathbb{N}\cup\{0\}$, must be zeros of $T(z+c)$. Otherwise, considering (3.6), we would obtain a contradiction. Clearly, number of zeros of both $T(z)$ and $T(z+c)$ are infinite also. On the other hand, $q(z_0+mc)\neq 0$ for $m\in\mathbb{N}\cup\{0\}$. In fact, $z_0$ being a zero of $G(z)$, hence $z_0$ must be a zero of $T(z)$ as well as $T(z+c)$, otherwise by (3.6), we get a contradiction.

Let $z_0$ be a zero of $G(z)$ with multiplicity $k\geq 1$. Then, as $q(z_0+c)\neq 0$, in view of (3.6), we see that $T(z_0+c)=0 i.e,$ $z_0+c$ must be a zero of $T(z)$. The multiplicity of $z_0+c$ cannot be exactly equal to $k$, otherwise, we will get a contradiction in view of $n > p$. Thus, we see that the multiplicity of the zero $z_1 = z_0 + c$ of $T(z)$ must be $nk = pr$, for some $r$ $(\neq k)$. By the similar argument, $z_2 = z_0 + 2c$ will be a zero of $T(z)$ with multiplicity $n(nk) = n^2k$. Continuing inductively, it can be easily shown that $z_m = z_0 + mc$ will be a zero of $T(z)$ with multiplicity $n^m k$. Since $n \geq 2$ and $c \neq 0$, the sequence of zeros (counting multiplicities) is of infinite convergence exponent, this is a contradiction. Therefore, $T(z)$ and hence $T(z+c)$ have finite number of zeros, that is, they are polynomials. Consequently, $G$ has finite number of zeros and we see that $\lambda(f) = 0 < \sigma(f) = \deg \beta(z)$.

Let $z_*$ be a zero of the polynomial $T(z+c)$. Then, $z_*$ must be a zero of the polynomial $T(z)$, otherwise, from (3.6), we obtain an immediate contradiction. Hence, all the zeros of the polynomial $T(z+c)$ are also the zeros of the polynomial $T(z)$, and this is possible only when $T(z+c)$ and $T(z)$ are constant multiple of each other, since $T(z)$ and $T(z+c)$ both are polynomials. We suppose that $T(z+c) = \eta T(z)$, where $\eta \neq 0$. A simple computation shows that $\eta = 1$, and hence $T(z+c) = T(z)$ which turns out that the polynomial $T(z)$ is periodic of period $c$ and this leads to the fact that $T(z)$ must be a constant, say $A_0$. Now it follows from (3.6) that

$$\frac{A_0^{n-p}\left(1 - e^{\beta(z+c)-\beta(z)}\right)}{q(z)} = e^{Q(z)+p\beta(z+c)-n\beta(z)}. \tag{3.7}$$

Finally, it is easy to see that $1 - e^{\beta(z+c)-\beta(z)}$ has infinitely many zeros whereas $q(z)$ has only finite number of zeros. Thus, in view of (3.7), we get a contradiction, unless the polynomial $\beta(z+c) - \beta(z)$ is a constant $(\neq 2k\pi i,\ k \in \mathbb{Z})$ and $q(z)$ is a constant $(\neq 0)$. Hence, $\beta(z)$, being a polynomial, it must be linear, that is of the form $\beta(z) = a_1 z + a_0$, where $a_1(\neq 0)$, $a_0 \in \mathbb{C}$. Thus, we conclude that $f$ assumes the form $f(z) = Ce^{a_1 z + a_0}$, where $C$, $a_1(\neq 0)$, $a_0 \in \mathbb{C}$. This completes the proof. □

## References


[1] I. N. Baker, *Multiply connected domains of normality in iteration theory*, Math. Z. **81** (1963), 206–214, DOI 10.1007/BF01111543. MR153842

[2] I. N. Baker, *An entire function which has wandering domains*, J. Austral. Math. Soc. Ser. A **22** (1976), no. 2, 173–176, DOI 10.1017/s1446788700015287. MR419759

[3] I. N. Baker, *Wandering domains in the iteration of entire functions*, Proc. London Math. Soc. (3) **49** (1984), no. 3, 563–576, DOI 10.1112/plms/s3-49.3.563. MR759304

[4] I. N. Baker, *Some entire functions with multiply-connected wandering domains*, Ergodic Theory Dynam. Systems **5** (1985), no. 2, 163–169, DOI 10.1017/S0143385700002832. MR796748

[5] I. N. Baker and A. P. Singh, *Wandering domains in the iteration of compositions of entire functions*, Ann. Acad. Sci. Fenn. Ser. A I Math. **20** (1995), no. 1, 149–153. MR1304112

[6] A. F. Bearden, *Iteration of rational functions*, Springer-Verlag, Berlin, Heidelberg, New York, 1991.

[7] W. Bergweiler, *Iteration of meromorphic functions*, Bull. Amer. Math. Soc. (N.S.) **29** (1993), no. 2, 151–188, DOI 10.1090/S0273-0979-1993-00432-4. MR1216719

[8] W. Bergweiler, *Newton's method and a class of meromorphic functions without wandering domains*, Ergodic Theory Dynam. Systems **13** (1993), no. 2, 231–247, DOI 10.1017/S0143385700007343. MR1235471

[9] W. Bergweiler and N. Terglane, *Weakly repelling fixpoints and the connectivity of wandering domains*, Trans. Amer. Math. Soc. **348** (1996), no. 1, 1–12, DOI 10.1090/S0002-9947-96-01511-5. MR1327252

[10] W. Bergweiler, M. Haruta, H. Kriete, H.-G. Meier, and N. Terglane, *On the limit functions of iterates in wandering domains*, Ann. Acad. Sci. Fenn. Ser. A I Math. **18** (1993), no. 2, 369–375. MR1234740

[11] L. Carlson and T. Gamelin, *Complex dynamics*, Springer-Verlag, Berlin, Heidelberg, New York, 1993.

[12] Y.-M. Chiang and S.-J. Feng, *On the Nevanlinna characteristic of $f(z+\eta)$ and difference equations in the complex plane*, Ramanujan J. **16** (2008), no. 1, 105–129, DOI 10.1007/s11139-007-9101-1. MR2407244

[13] A. È. Erëmenko and M. Ju. Ljubich, *Examples of entire functions with pathological dynamics*, J. London Math. Soc. (2) **36** (1987), no. 3, 458–468, DOI 10.1112/jlms/s2-36.3.458. MR918638

[14] A. È. Erëmenko and M. Yu. Lyubich, *The dynamics of analytic transformations* (Russian), Algebra i Analiz **1** (1989), no. 3, 1–70; English transl., Leningrad Math. J. **1** (1990), no. 3, 563–634. MR1015124

[15] S. Fan and Y. Wang, *Dynamics of transcendental entire maps on Berkovich affine line*, J. Dynam. Differential Equations **25** (2013), no. 1, 217–229, DOI 10.1007/s10884-012-9281-2. MR3027640

[16] P. Fatou, *Sur les équations fonctionnelles* (French), Bull. Soc. Math. France **47** (1919), 161–271. MR1504787

[17] P. Fatou, *Sur l'itération des fonctions transcendantes Entières* (French), Acta Math. **47** (1926), no. 4, 337–370, DOI 10.1007/BF02559517. MR1555220

[18] S. Ghora, T. Nayak, and S. Sahoo, *On Fatou sets containing Baker omitted value*, J. Dynam. Differential Equations **35** (2023), no. 3, 2621–2639, DOI 10.1007/s10884-021-10060-y. MR4627826

[19] A. A. Gol′dberg, *On one-valued integrals of differential equations of the first order* (Russian), Ukrain. Mat. Ž. **8** (1956), 254–261. MR85396

[20] L. R. Goldberg and L. Keen, *A finiteness theorem for a dynamical class of entire functions*, Ergodic Theory Dynam. Systems **6** (1986), no. 2, 183–192, DOI 10.1017/S0143385700003394. MR857196

[21] R. G. Halburd and R. J. Korhonen, *Difference analogue of the lemma on the logarithmic derivative with applications to difference equations*, J. Math. Anal. Appl. **314** (2006), no. 2, 477–487, DOI 10.1016/j.jmaa.2005.04.010. MR2185244

[22] R. G. Halburd and R. J. Korhonen, *Nevanlinna theory for the difference operator*, Ann. Acad. Sci. Fenn. Math. **31** (2006), no. 2, 463–478. MR2248826

[23] W. K. Hayman, *Meromorphic functions*, Oxford Mathematical Monographs, Clarendon Press, Oxford, 1964. MR164038

[24] J. Heittokangas, R. Korhonen, I. Laine, and J. Rieppo, *Uniqueness of meromorphic functions sharing values with their shifts*, Complex Var. Elliptic Equ. **56** (2011), no. 1-4, 81–92, DOI 10.1080/17476930903394770. MR2774582

[25] J. Heittokangas, R. Korhonen, I. Laine, J. Rieppo, and J. Zhang, *Value sharing results for shifts of meromorphic functions, and sufficient conditions for periodicity*, J. Math. Anal. Appl. **355** (2009), no. 1, 352–363, DOI 10.1016/j.jmaa.2009.01.053. MR2514472

[26] J. Heittokangas, H. Yu, and M. A. Zemirni, *On the number of linearly independent admissible solutions to linear differential and linear difference equations*, Canad. J. Math. **73** (2021), no. 6, 1556–1591, DOI 10.4153/S0008414X20000607. MR4350553

[27] J.-H. Zheng, *On the non-existence of unbounded domains of normality of meromorphic functions*, J. Math. Anal. Appl. **264** (2001), no. 2, 479–494, DOI 10.1006/jmaa.2001.7681. MR1876745

[28] Z.-G. Huang and J. Wang, *Fatou sets of entire solutions of linear differential equations*, J. Math. Anal. Appl. **409** (2014), no. 1, 275–281, DOI 10.1016/j.jmaa.2013.07.010. MR3095038

[29] G. Julia, *Sur l'itération des fonctions rationelles*, J. Math. Pure. Appl. **4** (1918), 47–245.

[30] I. Laine, *Nevanlinna theory and complex differential equations*, De Gruyter Studies in Mathematics, vol. 15, Walter de Gruyter & Co., Berlin, 1993, DOI 10.1515/9783110863147. MR1207139

[31] I. Laine and C.-C. Yang, *Clunie theorems for difference and q-difference polynomials*, J. Lond. Math. Soc. (2) **76** (2007), no. 3, 556–566, DOI 10.1112/jlms/jdm073. MR2377111

[32] R. Korhonen, J. Wang, and Z. Ye, *Lower order and Baker wandering domains of solutions to differential equations with coefficients of exponential growth*, J. Math. Anal. Appl. **479** (2019), no. 2, 1475–1489, DOI 10.1016/j.jmaa.2019.07.007. MR3987918

[33] A. A. Mohon'ko and V. D. Mohon'ko, *Estimates of the Nevanlinna characteristics of certain classes of meromorphic functions, and their applications to differential equations* (Russian), Sibirsk. Mat. Ž. **15** (1974), 1305–1322, 1431. MR404625

[34] G. M. Stallard, *A class of meromorphic functions with no wandering domains*, Ann. Acad. Sci. Fenn. Ser. A I Math. **16** (1991), no. 2, 211–226, DOI 10.5186/aasfm.1991.1626. MR1139794

[35] N. Steinmetz, *Rational iteration*, De Gruyter Studies in Mathematics, vol. 16, Walter de Gruyter & Co., Berlin, 1993. Complex analytic dynamical systems, DOI 10.1515/9783110889314. MR1224235

[36] D. Sullivan, *Quasiconformal homeomorphisms and dynamics. I. Solution of the Fatou-Julia problem on wandering domains*, Ann. of Math. (2) **122** (1985), no. 3, 401–418, DOI 10.2307/1971308. MR819553

[37] Y. Wang, *Wandering domains in the dynamics of certain meromorphic functions*, Bull. Austral. Math. Soc. **59** (1999), no. 1, 99–104, DOI 10.1017/S0004972700032639. MR1672803

[38] J. Wang and Z. Chen, *Limiting direction and Baker wandering domain of entire solutions of differential equations*, Acta Math. Sci. Ser. B (Engl. Ed.) **36** (2016), no. 5, 1331–1342, DOI 10.1016/S0252-9602(16)30072-8. MR3529620

[39] Z. T. Wen, J. Heittokangas, and I. Laine, *Exponential polynomials as solutions of certain nonlinear difference equations*, Acta Math. Sin. (Engl. Ser.) **28** (2012), no. 7, 1295–1306, DOI 10.1007/s10114-012-1484-2. MR2928479

[40] H. X. Yi and C. C. Yang, *Uniqueness theory of meromorphic functions*, Science Press, Beijing, (1995) (In Chinese).

[41] G. Zhang, *Note on wandering domains in the dynamics of solutions of certain difference equations*, Adv. Difference Equ., posted on 2015, 2015:25, 7, DOI 10.1186/s13662-015-0359-0. MR3304953

Department of Mathematics, Jadavpur University, Kolkata-700032, West Bengal, India

*Email address*: mbahamed.math@jadavpuruniversity.in

Discipline of Mathematics, School of Basic Sciences, Indian Institute of Technology Bhubaneswar, Bhubaneswar-752050, Odisha, India

*Email address*: avrao@iitbbs.ac.in

SELECTED PUBLISHED TITLES IN THIS SERIES



For a complete list of titles in this series, visit the
AMS Bookstore at **www.ams.org/bookstore/conmseries/**.